PHILOSOPHY, HISTORY AND SOCIAL ACTION

BOSTON STUDIES IN THE PHILOSOPHY OF SCIENCE

VOLUME 107

PHILOSOPHY, HISTORY AND SOCIAL ACTION

Essays in Honor of Lewis Feuer

With an autobiographical essay by Lewis Feuer

Edited by

SIDNEY HOOK

Dept. of Philosophy, New York University, and the Hoover Institution

WILLIAM L. O'NEILL

Dept. of History, Rutgers University

and

ROGER O'TOOLE

Dept. of Sociology, University of Toronto

KLUWER ACADEMIC PUBLISHERS

DORDRECHT / BOSTON / LONDON

Library of Congress Cataloging-in-Publication Data CIP

Philosophy, history, and social action : essays in honor of Lewis Feuer : with an autobiographical essay by Lewis Feuer / edited by Sidney Hook, William L. O'Neill, and Roger O'Toole.
p. cm. — (Boston studies in the philosophy of science : v. 107)
Bibliography: p.
Includes index.
ISBN 9027726442
1. Philosophy. 2. History — Philosophy. 3. Social sciences—–Philosophy. 4. Social action. 5. Feuer, Lewis Samuel, 1912–
I. Feuer, Lewis Samuel, 1912– . II. Hook, Sidney, 1902–
III. O'Neill, William L. IV. O'Toole, Roger, 1942– . V. Series.
Q174.B67 vol. 107
[B29]
320.5—dc19 87–36905
CIP

Published by Kluwer Academic Publishers
P.O. Box 17, 3300 AA Dordrecht, The Netherlands.

Kluwer Academic Publishers incorporates the publishing programmes of
D. Reidel, Martinus Nijhoff, Dr. W. Junk and MTP Press

Sold and distributed in the U.S.A. and Canada
by Kluwer Academic Publishers,
101 Philip Drive, Norwell, MA 02061, U.S.A.

In all other countries, sold and distributed
by Kluwer Academic Publishers Group,
P.O. Box 322, 3300 AH Dordrecht, The Netherlands.

Printed in The Netherlands

TABLE OF CONTENTS

EDITORIAL PREFACE

Two articles by Lewis Feuer caught my attention in the '40s when I was wondering, as a student physicist, about the relations of physics to philosophy and to the world in turmoil. One was his essay on 'The Development of Logical Empiricism' (1941), and the other his critical review of Philipp Frank's biography of Einstein, 'Philosophy and the Theory of Relativity' (1947). How extraordinary it was to find so intelligent, independent, critical, and humane a mind; and furthermore he went further, as I soon realized when I looked for his name on other publications. I recall arguing with myself over his exploration of 'Indeterminacy and Economic Development' (1948), and even more when I read his 'Dialectical Materialism and Soviet Science' (1949). More papers, and then the fascinating, sometimes irritating, always insightful, books.

His monograph on *Psychoanalysis and Ethics* 1955, the beautiful sociological and humanist study of *Spinoza and the Rise of Liberalism* (1958), his essays on 'The Social Roots of Einstein's Theory of Relativity' (1971) together with the book on *Einstein and the Generations of Science* (1974), the splendid reader from the works of Marx and Engels, *Basic Writings on Politics and Philosophy* (1959) which was a major text of the '60s, the stimulating essays on the social formation which seems to have been required for a modern scientific movement to develop, set forth most convincingly in *The Scientific Intellectual* (1963). These are my choices; the reader will want to read and meditate on many others from the impressive bibliography in this book. Lewis Feuer has never avoided controversy nor abandoned either intelligence or courage. Agree with him or not, every reader will find his writings not only pungent but also powerful, critical but also admirably steeped in humane values. As one of his admirers, Sidney Hook, has written of Feuer's article 'Noumenalism and Einstein's Argument for the Existence of God' (1983): "I believe Einstein would have welcomed Feuer's argument which will impress even those who like myself remain unconvinced by it but in consequence are less sure of their own negations".

Centre for Philosophy and History of Science ROBERT S. COHEN
Boston University
April 1988

LEWIS S. FEUER

EDITORS' FOREWORD

For more than half a century the writings of Lewis S. Feuer have made a unique and provocative contribution to many branches of scholarship. Resistant to the tides of academic over-specialization and disciplinary isolation, Feuer has been an intrepid, astute, and often iconoclastic explorer of such topics as intellectual authoritarianism, the ideas of Spinoza and Dewey, the life and work of Karl Marx, the religious context of seventeenth century science, the generational basis of student revolt, the social roots of Einstein's theories, Jewish contributions to American life, and the social psychology of imperialism. Writing with verve, simplicity and impeccable scholarship, he has sought to relate and properly delimit the psychological and sociological components of discovery in a broad range of intellectual enterprises.

The contributions assembled in this volume have been especially written by many distinguished scholars in order to pay tribute to the life of thought and action recounted by Professor Feuer in his autobiographical essay. The collection honors Feuer both by exemplifying qualities characteristic of his work and by reflecting the immense variety of his scientific and literary interests. Besides admiring Lewis Feuer, most of the contributors are his friends too and have welcomed this opportunity to acknowledge his warmth, generosity and many acts of kindness rendered by him over the years.

We are grateful to Kathryn Feuer for generously responding to all our requests for information. In addition to writing articles some contributors assisted us in other ways, and we wish especially to thank Irving Louis Horowitz, Paul Kurtz and Sidney Ratner for their aid and counsel. This book would not exist save for the efforts of Robert Cohen, to whom we extend our gratitude, as also to Mrs. J. C. Kuipers of Kluwer Academic Publishers for her editorial support. Most of all we would like to thank the contributors for what has been an impressive display of patience on their part.

SIDNEY HOOK
WILLIAM L. O'NEILL
ROGER O'TOOLE

LEWIS S. FEUER

A NARRATIVE OF PERSONAL EVENTS AND IDEAS

I was born in New York City's Lower East Side in December 1912, in a tenement house on 5th Street off Avenue C; until I was almost thirteen years of age, my family lived longest in a house on that Avenue between 8th and 9th Streets. I still remember riding on the horse-drawn street car through 3rd Street toward Avenue C. Several cultures competed and commingled in the setting of my formative years; the earliest was the Orthodox Jewish. At home, we observed strictly the dietary laws; our mother lit and blessed the Friday night candles; we knew Saturday's taboo against writing, we were happy at Passover feasts, and gloomy on the Day of Atonement. For seven years I attended classes during the late afternoon or evening at the Downtown Talmud Torah at Houston Street, and later the Hebrew High School. The peddler's push carts lined our avenue, and five days a week I woke to the sounds of peddlers screeching their wares in Yiddish, beseeching, supplicating, uttering challenges, insults, flatteries; on wintry days, they would burn wood and oil lamps to keep warm, and they feared the policemen's clubs; they were mostly bearded, and wore strange hats. This Yiddish culture was quite different from the Hebraic one that I received at the Downtown Talmud Torah, where a group of young gifted teachers, in their early twenties, and working their way through college, read with us the books of the Bible; several of them later became noted scholars and public personalities: Oscar Janowsky, Frederic Ewen, Simon Rifkind.

During the days, however, I was a cheerful pupil at Public School 64, a school famed for its principals who encouraged some daring experimentation in education (Irwin and Marks, 1924). At P.S. 64, an experiment had been inaugurated, the 'Terman classes' they were called, thus named after the famous Stanford psychologist. Through psychological and physical tests, given throughout the East Side's schools, several small groups of boys and girls were selected to enjoy special curricula with specially chosen able teachers. That is how it happened that I was advanced rapidly, so that I entered college at the age of fourteen. The 'skippings', however, were an ordeal especially when I was jumped past a year of arithmetic; also there was a later social

S. Hook, W.L. O'Neill and R. O'Toole (eds.), Philosophy, History and Social Action, 1–85.

imbalance with regard to older classmates at college. On the other hand, I had one remarkable teacher, a Mr Rosenzweig, who read to us Van Loon's *The Story of Mankind* which had just appeared when I was in the sixth grade, and kindled in me an interest in the larger historical questions of man; we also began early learning French, a language to which I became much attached. Our classes were the only co-educational ones in a large boys' school.

The American culture we experienced was more cheerful, joyous, optimistic than the sombre one we imbibed at the Talmud Torah. I revelled in American history, having devoured the various series of novels by Joseph Altsheler, and I read all the books on the Civil War I could find at the Tompkins Square and Ottendorfer Public Libraries. A gloom however, diffused from the Books of Samuel and Kings that we read at the Hebrew school, and I never liked their endless scolding of the monarchs that included Solomon himself; I did enjoy the legends of the Talmud and its pithy sayings, but found the legalistic wranglings boring. Apart from a few passages, the prophets Isaiah and Amos seemed to me monotonous killjoys, always pronouncing their forecasts of ruination. Above all, I disliked the aridity of the long synagogal services, the mumbling races in quick reading, and managed to avoid them. Finally, shortly after we moved to the East Bronx in 1925, I refused, despite my father's anger, to continue at the Hebrew High School.

My father, a garment workingman and a trade unionist, had read and marvelled at Edward Bellamy's *Looking Backward* when he was young, and was briefly around 1910 a member of the Socialist Party; having cast his vote for the imprisoned Eugene V. Debs as President of the United States, he transmitted to us in a vague way the socialistic component in the East Side culture. In one respect, my father was unusual; he spoke English rather well, and with a slight Canadian accent, even using expressions like 'constable' and 'pon my honor'. He had attended public school in Toronto, Canada, for several years, because his father, a rigid orthodox Jew, had when he migrated from a hamlet in Austrian Poland gone directly to Canada around 1893. My father, a young boy, followed with the rest of the family seven years later; thus his first glimpse of the New World was at the port of Saint John, New Brunswick, from the deck of the ship *Champlain*. Curiously, in later years, I taught at the University of Vermont beside Lake Champlain. My father quarrelled bitterly with my grandfather one day because he neglected wearing a hat indoors; he ran away from home to New York City, where he later married my mother, whom he had met at a social event of their village mutual aid society.

My father was a taciturn man; I cannot remember that during all my school years he ever asked me once a question concerning my work, teachers, or friends. He was, of course, weary at the end of the day's work, but he was equally taciturn during the long 'slack' seasons. My mother was very bright and beautiful, but melancholy. She felt keenly the fact that she could neither read nor write. Since her widowed mother in Czortkow had been too poor to allow her to go to school, my mother began helping to support the family from the age of six or seven. In her teens, she emigrated with an aunt to America, and worked in a garment factory to earn the passage-money for her mother and the younger children. Soon after she enrolled me in the first grade, she joined afternoon classes for foreigners meeting at public schools and the center of the National Council of Jewish Women; she enjoyed such classes for several years and their social activities as well; a school excursion to Washington, D.C., and the White House brought her much pleasure; she learned to write in a rather large script, and would send me letters at Divinity Hall, Harvard University. My fellow residents at the morning mail table would be puzzled by the unusual child-like orthography. I should add that my kindergarten teachers at P.S. 36 after a year thought that I was somewhat backward, and since I was not quite six, that I should remain another term before commencing the first grade. My mother intervened, and somehow conveyed to the teachers that I was a big boy, and that it was a shame to keep me among the little ones. Her eloquence won the day.

When political campaigns were under way, the Lower East Side seethed with color and excitement; Election Day was a glorious holiday, and we children would blow our toy horns in the streets like a chorus calling the citizens to exercise their franchise. We wore the button pictures of all the candidates with a magnificent impartiality, and walked on the evenings down the Avenue listening at each corner to the platform orators. We listened to the Republicans and Democrats debating about immigration, the bus lines, and Teapot Dome, and we laughed when the socialist stumpers became picturesquely vituperative. Once, however, it must have been in 1922 or 1924, at the 4th Street corner, I heard a fine looking socialist speaker referring courteously to his opponent for the governorship of New York; the speaker was

Norman Thomas. We were all however partisans of Alfred E. Smith when he tried hard in 1924 to secure the Democratic nomination for the presidency. My brother and I walked to the Madison Square Park, next to the Garden where the endless roll-calls of the convention were being taken. I imbibed the editorials of *The New York World* that Walter Lippmann was writing. At home we listened to our father flaring into argument with Republican canvassers, and with a heat that surprised me assailing the silent Calvin Coolidge: "Roosevelt wouldn't have let the coal strike last so long."

Every year or so, my father's local in the International Ladies Garment Workers Union would call or be called to strike. An atmosphere of some anxiety touched the household at those times, and my mother would say gravely that we would have to watch our spending because the strike might last a long time. Once there was a stormy scene in the house when it transpired that a ne'er-do-well relative, a fur worker, had acted as 'scab', a strike-breaker; it seemed about the lowest act a man could perform, and his name was not mentioned with respect.

In later years, when I would hear Harvard Marxists talking airily and altruistically of strikes and militancy, I would recall vaguely my mother worrying about the food, and my father, a periodic member of the local's executive board, returning home glumly after having acted as a roving picket during the night, searching the windows in the high buildings to check whether lights from the shops betrayed that 'scabs' might be working. Class struggle was sombre, joyless, even though we children, with irrepressible cheer, came to poke fun at the ritual of the annual strike.

To be a Jew, and to be an American, really signified different things. To be a Jew was to hear stories at family gatherings from the age of five onwards of pogroms carried out by Russian soldiers and Cossack horsemen invading my mother's village, killing my grand-uncle and defiling his daughters; one learned that to be a Jew was to live by a higher code of manhood; if at the edge of the 7th Street Park, the border of the Slavic neighborhood, I saw miserable drunkards thrown out bodily upon the sidewalk by a saloon-keeper, I knew that a Jew would not deport himself in ways thus unworthy of men. To be a Jew was to know that life was beleaguered with uncertainties, and that to retain one's place among a mankind whose grasp upon civilization was precarious would require all one's forces and stability in judgment. When the Prohibition Amendment was enacted, I was among its child stalwarts, and I have adhered to the legislation permanently despite its repeal.

To be an American, however, signified a world of hope, where one was judged for what one was; a land of security with parks and forests where one could roam and fear no one, of schools and summer camps and the Boy's Club where we played in game rooms and attended weekly movies, of libraries of wonderful books where you could sit into the late afternoon and evening and where the librarian carefully wrote your name on a card and called you 'Master', of vacation time when all the P.S. 64 boys were taken to the Polo Grounds and we cheered Babe Ruth. America, where men played games, and were friends even when they lost; a land which asked people not to hate each other.

How did books first come to our bookless home? There was indeed only one book in our household to begin with – an old tattered prayer-book, not even a Bible; then the persuasive ways of American society intervened, and wrought their transformation. In the garment factories, the workingmen would enlighten each other that there were collections of books called 'encyclopaedias' that told all the knowledge in the world. Some had heard that with such books your children could answer all the questions that their teachers gave for homework. The motive of emulation was happily strong; soon many workingmen with sons decided that an encyclopaedia might be as necessary as a Singer sewing machine. Thus, in due course, sets of *The Book of Knowledge* and *The Outline of Science* were delivered to our house, to the utter confusion of my mother who didn't know where to put them. I was entranced by the illustrations, memorized poems, and for several Saturday afternoons tried to master a chapter on the theory of relativity.

Meanwhile, I had become much interested in what later would be called 'the sociology of the intellectual class'. It all began, I think, when my teacher, Mr Rosenzweig, in his last reading from *The Story of Mankind*, told of the Marquis de Condorcet, that noblest of characters, a writer for the encyclopaedia, devoted to the cause of the poor, and a leader of the Revolution; giving all and receiving nothing, he had nonetheless refused to despair, and affirmed, though as he was to die, the philosophy of progress. I began to wonder why the people who understand such things weren't the rulers of society. A year or so afterwards, I would walk on my weekend evenings across the East Side to the Cooper Union, where all sorts of unusual lecturers held forth under the auspices of the People's Institute. The audience was composed, apart from some of the Bowery's denizens, mainly of middle-aged working men and culture-seeking clerks who had never had the chance to become educated, but who read and argued earnestly. One evening in 1923, I heard a woman lecturer depart from her announced subject in order to speak of the newly elected first Labor Government of Britain, and its 'intellectuals'. The speaker and questioners talked of the 'laws of economics', and the mysterious 'laws of supply and demand'. Again I wondered why the people who knew all these laws weren't ruling society scientifically. They all seemed such noble-minded persons. One evening I heard a Hindu lecturer, Dhan Gopal Mukerji, whose children's books I had read. I was awed by actually listening to and seeing an author in the flesh, a man who had written books. He lectured upon the life and teachings of Mahatma Gandhi. When he was done, he was most discourteously heckled, yet he responded with such gentle courtesy that he won the applause of the audience. Of such I thought are the intellectuals. Even the textbooks of history that we read at the De Witt Clinton High School, written by James Harvey Robinson and Charles A. Beard, made the intellectual class of Western Europe the bearers of mankind's hopes, the true heroes of history. Shortly after we moved to the Bronx, the agitation to save the lives of the two anarchists, Sacco and Vanzetti, from execution in Boston, reached a desperate crescendo in the appeals of the intellectuals. The editorials in *The New York World* declared their last hour's appeals to reason; the columnist Heywood Broun, a Harvard alumnus, indicted his alma mater as 'Hangman's House'. And on the eve of the fatal day, on Tremont Avenue, Norman Thomas, standing before a huge crowd, called for an avalanche of telegrams to the Governor of Massachusetts to stay the execution. His voice, reverberating in the summer's air, seemed to subdue the clangs and rumbles of the street-cars: "Better to be a Jesus than a Pontius Pilate; better to be a Bartolomeo Vanzetti than a Judge Webster Thayer; better to be a shoemaker Nicola Sacco than a Harvard President A. Lawrence Lowell, . . ."

My interest in Spinoza was first stirred in early adolescence when I read articles in Jewish magazines describing him as a great heretic who succeeded in combining science with religion. In the high school lunch room I argued much with a militant atheist whose unpleasant personality seemed to me not unrelated to his atheism. One of our teachers was cited as commending Spinoza for his rigorous mathematical method. In the worker's enclave in Crotona Park, where on summer nights they argued nightly upon the existence of God, and of Left versus Right, Spinoza was also cited. Finally I went to the Public Library, and borrowed a newly published edition of Spinoza's *Ethics*. I worked hard for several days on its First Book, and then told my father sadly that I found Spinoza's proof for God's existence unconvincing, that Spinoza was really trying to demonstrate that 'if God exists, then he exists necessarily', but that he hadn't proved that God does actually exist. My father shook his head with a slight smile; it was the only philosophic discussion we ever had. Years later I discovered that though he had sent all his sons to Hebrew school, he was an atheist himself, and adhered to the rules for Kosher food only at home. My mother berated him when she discovered he had partaken of a ham sandwich for lunch: "Your son who went to Harvard wouldn't do a thing like that."

My first political publication took place too in high school when an essay of mine on 'The Matteotti Trial' won a third prize in *The New York World*'s 'Biggest News of the Week' contest. It dealt with the failure of Mussolini's fascist regime to punish the murderers of the brave socialist deputy Giuseppe Matteotti. I received a check that week for $7.50, and felt very rich for a while.

My teachers at the De Witt Clinton High School were mostly about as fine a group, I think, as could have been assembled. My senior class French teacher, Edouard de Champlon, lectured to us in French about Molière and Stendhal, and the Battle of the Marne in which he had fought in 1914, and been wounded. He was the first European intellectual I knew, and he lived in the world of ideas. Sylvia Levinson in history was a disciple of James Harvey Robinson, and Samuel D. Moskowitz was a loyal adherent to Charles Beard's wisdom on the economic basis of American political conflicts. Rupert H. Hopkins introduced us logically to the formulae of electricity, and even talked of Einstein. We memorized Burns's love poems in Larry Naumann's class, but he, a true romantic, later committed suicide. And Daniel G. Krane, once the metropolitan wrestling champion and a veteran of the Argonne Forest, read *Hamlet* with us, and secured for us tickets (at reduced rates) for the productions of the Theatre Guild; from the back seats of the second balcony, I listened to Werfel's *Juarez and Maximilian*, and became a devotee of drama. Later I secured a job as assistant on Saturday afternoons in the hat-checking room of the Theatre Guild; my wage consisted in being able to see the play free of charge. Very soon I was an admirer of Bernard Shaw whose plays were full of fun, though at that time I was puzzled as to what he was trying to say.

I became a member of the Library Squad at high school; it was responsible for the shelving and circulation of books in the school's large library and in my last term I was elected captain of the squad. The motion was made by the retiring captain, Joseph P. Lash, later most well known for his books on Eleanor Roosevelt, and seconded by Benjamin N. Nelson, later a well-known scholar of Max Weber's sociological ideas. One Friday afternoon, as the week ended, I read absorbedly Jack London's novel *Martin Eden*.

I also worked while I was at high school at whatever jobs I found. One summer I was the delivery boy for a fine bookstore near Wall Street, *Ball and Wilde*, and had the chance to glimpse such offices as that of the editor of the *The Wall Street Journal*. For a while my Sundays were spent till midnight working on the subway station of the Hudson Tubes at Grove Street as the clerk of the news-stand. Thirteen years old, very naïve, I was rendered surprised and speechless when a woman asked to buy a package of cigarettes. I had never before encountered a woman who smoked.

At the City College, my professor in intellectual history and the history of socialism, J. Salwyn Schapiro, had already written an article entitled 'The Revolutionary Intellectual' that coped with the problem of their motivations and aims. One day in class, he asked in his simple, unassuming way: "Why do they do it?" The question continued to haunt me. It was in his classes that I read Peter Kropotkin's *Memoirs of a Revolutionist* (I already owned the *Revolutionary Pamphlets*) and G. D. H. Cole's *Short History of the British Working Class Movement*, two books that I found utterly absorbing and that seemed to foreshadow the creative alliance between the working class and the intellectuals. (Years later I got to know Kropotkin's ageing daughter, the still beautiful Princess Alexandra; enduring her poverty stoically, she never seemed to take her father's political ideas seriously.) It was in Schapiro's course on the nineteenth century, furthermore, that I read John Stuart Mill's *Autobiography* and *The Subjection of Women*; the New York Public Library at 42nd Street gave us the reading room where my friends and I interspersed our studies with shy conversation with the extremely intelligent and high-minded girls from Hunter College. Reading with Schapiro I acquired an abiding interest in John Stuart Mill who remains for me to this day the model of the dispassionate and sincerely scientific social thinker. In later years, at Toronto, I devoted much time to Mill in the course I gave every year on the history of sociological theory, and the essay I read in 1973 at the Mill Centenary was, I think, the best I ever wrote in the domain of intellectual history. When I was an undergraduate, I put much emphasis during an oral report on Mill's socialist avowal, and Schapiro gently criticized me for making too much of it; years later I learned from Mill's last writings of his basic criticism of the socialist program as well as of the revolutionary socialist character.

The first historian of science I knew was Benjamin Harrow, my professor of chemistry. I read all his books. Once he was much interested in an alternative theory of the ionization of amphoteric hydroxides that I suggested, and was disappointed that I did not continue in chemistry.

At college, fortunately, I was appointed by Professor Charles Alfred Downer to be the student librarian of the French Department. Apart from the pleasant work and wages, the French Library became the meeting-place for a whole circle of intellectuals. My superior, a Frenchman, Gaston Laffargue, was a kindly, realistic veteran of the colonial war in 1885 in Indo-China. Never did I dream that I too would find myself a soldier among Indo-Chinese. Both men died during my senior year, and I missed their presence greatly.

My political involvement too became more active at college. In the fall of 1930, I was elected president of the Student Forum, the liberal socialist student club, and during the electoral campaign invited Heywood Broun and Morris Hillquit to address us. Broun, a famed humorist, was a socialist candidate for Congress, and spoke whimsically that he was an old man trying at last to do the right thing. Hillquit in 1930 was the first man to address students on the menace of the Nazis, what they signified, and their portent for the Jews. I remember how he cited August Bebel's famous remark that 'anti-Semitism is the socialism of the damned fool'. Few people were then taking the Nazis seriously, but Hillquit, America's greatest figure in the Second International, had glimpsed something of the horror that was to deploy its shadow upon many lives during the coming decade.

During my senior year, several friends and I revived a long-dormant literary society *Phrenocosmia*. It fulfilled an important need, for there were no seminars at the college in which students could read and discuss papers of their own. We met at each other's homes chiefly, and our papers were youthful, and untempered in aspiration. Our titles were large-intentioned; 'Essence, Existence, and Deity' was the title of one I read, obviously affected by Samuel Alexander's big book. We were joined at our meetings by several young women intellectuals who were denominated the 'Hunter Epistemological Society'. We solved the problem of political ambition in an unusual fashion; every member was authorized to call himself President of Phrenocosmia. At the same time, I took the initiative in organizing a political reform movement for our senior class. As in the Tammany-Mayor Walker politics of the time, there was a good deal of corruption in City College student politics. Not only were ballot boxes stuffed in student elections, but a considerable 'graft' prevailed in such departments as the cap and gown committee that negotiated the contract for the renting of that ceremonial paraphernalia. Although by present standards it was a small total sum, it meant a lot more during the onset of the depression. We organized a 'Left Wing' to contest the student elections; I persuaded Abe H. Raskin to be our presidential candidate, while I ran for secretary. We won a majority, and at our first class meeting we proceeded to enact our program. A hold-over, however, of the previous corrupt regime was already engaged in negotiating an unauthorized corrupt contract. When I explored the matter, he introduced a motion declaring I was no gentleman; I abstained from voting, but the Class Council by a vote of 3 to 2 defeated the motion. As the headline stated the next day, I was pronounced a gentleman by one vote. Our class rogue then resigned, dropping his one credit class at the college. I always found him a friendly, human person, and though a rogue in financial dealings, he was devoid of the sadistic fanaticism of the class ideologists in the Social Problems Club who became Communist Party functionaries. Forty years later our class rogue visited me at Toronto, and we talked of the current irrationalities. I was saddened later to learn at a class reunion that he had been disbarred, and had died broken-hearted.

My last month of college was marred when I was suspended from classes by a vote of the faculty that was enraged by a letter I had written that criticized their standpoint as 'authoritarian'. It all started with an incident that occurred during my term of office as the chosen secretary of the Faculty-Student Discipline Committee. Several students of the communistic Social Problems Club who had distributed a leaflet were punished unilaterally by the college president or dean, without the latter's having submitted the case through our regular channels of procedure. When I joined in pressing a criticism of their action, though finally alone in so doing, my suspension was voted. Two weeks later, the suspension was, however, lifted by a vote of 43–38. Some friends, especially Abraham H. Raskin, later an editor of *The New York Times*, were especially active on my behalf. When I heard that the case was going to reach the newspapers, I was obliged to forewarn my parents; that was the hardest thing of all.

At graduation, nonetheless, I was awarded the Cromwell Medal in History; I had previously been the recipient of the Ward Medal in Logic. I had always expected to go on to graduate studies in history, but I was enthralled with the problems that Morris R. Cohen raised in his fall course on 'The Philosophy of Civilization', that was actually the nearest thing that the college had to a course on sociological theory. To some extent also I felt the desire to find in philosophy an equivalent, intellectually satisfactory, for the vulnerable religious standpoint. A graduate scholarship in philosophy was available at Harvard. I was drawn to the idea of leaving New York City, and living at a university. In view of the fact that I later came to the conclusion that philosophical ideas are best conceived and cultivated as an outgrowth of a concern with basic questions in some empirical science, this decision of mine was mistaken. As a student of philosophy, moreover, I missed that coping with factual problems and realities that I so much enjoyed in historical and political research. I was also to find that a Ph.D. in philosophy was singularly out of place during the depression years.

Transplanted to the philosophers' classes at Harvard brought a puzzling experience. I gradually discovered that each one thought that the others were teaching either trivia or nonsense, but there seemed to be an accord not to say so too publicly. Whitehead was expounding his panpsychist world-scheme, and occasionally indicated by his tepid interest that C. I. Lewis was repeating a tedious solipsism. In Lewis's seminar, a hapless Whiteheadian once gave a paper using Whitehead's notion of feeling as his basic category; to my surprise, Lewis lost his composure, and called it 'shamanism' and *schwärmerei*; at the next meeting, he apologized warmly, but I thought that enquiring into this deep divergence of opinion might have been more helpful. Ralph Barton Perry was teaching his 'theory of value' energetically, but Whitehead told our seminar one evening that as an undergraduate he had been compelled to read Sidgwick's *Methods of Ethics*, and found it dull beyond words. He thought that 'orthodox' American idealists such as Hocking were doing poorly what F. H. Bradley had long before done better, though he was amiably proud that he himself had never read more than a page of Hegel, and a 'handbook' or two. D. W. Prall told me some years later that he regarded philosophy as 'rubbish', though he somehow managed to give courses on Leibniz and Spinoza that were enthusiastic. H. M. Sheffer was acting the role of the genius who had a great secret masterpiece in the making; meanwhile, he perpetually warned the class against plagiarizing his results, though no one ever knew what these were; he also insisted that logic had nothing to do with philosophy, which was disenchanting because *Principia Mathematica* was revered as our Tables of Truth from the Cambridge Heights, and their bearer, Bertrand Russell, had proclaimed that logic was the essence of philosophy. Sheffer's character reminded once again of Balzac's story *The Unknown Masterpiece* that tells of a master painter who is presumably re-working his greatest picture all his life; then when his disciples finally uncover it, they find a canvas painted formlessly white. Harry Austryn Wolfson, who was completing in those years his book on Spinoza, used to say of his friend Sheffer: "There but for the grace of God go I." He felt that Sheffer had succumbed to the strains that were then the lot of a Jew seeking to enter the American university world; the salaries of both men were for a number of years met by grants from Jewish donors, and Wolfson regarded himself as a kind of symbolic representative of Jewish thought in a Gentile setting. He told me after several years that he could not understand why anyone should study philosophy in our time, that it was superseded by science; he himself had become a professor of Jewish philosophy and history, he said, because he had been so poorly educated in his childhood and youth, when he had perforce studied only Jewish learning; his profession, he said, was the consequence of his bad education.

Whitehead's speculative construction seemed to me arbitrary and somewhat artificial. He invoked opposed principles of continuity and emergence at his convenience. Though an emergent evolutionist, he did not regard sentient experience as one such emergent; to refute scientific materialism, he thought the material world itself had to be taken as composed of 'occasions of experience', each grasping ('prehending') its items selectively for its subjective formation. It seemed to me the argument for this panpsychist extrapolation was weak. Furthermore, his projecting a Victorian liberalism on the universe as a whole with the cosmos seeking new epochs of creative adventure, seemed to me a provincio-morphism, which won people's attention because Whitehead was the author of more than half the *Principia Mathematica*. The *argumentum ad Principia Mathematica* was Harvard's characteristic and unspoken fallacy. Whitehead seemed to set aside all the numerous evidences of nature's lines of regressive evolution. C. I. Lewis, on the contrary, had difficulty even in holding on to a real, external world; he strongly denied that his 'experience' was self-enclosed, or solipsistic, for he argued, "I do not will my own experience." His assumption seemed to be that since he did not will the contents of his experience, they obviously had an external source. But this causal principle, 'whatever I do not will in my experience has a causal source external to me' itself could never be validated in Lewis's enclosed selfhood. Thus Lewis's sense of reality struggled with the constraints of his own formal philosophy.

When the first graduate students who had been pupils of Wittgenstein began arriving in 1931–1933, the drift to logical positivism became strong. Lovejoy, as a visiting professor, staunchly and acutely analyzed its ambiguities, and its alternating oscillation between solipsism and behaviorism. With every philosopher, however, regarding most of his colleagues as purveyors of nonsense, one can understand why the doctrine that all philosophy was nonsense, was soon going to spread among graduate students. Lovejoy tried to engage in salutary interchanges but was singularly unsuccessful in this effort; he was usually a solitary faculty personage at meetings of the Harvard Philosophical Club. Moreover, it seemed a paradox (unnamed) that young graduate students after militantly and stridently declaring all metaphysics, epistemology, and ethics to be nonsense, still averred their intention to take doctorates in what was presumably nonsense, and to plan to spend the rest of their lives in the rather arid pursuit of convincing students that the course they had elected was devoted to the types of nonsense that had been written by persons who called themselves philosophers. Also a strange aura of nonsense and irrationality hovered about the logical positivists themselves. Lovejoy struggled hard to demonstrate that they did not believe what they were saying when they asserted they were 'methodological solipsists'; none of his solid grounds, however, could ever possibly constitute a crucial experiment for the willfully fixated solipsist who swallowed every consequence. More and more I came to the conclusion that genetic analysis, even psychoanalytical considerations, could alone penetrate and disperse the ultimately emotional blockages that gave rise to irrationalities. But then, how to show that such genetic considerations were not always excluded by the alleged 'genetic fallacy'?

My study of Freud's writings began during my third year as a graduate student when I had many discussions with a fellow graduate student,

Martha Wolfenstein, later professor of clinical psychiatry at the Einstein Medical College. Prior to meeting her I had tended like almost every City College student who had taken Morris Cohen's courses to follow their teacher's example, and to dismiss Freud without having read him. Martha, however, knew Freud's writings thoroughly, and had a wonderful knack for applying his ideas suggestively to philosophical doctrines. She read philosophers less with an interest in whether what they were saying was true or false but rather for the causation of their ideas in the structures of their unconscious. Martha was the only graduate student I met who had an interest in psychoanalytical ideas, and she stood isolated from the Wittgensteinian and behaviorist currents that had affected almost all the able students. Wittgenstein's critique of philosophy seemed to me a linguistic echo of what Freud had done much more powerfully on a psychological level while remaining free from the grandiose solipsist-sounding pronouncements that made the *Tractatus Logico-Philosophicus* read at times like a *Tractatus Logico-Neuroticus*. For all his logical apparatus, and his Austrian background, I thought Wittgenstein exhibited the traits that Santayana had described in *Egotism in German Philosophy*, and I agreed when I heard Whitehead several times speak of the appeal that Wittgenstein was exerting upon 'impressionable young people'.

Lovejoy was the only philosopher I met at Harvard who imparted a sense of wisdom, and I used to enjoy greatly the discussions that would take place at his rooms at Adams House. His arguments failed to make any dent upon the young positivists, because a blend of the solipsistic and behavioristic standpoints was precisely what was emotively congenial to them; in short, non-logical emotions rather than logical grounds were the basic determinants of philosophies. The final arguments for philosophies rested on the appeal to their emotional mainsprings. If the adherent to solipsism clung to it unyieldingly, then no matter how much its consequences were inconsistent with our assumptions in daily life, its fixated scheme remained impregnable. Even in the New York Philosophical Circle, of which I became a member in 1935, any person who accepted the reality of the external world was regarded by its positivist members as close to being philosophically non-sane.

My sense of reality was more satisfied in the year's course I took in 1932–1933 in the history of science with L. J. Henderson and George Sarton. The chairman of the Department of Philosophy interposed some objections to my taking this course, and misgivings were expressed about my electing the history of science as one of my two minor fields for my doctoral examination. Henderson, who had recently been converted to the sociological standpoint of Vilfredo Pareto, approached the history of science in that spirit; once a close co-worker of the idealist master, Josiah Royce, Henderson was now militantly anti-philosophical, and his lectures were filled with such terms as 'conceptual scheme', 'first approximation', and above all, 'equilibrium'. Sarton's lectures were the best I heard at Harvard – filled with facts yet sensitive to the human character of the scientists; his own philosophy was that of a staunch Comtist, and he would actually blush as he would affirm his faith in progress. He was the only professor I ever had who actually one day referred to Marxism, and even affirmed its importance for scholarly study; that was in 1933 when men's spirits were diminishing, jobs vanishing, and when the first ominous signs appeared that the very existence of Jews would be threatened, Sarton told us one day of 'the strong delegation' that the Soviet Union had sent in 1931 to the International Congress of the History of Science and Technology in London, and how they argued for the Marxist, economic interpretation of the history of science. Although much impressed by their papers, he nonetheless had remained unconvinced of their basic hypothesis. The problem stirred me; I managed to get hold of a volume of the papers, but found myself lacking in the historical and scientific backgrounds for evaluating a theory which, if true, had such profound significance for the philosophy of science.

My doctoral thesis attempted to show that an absolutist conception of space-time was still underlying to our scientific, operational notions; its opening chapter won a Bowdoin Prize and Medal, but I soon decided there were basic defects to my approach, and I made no effort to publish any parts of it in journals. Some of its chapters, I think, had value; in one, I attempted to show that the calculus of propositions presupposed a world of spatio-temporal regions to which contents (predicates) were assigned; another chapter tried to show the flaw in such arguments as McTaggart's against the reality of time. Meanwhile too, my first article in philosophy was published in *Mind* in 1934, a note in answer to a challenge by F. C. S. Schiller with respect to the meaningfulness of the concept 'universe'. Later, I became uneasy with this sort of 'philosophizing'; this kind of writing about words was not difficult to do, but gave one no feeling of genuine knowledge or work. During my next two years at Harvard I was an assistant in the metaphysics course, but most of my time I was auditing courses in physics, economics, anthropology, and especially Henry A. Murray, Jr's in abnormal psychology. Also much of my time was devoted to helping organize what was named the Cambridge Union of University Teachers.

In the summer of 1935 I had attended the convention of the American Federation of Teachers at Cleveland, and discussed with its officers the founding of a 'local' at Harvard. There I also met two brothers, both economists, who were beginning their careers at Harvard, and I raised with them the possibility of an organization there. Returned that fall to Harvard, I spoke to all the assistants I knew, while the economists' group burgeoned at the Harvard houses to include a number of outstanding members of the departments of history and literature. Thus, the Cambridge Union of University Teachers was founded; I became its secretary-treasurer, and a delegate as well to the Boston Central Labor Union (A.F. of L.) meeting every two weeks at the Catholic Trade Union Hall. There we voted, after the debates on such questions as whether to endorse the corrupt Governor Curley or not, or whether to found a Labor Party; sometimes angry words were exchanged and once even some physical threats were made by the president of the Bartenders' Union. My own ideas of the American labor movement had been largely based on what I'd seen of my father's fellow-workers in their self-organized credit union, which my father served as secretary for many years. Consequently, there was a kind of 'Jewish fallacy' in my observations and reflections, that is, I tended to think that the idealism and cultural aspiration of the Jewish working class were typical of all sections of the working class. Actually I was wrong, for the Jewish working class in America lasted little more than a generation; this was a working class with a bourgeois teleology; their consciousness always transcended their existence, even when it spoke in the accents of Marxist class struggle.

Several of the men I met through the Cambridge Union of University Teachers became life-long friends; I always valued much the association with its Recording Secretary, Ernest J. Simmons, a scholar of Slavic literature, and always generous and loyal, in Royce's sense. On the whole, however, the impressions I had of the Harvard trade unionists gradually confirmed the generalizations concerning intellectuals that I found expressed, for instance, among such writers as Max Nomad and Selig Perlman; socialism and the labor movement provided the rationale and the occasion for an emerging university intellectual class that was ambitious for political power. That aim circumscribed their conception as to what activities were appropriate for their union. It thus never crossed their heads that the anti-Semitism that pervaded the Harvard system was an issue worth raising. One episode illumined this reluctance vividly. I had been informed by William Ernest Hocking, the chairman of Harvard's Philosophy Department, that he would be able to recommend me for a post only in a large state university where the ratio of 6 Gentiles to 1 Jew had not been exceeded. When the Union's Executive Council (of which I was a member) met, I brought the incident to their attention. I expected some stand would be taken, however minimal, to defend the rights of its members. To my surprise, there was none whatsoever. Instead, they engaged in much handwringing, and made allusions to the Jewish quota they themselves helped to administer in their Houses, and told of extraordinary graduate students in economics, history, and political science who could nevertheless not hope for appointments at Harvard because they were Jews. The notion in 1937 of speaking as a Union against anti-Semitic practices simply did not occur to the Union's Executive Council. They were obviously (and in all conscious sincerity) concerned with cooperating with Boston labor leaders, with raising money for various committees ranging from that on behalf of the Spanish republic to one for the Vermont quarry workers; they sought in spirit to merge themselves with the cause of the working class, which they hoped to serve as future economic planners and political guides. I shared their hope that a labor-intellectual alliance might bode a better future, but a certain doubt as to the depth of the idealism of the intellectual class, and their professed mission, also began to grow.

The mid-thirties at Harvard were an age of discussions of Marx, Weber, and Keynes; all sorts of discussion groups flourished. From the outset, I felt a certain unreality about Weber's arguments concerning the influence of the Calvinist ethic of asceticism and guilt on the rise of capitalism, and I had similar doubts when this notion was extended to a causal thesis concerning the rise of modern science. Lecky and Mill had both linked the philosophy of industrial and scientific advance to a hedonistic utilitarianism, while Morris Cohen, like Ernest Renan before him, had observed how the centers of trade such as Venice also tended to become the centers of science. When L. J. Henderson queried me on my Ph.D. examination as to the sociological factors involved in the rise of science, I answered along these lines, but it was not for some years that I was able to develop the theme. At Harvard I also had long discussions with a young economist who later published an important theorem in welfare economics. I outlined that a kind of social contract axiom was required to specify the degree of inequality that a lower class would accept in exchange for the heightened productivity and standard of living associated with the increased rewards to a functioning upper class. This was consonant with a principle that Engels had already advanced as a historical justification for class systems in the past. I was much intrigued, however, by the notion that human jealousy might be so strong that some lower classes might prefer an egalitarian system even if it meant a lower standard of living, and at the meeting of the Unity of Science Congress in 1941, I discussed such ideas, which seemed to explain the perennial appeal of such ideas of absolute equality as those that Bernard Shaw advanced. But in the latter thirties, the basic problem was still the basis of value-judgments generally: were they founded on generic intuitions, or could scientific method in some way resolve basic ethical differences, or were they ultimately altogether emotive and non-rational?

In 1939, I was appointed an instructor in the Department of Philosophy at the City College principally through the effort of the retiring chairman, Harry Allen Overstreet; it was his last official act, and I remember how his face was pale and drawn after the Faculty Committee meeting at which he had encountered some angry opposition. Among the courses I was assigned was one called 'Philosophy of Civilization' that Morris R. Cohen had given. I used it to discuss the theoretical ideas of Marx, Freud, Veblen, Pareto, and Spengler, and thus came to write during the next three years several articles on historical materialism that were published in the Marxist magazine *Science and Society*. My aim was to state historical materialism in a rigorous scientific fashion so as to divest it of all association with metaphysics, Hegelian or otherwise, and to present its propositions so that they could be verified or contravened. The usual critiques by Dewey and Russell such as that historical materialism was a 'monistic' theory, seemed to me about as significant as philosophical arguments against Einstein's attempt to establish a unitary field theory. If Marx's historical and economic theories were to be falsified, it would have to be on straightforward empirical grounds. Thus, likewise, the 'dialectical' approach to economic systems was independent of any metaphysical theory of being; it was rather a first effort at stating that there were systems in which unstable equilibria, far from being resolvable within the given framework, evolved as a consequence of their internal workings in a cumulative disequilibrium; the latter was then resolved only with the emergence of novel forms of equilibrium, with novel systemic states, or conditions. I regarded this 'disequilibrium' analysis as defining a 'form of law' (in Wittgenstein's usage) comparable to such forms as those of least action, or conservation laws; this was what later was called in Oxonian Latin school fashion a new 'paradigm'; there was no metaphysical, or *a priori* mandate, that all systems should take such a disequilibrium form; that was a question again that could only be decided on the basis of empirical findings, predictions, and verifications.

Both these papers evoked strong criticisms from rigid communist theoreticians in whom such independent analysis elicited all sorts of anxieties; these objections reached the proportions of a barrage when an essay of mine entitled 'Ethical Theories and Historical Materialism' was published shortly after I entered the army in August, 1942. A squad of doctrinaires then advanced notions deemed more acceptable to official Marxism than my own views that drew on Mill, Locke, and Freud. At this juncture, however, I must say a word as to how these views had developed.

As I reflected upon the inarticulate, non-logical premises that constituted the psychological bases of philosophies, it struck me one day that G. E. Moore's famous argument was, in effect, a resistance mechanism to the psychological analysis of ethical concepts. I had often read with disbelief Moore's argument that any proposed naturalistic equivalent of 'good' such as 'good is pleasure' would fail because otherwise the latter would mean the same as 'pleasure is pleasure', which is obviously not the case. It appeared to me that the curious emotional associations attached to the word 'good' derived from all those childhood experiences that gave rise in our unconscious to what Freud called the 'super-ego'. Hence, when we act in ways that contravene its admonitions as to what is 'good', we have experiences of guilt. As long as these unconscious mechanisms are ignited by the word 'good', and as long as these mechanisms are not rendered completely conscious and their emotive overtones dispelled, any proposed naturalistic definition of 'good' will be felt to be unsatisfactory precisely because the unconscious emotive overtones evoked by 'good' will be lacking. When the origins of the latter overtones are explored and made conscious, hedonistic-utilitarian equivalents to ethical terms emerged as their irreducible central meanings. The doctrine, moreover, that good is 'indefinable', by providing a philosophical resistance to psychological analysis, also served to underwrite the self-regarding superiority of the moral, social, and aesthetic judgments of Englishmen in a Cantabrigian cultural circle. Since all the overtones that led people to believe 'good' was indefinable, non-natural, and intuited, could be explained through psychological causation, they ceased to provide a philosophical ground for arguing that such an intuition existed.

When I presented a paper with these ideas in 1941 to the Second International Congress of the Unity of Science at Chicago, it received a warm reception from my discussant, Charles Leslie Stevenson, and from Herbert Feigl and Gustav Bergmann. But when I submitted the paper for publication to *The Journal of Philosophy*, its senior editor informed me that, in his judgment, any article using psychoanalytical categories would so much damage my prospects for an academic career that he could not allow its publication. A junior colleague of his, a philosopher of science, concurred, though his judgment seemed more founded on personal anxieties that were evoked when the word 'unconscious' was used. Thereupon, I added to my essay several pages that related the diverse psychological functionings of ethical notions to corresponding diversities in the socio-economic systems, and submitted it to *Science and Society*, whose editorial board I had recently joined. When the political criticism of my paper became intense, the editors planned to disavow my essay by having the concluding response written by a spokesman enjoying political approval. Whereupon I telephoned from my army post that unless I received the customary courtesy of the author's last response, I would resign from the editorial board. The editors then withdrew their plan, but instead made all sorts of deletions in my manuscript that I was not at hand to prevent. Years later I discovered that the British political scientist, Harold J. Laski, had in his last, posthumous book the *The Dilemma of Our Times* warmly praised my essay as a 'brilliant article', and used it to found his critique of a dominant strain in British philosophy; and indeed, G. E. Moore queried a house guest to learn some personal details of the essay's author. Meanwhile, however, the war had cast me upon the shores of the South Pacific island of New Caledonia, where experiences posed problems I had not encountered before in their immediacy.

Toward the end of 1943, a battered old troop-carrier took its swollen cargo of soldiers through the Panama Canal to the island of New

Caledonia in the South Pacific. The island was known to me through the annals of French revolutionary movements, for in 1871, when the Paris Commune was crushed and many thousands slain in their wildly passionate and irrational uprising, the surviving Communards were mostly sent into its semi-tropical, lonely exile. Now, however, New Caledonia was providing the primary base for American campaigns in the South and Southwestern Pacific; also, the island's nickel and chrome mines were being exploited vigorously for their strategic minerals; several thousand 'coolies' – that is, indentured laborers – recruited from their native Java and Tonkin (Indonesians and Vietnamese we would now call them) to work for five years.

While we were encamped at the Sixth Replacement Depot, I heard one evening that the Personnel Colonel was looking for a man who could speak French, and handle economic data. The next day I sought out the colonel, and volunteered for the post. Although I was classified as a radio and radar mechanic, I was sceptical about my technological skills. The army officers at the Nickel Production Office, Lend-Lease, and Liaison Office of the Foreign Economic Administration at the South Pacific Base Command, after talking with me, asked that I be transferred to their staff. Soon I was studying at my suggestion, the conditions of the Asian coolie laborers, and enquiring as to why there was so much labor unrest, and why production in the mines was falling.

The coolie labor system in New Caledonia had indeed, I found, become outrightly one of forced labor. During the war years, the contracts of indenture had expired; the French Administration had, however, 'unilaterally' renewed the terms of contract, thereby in effect, making industrial serfs out of coolies. In ugly, primitive, unsanitary barracks, mostly in the mountains far from any village, hundreds of Javanese and Tonkinese were crowded together, struggling to preserve a vestige of privacy for their families with thin makeshift partitions. The rations would have provoked a mutiny at sea, and the wages were about five dollars a month. I climbed to the mountain mines and barracks at Thio where near the ruins of an old Communard structure, I found the coolies in their quarters. Children, men, and women clustered around me. "Fini la guerre? Fini la guerre?", they asked animatedly; they were completely cut off from news. One coolie, especially excited by my wrist watch, listened closely to its rhythm, then ran into his barrack, and emerged with an alarm clock. He asked if we could exchange. With regrets, I told him that as I was climbing to the 'plateau', I could not carry a heavier clock. Each day, they told me, they would mount *doucement* to the nickel ore face, and chip away with their pick-axes. Though they assured me they disliked *les japonais*, and spoke of *le président Roosevelt* with reverence, they had no more concern for their 'productivity' then any slaves would have had. Every month some coolies would escape the mines or the smelter, and hide in Noumea, or seek the sanctuary of an American camp. If they were caught, as they usually were, the penalty customarily included a flogging. The Army of the United States was in the unenviable position of having agreed to return runaway coolie laborers to the local French authorities. The mines were managed by men whose sympathies had been with Vichy rather than with De Gaulle's Free French.

In Noumea, I became friends with the two persons on the island who were most closely concerned with the well-being of the coolies. One was Father Zimmermann, a Dutch Catholic priest, who conducted a mission for *les javanais* where they composed their own prayers, and on Sundays gave long, interminable dramatic dances re-enacting, to the gongs and gamelans, various episodes of Indonesian history. Father Zimmermann intervened on behalf of his coolies with the authorities when conditions became near intolerable. The second friend was Madame Jeanne Tunica; married to an exiled Spanish syndicalist, and in her early fifties, she was the island's only communist. Many years later when Bertram D. Wolfe published a book entitled *Strange Communists I Have Known*, I thought Madame Tunica was stranger than any in the collection that Bert had assembled; in fact, I believe she was the strangest communist in the history of Communism. Some time during the mid-thirties she had gotten herself a membership card from the Paris office, but as she was the only communist member in all New Caledonia, her soliloquies were the equivalent of a caucus, politburo, and membership meeting all rolled up in one. Her communist doctrine was a mélange of Jean Jaurès, Anatole France (*le grand maître*), Lénine, and Louise Michel, the anarchist *vierge rouge* of the Paris Commune who had spent eight years in New Caledonia. She was the despair of visiting Australian and American Communists who chided her for not following the party line. "On me dit que je ne suis pas sur la ligne," she said wistfully, but then explained she hadn't had a directive for many years. She had acquired some fortune of her own, because while her husband operated an automobile repair shop, she ran a bar for American servicemen that dispensed drinks (low-grade, I was told). Meanwhile, however, she befriended the Tonkinese coolies, talked with them, instructed them, advised them. She provided me with much information that I used in a report I wrote for the Army Command which concluded with a recommendation for appropriate steps that might be taken for the replacing of the coolie labor system with one of free, mobile workingmen, or alternatively, an army labor corps for the operation of the mines and smelter. I felt that for the United States, this was a fateful first encounter with Indo-Chinese and Indonesian working people, and that on this island, a precedent should be set which would win the confidence of Asian laborers, and establish American social leadership in the Pacific. My report was shelved, stamped with a highly secret classification, and I was assigned to more operational duties in the army's lend-lease relations with the French army and colonists. The French Chief of Economic Affairs, Henri Lemonnier (a socialist and a literary man) and I became good friends, and I saw much of the local French naval commander Pierre Vilbois, a man of heroic deeds, veteran of Dunkirk, and of several secret missions to organize resistance against the Nazis.

At Madame Tunica's house, the Indo-Chinese coolies, with whom I talked occasionally, finally however began to plan a strike, and called themselves communists. Madame Tunica literally had to instruct them not to use violence against any of their fellow-coolies opposed to such a step. I advised the smelter laborers to inform the United States Command that they planned to strike against the coolie labor system, and to ask for American mediation with the French authorities. I emphasized that the mere awareness on the part of the French mining managers that the American Army had its eye on what was going on would diminish their use of corporal punishment against the laborers. The coolies did possess a single Romanized Vietnamese typewriter (its owner was therefore regarded as *un intellectuel*), but they were puzzled as to how to write or address such a letter. Therefore I wrote a draft of one, that appealed, in their name, to the tradition of Abraham Lincoln. Madame Tunica then made a French version though she could not forebear adding some flamboyant phrases and lines that were distinctively hers. On my suggestion, the coolies transmitted their letter through the lawyer, Maître Gervais Bourdinat, grandson of a Paris Communard, who venerated his ancestor's idealism, *plein de volonté*. At the army headquarters I saw the letter arrive, and the general's answer leave, with its statement of regret that the United States Army had no authority to intervene in the internal affairs of New Caledonia.

Events now leaped ahead, as they so often have a way of doing. The

French authorities raided the coolie barracks, arrested their leaders, and confiscated their typewriter. A strike began spontaneously and energetically. Then a trial took place that was a colorful spectacle, as Indo-Chinese women bedecked in all their costumery thronged in the courtroom. Maître Bourdinat cross-examined the smelter managers with a magisterial bearing and power, and then delivered a stirring address on *les Droits de l'Homme*. In later years, I thought that the Vietnamese Revolution began with the Rights of Man which were then obliterated into Soviet nullities. To the joy of the courtroom crowd, the Tonkinese union leaders were acquitted. A few months later, the De Gaulle regime in Paris, learning of what was taking place, pressed the New Caledonian officials to enact the abolition of the coolie labor system. Thereupon the local General Council was persuaded to take this action in patriotic loyalty, with some safeguards against a big influx of free Asians into the town of Noumea. Everywhere the coolies celebrated their freedom; at one of the celebrations, they said they should present a gift to the American sergeant. One of their leaders, Soan, painted a picture of his Vietnamese coastal birthplace, and through Madame Tunica presented it to me. Some American intelligence operatives overheard the coolies talking about the American sergeant who had helped them, and duly reported their information to the colonel of intelligence. The intelligence section had indeed opposed the freeing of the coolies: now their colonel pressed for court-martial proceedings against me on the ground that I had not merely investigated the coolie labor system but had instigated the movement for their liberation. My own Camp Commandant, Arthur J. Solomon, a brave veteran of combat, refused, however, to sign the papers for my court-martial, and replied simply that all my actions had been honorable and in the highest traditions of the American people. Thereupon, the intelligence officer 'telexed' to the Central Pacific Command for authority to supersede the Commandant, and to initiate a court-martial against me. The grapevine among enlisted men informed me that this most serious trouble was in the making. At once I went to the colonel in charge of troop movements, and volunteered to be sent to the front. As he happened himself to be a social scientist, he was kindly disposed toward me, and agreed to my request. A few days later, I left at the Tontouta airport, a solitary soldier with his carbine, bearing on my person the order that wherever I arrived, I was to be sent forward by the next available transportation toward the front. I flew from one island to another, Guadalcanal,

Kwajalein, Saipan, and finally landed in Iwo Jima as hostilities ended. My field notes on New Caledonia and collection of documents had been confiscated before I left. Madame Tunica and I corresponded for several years: her house was bombed and largely destroyed. She returned to Paris, but after a brief taste of Communist Party life in the metropolis, she decided to return to the South Pacific, to the New Hebrides, to spend the last of her days *tranquillement*. A letter from a Javanese coolie reached me when I was in New York after being demobilized and seeking to resume to academic work. Often I wondered whether I might find some vocation in these new frontier regions of social development. The workings of imperialism always remained a central problem for me. As the years went on, I came to reject Hobson's, Lenin's and Schumpeter's theories of imperialism; I no longer regarded it as a phenomenon that was peculiar to some specific social structure, of the capitalist or pre-capitalist variety. Rather imperialism was a universal historical theme, recurrent under all modes of social systems; thus there have been feudal imperialisms, capitalist ones, nomadic imperialisms of the Tartar kind, piratical Viking imperialism, democratic imperialism under the Athenians, Roman senatorial and Caesarist imperialism, and in our own time, Soviet societist imperialism. There have been imperialisms associated with a surplus of capital to invest, and others as the Russian and Italian linked to a shortage of capital. The most significant question to ask concerning a mode of imperialism is whether it makes for progress or regress. British imperialism in the nineteenth century, the French, and the Dutch, raised the population and culture and happiness of the areas they administered; and so likewise has the American imperialism of the twentieth century. By contrast, the Soviet imperialism, like the Tartar and the Spanish and the Nazi, has brought regression to societies, and even planned genocide. I regarded it as noteworthy that the Jews had long ago welcomed the Macedonian imperialism of Alexander the Great for the opportunities it brought in commerce, industry, and trade throughout the Mediteranean region and Western Asia; indeed this role of the Jews as participatory imperialists was revived under the British Empire when men like Barney Barnato and Alfred Beit joined with Cecil Rhodes to develop the mining industry of South Africa, while Julius Vogel became the most daring and imaginative Premier of New Zealand, and Sir John Monash of Australia the allies' most resourceful general. I recalled New Caledonia where an Alsatian Jew, Lucien Bernheim, its most renowned mining prospector, bequeathed to the colony the endowment for its library, the Bibliothèque Bernheim. A century of propaganda by intellectuals had given to the word 'imperialism' a pejorative sense, but Disraeli and Churchill, it seemed to me, had been right as against Lenin and Hobson, when they remembered that Roman imperialism had brought peace and industrial and cultural advance for several centuries that almost all the Mediterranean peoples came to welcome; Christianity, which diminished the Roman valor and military loyalty, while acting as well as an emotional-ideological barrier against the rise of a scientific movement, was the prototype in its regressive role of Marxism in its Leninist form.

My activities on behalf of the Asian coolie laborers, strangely enough contributed to my estrangement from the American communists and their fellow-travellers in the army in the South Pacific. It was a notable sociological fact that throughout the army posts, installations, and companies in New Caledonia, the 'Information and Education' assignments fell largely into the hands of 'left-wingers'; they knew how to draw up lecture plans and outlines, gather the source materials, and operate the mimeograph machines. One of them, a former circulation manager of the *Daily Worker*, and later a screen writer, managed to assemble a group of them for weekly dinners at *Le Cercle* in Noumea. Transient soldiers of the leftist community would take dinner with the circle, among them, for instance, a former student leader who had been a communist. Once arguments took place concerning the help I was giving to the Javanese and Tonkinese, because the 'party line' envisaged no such activity; its members waited for directives from abroad. The debates grew heated. It was then rumored that my sympathies were of a Trotskyist order. The whole experience reinforced upon me the sociological law that political intelligence diminishes with a long-term commitment to the Communist organization; highly intelligent men, book and magazine writers, trade union organizers, college and high school teachers, had their capacity to respond autonomously to political situations impaired. Their original ethical sense had become enfeebled, and they could act only as a collective subservient to the discipline of authoritarian directives. Even in a South Pacific island, the nucleus of leftist intellectuals had the totalitarian potential.

Three and a half months after my return home from the army, I began teaching at Vassar College; my assigned courses were on the history of philosophy in the nineteenth century, and a seminar in the philosophy of science. After Pacific Army camps, where women were scarce and where fellow-soldiers were almost always direct, honest, and unpretentious, the hot-house environment of an upper-class women's college made the latter seem unreal; teas and cocktail parties were interspersed with genteel scholarship, mostly not too probing, and not too searching. I had not previously been to a cocktail party in my life, and I had never discovered what their purpose was. While I was there, Vassar College spent a large sum of money employing several psychiatrists to study its workings. Their chief figure, well known as the psychiatric expert employed by the legal defence for Alger Hiss, wrote a long document that characterized Vassar College as a 'matriarchy'. This was, however, actually a misnomer, for the chief officers of the college, its president, dean, and the members of their primary committee, were all unmarried, and presumably childless. Freud told his Viennese circle that he thought that homosexuals were the best teachers, but he never stated upon what evidence he based this opinion (Numberg and Federn, 1967). "The best teachers are the real homosexuals, . . . so the repressed homosexuals are the worst, and the slowest." Purported lesbian professors do seem to identify themselves on the manifest level more wholly with their students, and perhaps the master–disciple relationship is one in which homosexual academics find a congenial ablation. At the same time, such teachers, however, have seemed to me to impair their students' sense of reality, and actually to mis-prepare them for their own life's choices. Plato in the *Gorgias* saw a side to Socrates' character and role that was absent from the portraiture of the martyr's role in the *Apology*. " . . . Such a one, . . . becomes effeminate . . . He flies from the busy centre and the market-place, . . . and talks in a whisper with three or four admiring youths, but never speaks out like a freeman in a satisfactory manner" (Jowett, 1875). Apart from a current of deviant sexual behavior that evidently existed at Vassar College, there was a pronounced strain of fellow-travellerism in the faculty and administration, and that despite the fact that it aimed to attract for its student body the daughters of the financial and social elite. Its president had signed the appeal for the celebrated Waldorf 'peace' conference of 1949 at which the Soviet spokesman was cheered when he denied that there were any repressions of intellectuals in the Soviet Union. (When there was a public outcry of criticism, however, the president withdrew her name.) Vassar fellow-travellerism had its comical aspect. One Vassar full professor was the author of a book on Marxism. At festive occasions, he would boast that he was solidly in 'the Party'; in situations of sobriety, he would insist he was a fellow traveller only. I was reminded of the millionaire in the Charlie Chaplin film who regarded the Tramp as his friend only when he was inebriated. The effect of fellow-travellerism on intellectual integrity was, in any case, hardly salutary. At one collegiate conviviality, the full professor told how he had recently been taken to task by the head of the Communist International Publishers for having referred a trifle condescendingly to Lenin in the preface of his book; Lenin, the Vassar thinker had written, was among the 'presumably competent interpreters' of Marx. The Bolshevik functionary made it plain this was *lesé* communist majesty. The Vassar full professor boasted with Marxist piety how he had assured the communist chieftain that the word 'presumably' would be deleted in the next edition.

During the two years after the Second World War, my own separation from the American Left became complete. For a while, I thought the post-war spirit might tend to liberalize the editors of *Science and Society*, and I submitted an essay on Russell's philosophical evolution which, while analyzing the psychological causes of his several transitions, was sympathetic in tone, and did not denounce him for his critique of the Bolshevik Revolution or even his readiness to present the dictator Stalin with an atomic ultimatum. A heated discussion ensued at the editorial meeting. I was actually able after a long afternoon's argument to secure a majority, but the debate itself made plain that the communist imprint of conformity to the 'party line' was so powerful that it was quixotic to hope in these quarters for a genuinely creative journal. I withdrew the essay myself, and sent a letter of resignation to the editorial board. That was not accepted at first, and a book review of mine of Philipp Frank's biography of Einstein was indeed published though the chief editor tried to delete a statement of mine concerning the resistance of Soviet philosophers to the theory of relativity. Shortly afterwards, my resignation was accepted. In England, a Marxist journal later criticized me for having written that philosophical idealism had sometimes had a constructive influence on science, as when, for instance, Schelling's *Naturphilosophie* had stimulated Hans Christian Oersted in his discovery of electromagnetism.

For a while, I tried to work within the American Labor Party, and to keep that organization from coming under communist control. While I was on a journey in 1948 to the newly created state of Israel to study the workings of the communal settlements as a form of democratic socialism, I was indeed nominated in my absence for the state assembly by the

Dutchess County branch of the American Labor Party. When I returned however, I declared at several public meetings that if Henry Wallace lived in the Soviet Union, he would as a geneticist probably suffer the same fate as had his Soviet co-worker Nikolai Vavilov – death in exile or imprisonment. Thereupon the communist organizers warned their members against me, and I, for my part, dropped almost all criticism of Truman's foreign policies. I wrote an article on these matters for *The New Republic* that was published on November 8, 1948.

In my view, with the menace of Hitler and the Nazi totalitarians removed, the last possible justification for the maintenance of dictatorship in the Soviet Union was gone; the 'capitalist encirclement' had vanished. But instead of the advent of a wave of liberalization in the post-war Soviet Union, the opposite was taking place: an intensification of repressions especially against free-minded, objective scientists, artists, and workers. A crucial experiment had taken place, and its result was a negative one for contemporary Marxism. Moreover, Stalin's effort to exploit economically and crush the independent Yugoslavia of Tito showed that a Soviet socialist imperialism had become a fact of our time, one that could not be accommodated to Marxist theory. This essay of mine evoked some vituperative mail and among colleagues at Vassar some stoney faces. It drew, however, the attention of the newly arrived Yugoslav ambassador to the United Nations, Josef Vilfan, and his successor, Alex Bebler. We had several discussions; at first, Vilfan still objected to my use of the term 'imperialism' as applied to the Soviet Union, and he found it hard to believe the facts concerning 'Lysenkoism'. By 1950, however, the Yugoslav ambassadors both acknowledged the facts concerning what they called 'bureaucratic science' and 'great power chauvinism' in the Soviet Union, and were predicting that an enmity would evolve between Communist China and the Soviet Union.

In 1951, I struck, with relative mildness, a junior colleague (male) at Vassar College because he had made some personally insulting remarks. He rushed to the Marxist chairman, who rushed to the president; a colleague of the chairman, an aesthetician, was meanwhile agitating against me alternately as a communist and as an ex-communist; and in due course, I was dismissed from my post. A Vassar trustee made statements utterly false that were published in *The New York Times*. The 'victim' was described as former prisoner in a German concentration camp, and I as a Spencerian. Actually the former, about eight years younger than I, had never been inside a concentration camp, and I at that time had been little interested in Herbert Spencer, or even knew much about him, since at City College we were taught that he had been demolished by Morris Cohen and Franz Boas. I thought of suing Vassar College but the president of the University of Vermont, William S. Carlson, who was proposing my name to his trustees for appointment, advised me not to do so. It would indeed have made things difficult, especially since Vassar's defence would in all likelihood have been that there was not much difference between a former Marxist and a Spencerian; indeed, they both do usually share a common regard for individual liberty.

The seven Vassar women who voted to oust me from their college, one may observe, included not a single scholar or scientist who was truly eminent in her own right. From the president down, they were moderately competent people who enjoyed honors or were granted appointments beyond their merits simply because they were women. They felt comfortable with such male auxiliaries as were mirror images of their own characters. In later years, the women who have been truly honored are those like Golda Meir or Margaret Thatcher or Jeane Kirkpatrick who rose by dint of their own abilities in a purely competitive setting. It is an open question whether the militant feminist agitation (among others) with its administrative and legislative victories, has led to a diminution or enhancement in intellectual stature of the departments or institutions in which they have been most influential. Possibly this effect is a temporary one. In any case, if John Stuart Mill were re-editing *The Subjection of Women* for publication today, he would find himself considering some unforeseen problems.

The University of Vermont was the only one at which I taught for which my affection was whole-hearted. The people and the faculty at that time were simple and unpretentious; the university's philosophic tradition reached back to President James Marsh, who had inaugurated American transcendentalism, and inspired the young John Dewey. Among my friends and colleagues were Raul Hilberg, the pioneer and unflaggingly honest inquirer into the destruction of European Jewry; Robert V. Daniels, a foremost authority in the history of communism and Soviet politics; John Huden, professor of education and self-taught investigator into early Vermont Indian artefacts; Lester Wallman, the devoted, hard-working professor of neurology, and researcher into the history of medicine; and my warm friend, Paul D. Evans, a man blessed by sagacity, an artilleryman, as he said, of 'Mr Wilson's Army'; his lectures on the French Revolution would refresh me at mid-day. The university's president, William S. Carlson, had served his life's apprenticeship as a geologist-explorer in Greenland, and then spent several years during the war in an Alaskan outpost; his Viking's eyes had seen something of the immutable sorrow in human existence. He secured the unanimous approval of Vermont's Governor and Trustees to my appointment, despite the fact that an objection arose from a most unusual quarter. By a coincidence worthy of Thomas Hardy's novels, it transpired that the Colonel of Intelligence in New Caledonia had re-settled in Burlington, Vermont, and on reading that I was in process of appointment, made objections. President Carlson, however, held to his decision. Two years later, Carlson became the president of the State University of New York, but held office for only a few years because an undistinguished governor resented his independent efforts to develop the then backward conglomeration of public colleges.

Our schedules for teaching at the University of Vermont were heavy, but I enjoyed the courses that ranged from logic to American philosophy to social problems. I was especially pleased that half my work was in sociology; indeed, that fraction of my academic self constituted the entire sub-department of sociology in the University of Vermont. Its enrollment began to grow inordinately to the consternation of the classical dean who fought as doughtily against 'progress' as Horatio did at the Bridge. Among my first native Vermont students were a future professor of political science at Princeton, and another who became Ambassador to Jordan. There was also a large company of Jewish students from New York who mainly intended to study medicine. All the students in nursing took their prescribed course in sociology with me; they hatched plots for the day when I would be their patient at the hospital, but that retribution was happily not visited upon me.

I took an active part in Vermont's faculty and political meetings, and introduced in committee and guided through its passage by the Faculty Senate the resolution that terminated racial and religious clauses in the constitutions of fraternities at the university. Elected president of the Vermont chapter of the American Association of University Professors, I worked with some success to increase the University's commitment to scholarly research, and at some exciting sessions, posed the unusually low salaries at the College of Arts and Science as a question for the first time of public debate. The president of the university, Carl W. Borgmann, a chemist with a Cambridge degree, was a fair-minded man who, when the facts were made clear, undertook to achieve the necessary remedial measures. Something of the atmosphere of the exaggeratedly mis-named 'McCarthyist Era' still persisted in the middle 50s. When I undertook as president of the Vermont AAUP to arrange for a showing of an interview between Edward R. Murrow and J. Robert Oppenheimer, some junior colleagues warned that I was courting a public, governmental, or university reprisal. A local newspaper, the *Vermont Daily News*, published by a non-Vermonter, did indeed feature a fear-engendering story. The only consequence was that we had a large audience for what was in no way a propaganda film.

I began to come to the conclusion during this period that the strength of Senator McCarthy resided not in his own power, for he had no organization behind him and possessed neither eloquence nor political program, nor personal magnetism. What gave him his apparent strength was the weakness, guilt, and cowardice of the intellectuals. The so-called 'Fifth Amendment Communists' followed the behest of the Communist Party to plead the Fifth Amendment but not solely because of party pressure. More important was the fact that they were utterly confused and bewildered when it came to defending Stalin's policies of two decades – the repression of individual liberties, the spectacle of judicial mockery in the Moscow Trials, the persecutions of followers of Mendel and Einstein, the occupation of Czechoslovakia and Poland, the Berlin blockade, the provocation of the Korean War, and the virulent anti-Semitism. Senator McCarthy was the one who had intuitively grasped and made a theatre of the shallowness and hypocrisy of the left intellectuals. Faced at a committee hearing by a disrespectful questioner, the confidence of the cocktail party communist collapsed. He stuttered, grasped feebly for argument, tried lamely to posture as a hero, martyr, or tribune of the people, but then the play-acting withered, and he stood, not the moral accuser of the system, but its pathetic beneficiary, clutching at the Fifth Amendment that gave him the privilege of silence – he, who was supposed to be the fearless, irrepressible voice, proud of his deeds and actions, finally aligning himself with the Corrupt Knavery of the Fifth Amendment Pleader: the tax-evaders, the bribe-takers among policemen and judges, the dishonest machine politicians and embezzling bankers. Was there ever a group of self-anointed revolutionists who made a sorrier exit from the world-historical stage to which they aspired? Were these the inheritors of Karl Marx who had proclaimed that the communists did not conceal their aims, that they were no secret society? Were they the successors to Eugene V. Debs who had welcomed every court and committee as a forum in which to defend openly both his ideas and actions? In later years, it was intellectual kinfolk who tried to invent a mythology of the intellectuals' courage during the 'McCarthyist Era'. But Senator McCarthy, however much humbled and humiliated for his own personal failings and weaknesses, and political swashbuckling and irresponsibilities, had nonetheless inflicted defeat in the Battle of the Intellectuals, a defeat all the more acute and crushing because the antagonist seemed so unworthy a foe. If in England, France, and Italy, the Communist Parties grew and their influence was augmented among intellectuals, in the United States alone the Communist Party did not revive after Senator McCarthy triumphed in the Battle of the Intellectuals.

At Vermont, I set out to define my philosophic standpoint, and wrote an essay to explain why I rejected the then fashionable view that the forms of language tend to determine their users' philosophies. Entitled 'Sociological Aspects of the Relation between Language and Philosophy', it noted that the most diverse philosophies had arisen and flourished within the same linguistic setting; furthermore, that the absence of the verb 'to be' in Semitic languages (as Hebrew) had posed no insuperable obstacle to medieval Jewish philosophers who had adopted Aristotelian conceptions of being and proofs of the existence of God. Likewise the thesis of linguistic relativity – that different languages projected incommensurable universes – seemed to me decisively contravened by the fact that people of diverse languages do manage to communicate adequately with one another; finally there was no analogue among the alternative languages to the Michelson-Morley experiment that had required a principle of physical relativity; people could translate well from one language-system to another.

When George Boas in his volume of Carus Lectures *The Inquiring Mind* wrote that he accepted the "general point of view about the relation between language and philosophy" expressed in my article, I felt reassured that the historical evidence sustained my thesis.

Philosophies, in my view, were essentially answers to unanswerable questions that were shaped by the character-traits of the particular thinker in their interaction with the social, scientific, and cultural settings in which he lived. I had begun while teaching at City College to apply, or test, this approach by a study of the formation of Spinoza's philosophy, and during the intervening years kept gathering relevant facts. My friend and professor, Harry Austrya Wolfson, had written a powerful book in which Spinoza's philosophy seemed distilled from a set of reactions between medieval Jewish philosophical texts and Descartes' writings. I, on the other hand, linked Spinoza's most original tenets to his involvements with the political revolutionary, Franciscus van den Ende, his youthful emotional attachment to the mystic, communistic sect of Collegiants, and finally, his mature loyalty to the liberal party among the Dutch Republicans led by the statesman and mathematician, Jan de Witt. Spinoza's pantheistic mysticism was thus associated with the mystical current among the democratic religious sects of the seventeenth century; his hedonistic conception of the free man, however, was a point-by-point critique of the ethic of the Calvinist party, that bitterly opposed the liberal Republicans, while his enthusiasm for the mathematical method was one similarly shared by de Witt and his circle. Even the excommunication of Spinoza was probably far more related to his liberal and youthful radical political associations than it was with purely theological views. Spinoza's thoroughgoing determinism seemed to me, on the other hand, the outcome of a certain masochist strain in his own character, as he endeavored to take pleasure in regarding all his actions as altogether determined by causal laws, and in loving God intellectually as an all-inclusive determined whole. In later years, I have come to believe that among highly scientific-minded Jews, such as Einstein and Freud, the determinist propensity arises from a repressed desire to retaliate against irrational persecutors who, however, hold the dominant physical power. If they cannot retaliate physically, they do so intellectually; and their form of intellectual retaliation, or aggression, has taken the direction of explaining the behavior of such men as Hitler, or the mob that lynched de Witt, or the Viennese anti-Semites, as the determined consequence of causal laws; nothing so destroys the pretensions of human beasts as to subsume their actions under mathematical laws.

Because of the disparate emotional layers of motivation to his philosophy, Spinoza's system, however, remained basically disunified; his mysticism, for instance, made it impossible for him to formulate the laws of nature in which he believed, for the latter involved a language of numbers which, from his mystical standpoint, were unreal. And his belief in the eternity of the human mind was at odds with his psychophysical correspondence. Spinoza, the first philosopher in history to be explicitly an advocate of democracy, was similarly constricted when he tried to rear his democratic politics on a foundation that equated 'right' with 'power'. Moreover, since the masses in his view were irrational and volatile, he could scarcely indulge in Jeffersonian-like enthusiasm. He was especially baffled because he could not adduce empirical examples in the world's political history of democracies that had proved themselves stable. This is why his *Political Treatise*, in my opinion, remained an unfinished book. Yet one will always treasure his tribute in 1670 to Amsterdam as that city where all men could walk in freedom because the city, preoccupied with commerce and industry, valued men who contributed to its work, regardless of their religious opinions.

I myself during these studies had come to reject the determinist doctrine, and in an essay 'Causality in the Social Sciences' had drawn the distinction between those laws that are necessitarian and those are interventionist. Thus, laws of nature are necessitarian where the values to their independent variables cannot be affected by any human decision: on the other hand, interventionist laws are those whose values to their independent variables can be affected by human actions. Such a law as Hubble's, that the velocity of recession of galaxies is proportional to their distance from us, is an example of a necessitarian one because that distance can scarcely be affected by anything we do. Laws such as Boyle's law for the inverse relation of the pressure and volume of a given gas obviously allow for human intervention; we can decide to increase or decrease a given pressure-value. Among social phenomena, furthermore, outcomes of human interventions are often characterized by zones of indeterminacy. Such laws of social science as have been formulated or proposed always involve some allowance for freedom of human decision. Even Marx, who affirmed that his law for the evolution of capitalism was one of inevitability, left it to human choice to decide whether or not to cooperate with that alleged inevitability. Every determinist sociological law of order n generates an interventionist one of order $n + 1$.

Living in Vermont meanwhile enlivened my interest in the history of Vermonters who, under conditions of physical hardship, had made outstanding contributions to American thought. An essay of mine on James Marsh was followed by several studies in the life of John Dewey. As I got to know the often unexamined Vermont archives well, I was able to make some new archival discoveries. I found, for instance, the university librarian's list of every book that John Dewey had borrowed while he was an undergraduate student. The aged Edith Davidson

Harris, daughter of William T. Harris, founder of *The Journal of Speculative Philosophy*, told me of the young Dewey's letters to her father, which information I conveyed to a later chronicler, a teacher of ethics, who regrettably omitted acknowledgment. I thought seriously in later years of writing a biography of Dewey. That staid Vermonter posed an unusual psychological problem; all his life he was attracted to flamboyant, emotion-emanating personalities such as the educational reformers Ella Flagg Young and Colonel Francis Parker, the brother-promoters Franklin and Corydon Ford, the anarchists Emma Goldman and Carlo Tresca, the witty revolutionist and romantic Max Eastman, the Australian actor-therapist F. Matthias Alexander and the enthusiastic disciple Joseph Ratner, and the American Tolstoyan saint, Jane Addams. What Dewey meant by overcoming 'dualisms' had more to do with such emotional needs than it had with defects in logical theory. For Dewey, a proposition was 'true' if it tended to bring human beings together; he told the Harvard Tercentenary Conference in 1937 that the work of science was the greatest example of a cooperative mentality, rather forgetting not only the competitive drives among many scientists but also the impact of industrial and military competition upon the development of science.

The drama of Dewey as a representative American university intellectual seemed to me more significant than any originality in his ideas. In later years, while I was at California, I journeyed to have long interviews with Dewey's son, stepson, and youngest daughter, Lucy, and with such of Dewey's co-workers and friends as Max Eastman, William H. Kilpatrick, Bruce Bliven, Lucy Sprague Mitchell, George Counts, Roger Baldwin, and the second Mrs Dewey; in Japan, I talked with several of Dewey's old friends of 1919. Some generally accepted version of Dewey's background and life turned out to be partially mythologic. Burlington, the Vermont town in which Dewey grew up, far from having been a center of Protestant virtue, was embedded in dissoluteness; as an entrepot for the trans-shipment of lumber, it was also one for prostitution, an evil so rampant that according to a survey made when Dewey was a boy it touched almost every household. Part of Dewey's social democratic outlook may have been imbibed from his teacher H. A. P. Torrey, who idolized Abraham Lincoln and lectured to his classes on the theories of the socialist forerunner, Saint Simon, a daring theme at that time. A novel I read by the East Side Jewish writer, Anzia Yezierska, a member of Dewey's seminar in Philadelphia during 1916 on immigrant groups, was so transparently autobiographical that I informed Sidney Hook that it was evidently based on a romantic episode in Dewey's life, and I so stated in a published review a few years later of a biography of Dewey. Several months later the Curator of the Dewey Archives telephoned me in Toronto and Burlington to inform me excitedly that she had verified what I had written, and that a solution was now at hand as to the collection of Dewey's love poems that were housed in Southern Illinois University. A poem with which Ms Yezierska ended her novel was identical with one in Dewey's handwriting in his collection. The Curator asked me to collaborate in publishing her discovery but I declined, though I wrote her concerning still another poem at the end of Anzia Yezierska's formal autobiography (with an introduction by W. H. Auden); that too was found to be one of Dewey's. It had indeed been proposed a few years earlier that I go to Cleveland where the aged, blind novelist then lived; Sidney Hook had traced her whereabouts, and she had replied indicating that despite her political reservations, she was willing to tell of John Dewey. I was then, however, finishing *The Conflict of Generations* and did not look forward to eliciting the romantic reminiscences of an aged, unhappy woman, with a reputation among those who recalled her in Greenwich Village, as a highly irascible person.

Dewey probably affixed his name to more *ad hoc* political manifestoes, declarations of public conscience and witness, statements by committees undertaking to rectify wrongs done to individuals and peoples, and announcements for organizations of protest than any other person in the history of American civilization. Yet Franklin D. Roosevelt thought he was perhaps the 'worst' of the intellectuals. The philosopher in practical affairs may indeed be lacking in the qualities of wisdom; he may tend to succumb to the 'prophetic fallacy', that is, the ideological mode of thought. Dewey was scarcely the scientific observer when he travelled to the Soviet Union, a topic I later explored in an essay 'American Travellers to the Soviet Union'.

Another Vermont research of mine done during later vacational visits to Burlington had a significance for the study of American values and religion generally. I found in the archives of the Wilbur Library a copy of a house-to-house and church survey done in Burlington on a 'clear, crisp' Sunday on March 3, 1867; the study was painstaking and it had the exact figures and percentages of the population attending church, holding membership, and so on, for each congregation in Burlington. It occurred to me that we could replicate the study on a Sunday of that same month, March 8, 1964, almost a hundred years later. I secured the cooperation of a professor of psychology, Mervyn W. Perrine, and his students in social psychology, and then analyzed and compared the church attendances and membership for Burlington as they had changed during nearly a century's interval from 1867 to 1964. By and large, I found that the proportion of the Protestant community that attended church had diminished by almost a half. The liberal congregations, the Unitarian and the Congregationalist, had especially declined in their relative percentage of attenders; at the same time, however, the overall percentage of church members among Protestants had risen by 16 percent. The Roman Catholic church had grown significantly in its parishioners to possibly as much as 42 percent, and there were new religious groups – Hellenic Orthodox, Jewish, Christian Science and Pentecostal sects. In short, although church membership had become more customary among Protestant folk in the northern Vermont town of Burlington, the intensity of religious involvement had declined, as compared to the year 1867. And yet Burlington was far more religious in spirit than it had been in the revolutionary era of Ethan Allen, for in the year 1789 all Burlingtonians were militantly unchurched, regarded themselves as deists or 'proper-heathen', and mostly used the Sabbath to hunt and fish. A cyclical explanation for the fluctuations of religion and ideology seemed to fit the facts best.

In my last three years at Vermont I returned to studies in the logic and sociology of science. Professor Philipp Frank secured for me a small grant that enabled me to spend a summer at the Harvard Widener Library during which I wrote the first draft of my study of the ethic of the scientists who founded the Royal Society; I verified by careful biographical analyses that the overwhelming majority of both the founders and the members of the Royal Society of London had been hedonistic, not Calvinist, in their philosophy and indeed practice; mainly supporters of the royalist cause, not the parliamentarian, during the English Civil War, they had looked very much to the English aristocracy and monarchy for social support in committed membership. My analysis and factual studies contravened the standpoint of Max Weber's *Protestant Ethic*, a book which, in my view, had led to quite mistaken 'findings' on the part of such sociologists as Robert K. Merton. I presented a paper on my work in the spring of 1956 to the Eastern Sociological Society in New York; later it evolved into the major chapter of my book *The Scientific Intellectual*.

An essay on *The Principle of Simplicity* also drew a distinction between the methodological and metaphysical usages of simplicity. The latter is the kind of belief held by Einstein and Leibniz, for instance, when they affirmed that God in His Wisdom created a world that was governed by beautiful, simple mathematical laws. The methodological principle, on the other hand, is always applicable, even if the world were governed by complex laws, with multiple variables; for no matter how complicated the world, we would still choose to describe it with the simplest possible laws that would fit the complex data. This methodological, or economical, sense of simplicity is actually one form of the principle of verifiability; for if a person holding to his favored hypothesis were prepared to rebut the negative evidence of an experiment by inventing an auxiliary hypothesis to explain it away, and was indeed prepared to generate in principle an indefinitely large sequence of auxiliary hypotheses in order to cling to his primary hypothesis, we would judge him as psychologically fixated, and as logically the advocate of an infinite series of hypotheses with not a single verifiable member, and hence as a whole unverifiable.

I had the pleasure of a letter from Curt J. Ducasse telling me that it was the best treatment he had ever read of the principle of simplicity. A large number of psychiatrists wrote for reprints because evidently at one of their meetings they had been discussing how one symptom of mental illness was often provided in the persistent disregard of the principle of simplicity; the neurotic, in other words, tends to be, in Leibniz's term, a 'goropisor'.

In 1957 I was appointed Professor of Philosophy and chairman of the Social Science Integrated Course at the University of California, Berkeley. The latter was an interdisciplinary effort to provide a one-year course that would be given by lecturers and associates drawn from the various departments of social science. I was to give the lectures for the sociological part. The committee that chose me was most influenced, I gathered, by the young, gifted sociologist, Reinhard Bendix, and also, curiously, by the economist Andreas Papandreou, later the left socialist Prime Minister of Greece, who renounced his American citizenship to enter Greek politics, and seemed to hate the United States all the more because its intervention had probably once saved him from being executed by military rulers. I enjoyed the association with an able group of young scholars; the late Eugene Burdick, who became famous for his novel *The Ugly American* was our lecturer in political science, and among our associates were Harold Schiffrin, later Director of the Harry Truman Institute at the University of Jerusalem, and many others now occupying posts at California, Harvard, Johns Hopkins, Washington, Rutgers. During its last years, however, the teaching staff and students in the Social Sciences Course alike declined in their scientific character. My successor as chairman, a well-known writer on social questions, and a talented composer of articles in *Commentary*, tended to be quiescent to the left-wing atmosphere that grew heavy in 1965; one communist associate whom he had had chosen was allowed to give almost all his students the grade A as part of the Utopian euphoria that was characteristic of that interval; in the new élan, the new chairman had difficulty even in recognizing as his own, passages he had once written in an article on Senator Joseph McCarthy; ideology had replaced science. It was a period when a student's stand with respect to the Vietnamese War tended to determine his grade, and attendance at 'demonstrations' was accepted in lieu of class attendance. Also, the multiplication of fellowships and governmental assistance for research tended to limit the available associates to a lesser level of graduate student talent. Not long after I left Berkeley, the experiment was properly discontinued. Educational experiments of that kind thrive only in the most unusual circumstances, most notably, apart from the social situation, the presence of a group of talented scholar-researchers sharing enough of a common philosophy of scientific inquiry to make their cooperation fruitful. That was the case several generations ago, as Rexford Tugwell has narrated, when such an opportunity arose at Columbia University, and the famed *Contemporary Civilization* course was created. Otherwise human learning and teaching do best to follow traditional routes.

Meanwhile, at Berkeley, I pressed forward to complete my book on *The Scientific Intellectual*. What had begun with a study of the seventeenth-century English scientists was generalized into a sociological law that only where a hedonist-libertarian ethic prevails of the kind that leads to respect for the human body's senses and reason as an avenue to basic knowledge can a scientific movement arise. And such an ethic has prevailed in different societies in different times – in England at the time of the Restoration, in Catholic Venice and Padua in Galileo's time, among European Jews in the nineteenth century in Russia, Austria, Germany, and France, liberating themselves from Ghetto trammels, among Arab-speaking thinkers in early medieval Bagdad, detaching themselves from Moslem orthodoxy, among young Americans such as Benjamin Franklin who took Calvinist theology, original sin and predestination lightly, singing instead in praise of love, and renouncing money-making as soon as he had enough to enquire into nature, and to experiment with her phenomena. The ethic of science was Baconian, not Calvinist, for from the Calvinist standpoint every labor-saving device was a manoeuver to outwit God's curse upon man, and every experimental instrument was a way of evading the low estate of man's senses, tainted with his sinfulness. My view of the psychological origins of science seemed to me in accord with all that psychology has learned of the impact of guilt and repression on human thinking.

Thanks to the initiative of Irving Kristol, who had assumed the editorship of Basic Books, Inc., my book was published in 1963. Its reception was sharply divided into extremes. Followers in the United States of Weberian sociology as well as those historians who were offended that any discussion of the psychology of scientists should be tainted with psychoanalytical ideas, or even a sexual component, were hostile to my views. On the other hand, in England, Sir Harold Hartley, editor of the *Notes and Records* of the Royal Society and the volume *The Royal Society: Its Origins and Founders*, wrote that the critique of the 'Protestant asceticism' thesis had been 'most convincingly' demonstrated; A. Rupert Hall, the historian of seventeenth-century science, thought, despite his misgivings concerning my psychoanalytical conjectures, that my analyses of science in Scotland and New England were 'acute', and that 'most readers' would applaud my discussion of the Weber–Merton thesis. The vehemence with which the latter's proponents responded to my views suggested that some strong non-logical sensitivities were involved. In America, the outstanding intellectual historian of the seventeenth century, Marjorie Nicolson, concurred in my estimate of the role of the hedonistic coffee-houses in the life of the Royal Society, while Owsei Temkin, the Johns Hopkins scholar of medieval medical science, wrote that I had grasped the "psychological make-up of the men who created the new science". The anatomist Paul C. Roofe in *The Quarterly Review of Biology* found that my psychoanalytical hypothesis concerning Vesalius and the Anatomical Revolution was "probably the best yet offered concerning the behavior of this remarkable genius", and that my book's approach was "unique if not profound". And from the unexpected quarter of the president of the Venezuelan National Council of Scientific and Technological Research, M. Roche, came the opinion that my theory best explained the failure of the scientific movement in Spain. The *Encyclopaedia Britannica* article in 1971 on 'Inventions and Discoveries' took note of my theory of the role of the hedonist-libertarian ethic in the rise of science, and most recently in 1981. Michael Hunter wrote in his book *Science and Society in Restoration England* that the alleged statistical basis for "a significant 'puritan' strain among the Fellows . . . seems to have been mistaken", and that "more respect should be given to the arguments of L. S. Feuer". My own final reply to the Weberian 'Protestant ascetic' standpoint was published after I went to the University of Virginia. The highly emotional, personal, and prestigial investment in that standpoint, however, was one I never undertook to analyze.

As part of my work in the Social Science Integrated Course, I had taken the responsibility for a series of lectures every year on Soviet society. For several years therefore I studied the Russian language, and acquired a moderate reading knowledge. To acquire a more first-hand experience of Soviet economy and culture, I undertook to spend a semester in 1963 as a Visiting Scholar with the Soviet Academy of Science under the newly begun exchange program it had negotiated with the American Council of Learned Societies. My proposed subject of study was 'Recent Trends in Historical Materialism', and I went accompanied by our daughter, Robin, then 15 years old, and a few weeks later by my wife, Kathryn, who was at that time a Lecturer in Russian Literature at the University of California. I was attached to the Soviet Institute of Philosophy, which at that period included sociology; since the word 'sociology' had a bourgeois connotation to Stalinist-conditioned Soviet ears, it was at that time called something like 'concrete conditions of social existence'. The five months in the Soviet Union were about the most unpleasant I had spent in a long time. Although Soviet propaganda boasted about an English-speaking school that foreign children could attend, its existence turned out to be a figment of propaganda. With some threatening to leave, I finally got my daughter admitted to a middle Soviet school; her friends there were most helpful, but the pervasive dishonesty, the cheating on examinations, the non-payment of fares on the buses, the cynical attitude of the children to a proposed sojourn on a collective farm, the anti-Semitic remarks concerning the school's one Jewish teacher, the classroom anti-religious pressure, and the day her best friend, a Young Pioneer, confided to her that she had just learned that her grandfather had been murdered by Stalin, were also among her enduring impressions.

The saddest persons in the Soviet Union, however, we found, were those Americans who had emigrated to the Soviet Union during the thirties. Among such we became friends especially with two women; one had accompanied her parents to the Soviet Union when they, as idealistic communists and skilled workers, listened to the Soviet call for technical experts to come as industrial volunteers; the other, an American Young Communist went idealistically with her husband. The father of the first struggled to accept the 'justice' of the Moscow trials but then was himself arrested, and vanished in some Siberian camp. The daughter, arrested after the Second World War, was found to have written in her diary her doubts concerning Soviet justice; she spent years in a labor camp in Central Asia. The husbands of both women were tortured, and sent to labor camps as well. Both women, in the most bizarre conversation I have ever heard, discussed the 'friends' who had 'testified' against them. Should they forgive them or not? They argued as to whose labor camp had resorted to the greater cruelties, and they took comfort that as women they had not been subjected to the degrading tortures inflicted upon their men. Now, 'rehabilitated' under Khrushchev, one of them hoped her daughter would somehow manage to emigrate from the Soviet Union, while the other, allowed by the connivance of the regime to teach English privately, found her chief pleasure in reading American magazines that American friends left with her.

Though the winter of 1963 was a comparatively liberal period politically in the Soviet Union, at the Moscow Institute of Philosophy a semi-torpor prevailed. The head of the section on 'bourgeois philosophy' was a party hack who would not have passed a sophomore's examination in an American introductory course. A few brave persons were trying to do original or independent work, but their anxieties in their situation were obvious. When I had a discussion with the logician (and later satirist) Alexander Zinoviev, he sat tense, weighing each word as if it were a crucible of communication. That he deviated from party othodoxy was conveyed to me when he asserted that whether some theorem in mathematical logic were true or false had no bearing on dialectical materialism. The secretary of the Institute, Mshvenieradze, was obviously a corrupt tool of the KGB; at the reception for us the very first day, I predicted to my daughter that if ever I were assailed in the Soviet press, he would be the hireling. She was incredulous, but before a year was gone, my forecast was confirmed. The director of the Institute, Konstantinov, was a dull bureaucrat who had during the

Stalinist era obediently executed the party orders to denounce men who showed signs of honest thinking; he himself had written a book on historical materialism that was remarkable because it lacked a single fresh thought or observation. The Lenin Library, in which I enrolled as a reader, was the only one in my life's experience with libraries that asked on its application form whether one was a member of the Communist Party. A quick ruling was made that this question did not apply to foreigners, though it was indicated that communists were more apt to get access to special collections, that is, books 'isolated' from general readers.

Repeatedly I observed the readiness for bland lying that pervaded the Moscow Institute of Philosophy. Once, in the presence of two young, able sociologists, I was discussing with the head of their section, Mr Osipov, the lack of American sociological works available to them. We also got on to the question of freedom of publication. From his desk he produced a copy of a book of Kluckhohn's essays to show how free his choices were. I remarked that in the United States there was a Soviet bookshop right on 5th Avenue; why then didn't the Soviet Union allow an American bookshop in Moscow? Osipov claimed the difficulties were those of monetary exchange and the laws against private commerce. I said this difficulty could be surmounted, and added, that in the United States many books were published criticizing capitalism, and advocating socialism. Could he say that Soviet scholars had a free choice as long as not a single book could be published criticizing socialism, and advocating capitalism? Blandly, with that curious frozenness of feature that characterizes Soviet diplomats and spokesmen, Osipov replied that such a book could be published in the Soviet Union but that nobody wrote one. We all sat for a moment silently. We all knew that Osipov was lying; and the two young sociologists, never voicing an objection, were curiously rendered public sharers in the duplicity, partaking of its collective guilt. No society, apart from that of post-Nazi Germany, has been so penetrated by a social ether of collective guilt-feelings; it engenders a kind of cohesiveness that criminals have for one another, with a code for ostracizing the person who breaks the collective repression.

There were wonderful islands of aspiration to freedom, however, even within the interstices of the Soviet totalitarian system. In Tbilisi, the capital of Soviet Georgia, I found an Institute of Philosophy where free thinking was still valued. Its director, emerged from imprisonment under Stalin, had been 'rehabilitated'. Younger men were studying Charles Peirce's writings from photographed pages. The secretary, Chavchavadze, a kindly, thoughtful man with a rare taste in the arts, brought his two children one evening to take a walk with me along Rustaveli Boulevard. They whispered something to him. I asked whether they were worried by something. He replied sheepishly: "Both say they are a little disappointed; the older one expected that as an American, you'd be wearing a high hat like a capitalist, while the younger one thought you'd be wearing feathers on your head like an Indian." International relations improved when I treated them all to ice cream.

At my farewell party in Tbilisi, in response to the toasts, I said that though my visit to the Soviet Union had not encouraged me to think that their socialism was a system superior to that of the Americans, still if there were any country where the people's spirit might enable socialism to thrive, it was Georgia. That was the only republic in which persons even dared talk freely of the past; older members of the institute told me of their memories of Trotsky in Tiflis (as it was then called) when he still commanded the Red Army, and stood in his box at the Opera House at erect salute as he acknowledged the audience's cheers; they asked me questions concerning what had happened in later years to the exiled Georgian leaders such as Tseretelli; they were the only ones who recalled my early *Science and Society* articles, and would call me whimsically 'nashe Marxiste' (our Marxist), despite the fact that they, as well as I, had probably travelled far intellectually from Marxism.

My visit as a whole was no failure. Of that I had proof when articles criticizing my views, and assailing me personally, were published in Soviet newspapers and their philosophical journal after I left. The lectures I had given in Moscow, Tashkent, and Tbilisi had borne fruit. In Moscow, I gave a series on such subjects as American philosophy, the sociology of seventeenth-century science, and the relation between psychoanalysis and philosophy. At various junctures, the chairmen would become uneasy and intervene to assert that I was just giving my personal opinion. (Whose opinion could I have been giving?) When I noted that John Dewey had been shown to have been right in his defense of Leon Trotsky, for Khrushchev himself had acknowledged the Moscow trials were fraudulent, the chairman thus became much agitated. When I spoke on seventeenth-century science, I mentioned the pioneering volume of essays by Soviet scholars in 1931 entitled *Science at the Crossroads*; as nobody had heard of it, I began to tell them of the delegation led by Bukharin . . . whereupon the chairman, Mshvenieradze, jumped up to interrupt about something else. Everybody realized I had uttered a name that was meant to be repressed. When I spoke on psychoanalysis and philosophy, I was challenged to show of what conceivable use psychoanalytical ideas could be in the understanding of Soviet society. I replied that they themselves attributed the prevalence for twenty years of the 'cult of personality' to the irrational, paranoiac drives of Stalin rather than to the economic foundation of their social system, and that, therefore, even from their own standpoint, unless they wished to repress that whole period as an unproblem, they would find themselves confronted with psychoanalytical questions. That issue arose in Leningrad and Tbilisi as well, but the challengers made no effort to reply.

Usually at my first meeting with a department or institute in a Soviet city, the chiefs and aides would launch into a political attack on the United States. At Leningrad they dwelled on their sacrifices with an imputation that this was the fault of American capitalism. Finally, I replied that the Second World War began when Hitler and Stalin had both made a pact to invade Poland, and divide up its territory. The chairman replied that the working class in Leningrad would never stand for such a statement; I replied that there was no evidence that he spoke for the Leningrad workingmen. The next day the discussion was carefully confined to a sociological survey they were conducting on the causes of discontent in several factories.

I attended the synagogue services in almost every Soviet city I visited – Moscow, Leningrad, Tashkent, and Tbilisi. As I knew Hebrew and Yiddish sufficiently to establish a basis of confidence, I heard pitiful narratives of lifetimes spent under a variety of persecutions, petty and major. At Tashkent, the members in a weather-eroded, makeshift clay structure consisted chiefly of Polish Jews transplanted during the war to Central Asia. They put me to their security test by calling on me to read the week's portion from the Torah. A commotion ensued as the American visitor acquitted himself with all the skill of an alumnus of the Downtown Talmud Torah. Afterward they listened incredulously as I told them of Hebrew schools in American cities; they told me threats were being made to terminate their synagogue. Suddenly communist official arrived shouting angrily: "Here there is Soviet power! No discussions! Only praying!" A few months later I learned the synagogue had indeed been torn down, and the congregation dissolved.

At Moscow, meanwhile, I fortunately got to know a group of brave, 'dissident', graduate students drawn from several departments who assembled quietly, without public notice, in a seminar-office at the Moscow University. It all began one day when I visited the library-study of the Sections of Scientific Communism and Atheism, and an intense debate ensued between me and one of the lecturers. A bunch of students gathered around us; probably it was the first genuine public clash of basic ideas that they had ever seen. As my Russian was at times faltering, one student, with an excellent command of English, volunteered to be my interpreter. The association with him and his friends grew, and one day he told me that I would hear from those who followed 'real philosophy', and not the 'official philosophy'. A few days later, a voice, risking the hotel telephone and not identifying itself, suggested that I come to a certain place at a certain time at the University in the Lenin Hills where would I would be met for the promised philosophical discussion. I considered carefully whether to go, but since the discussion was to take place on university grounds and fitted into my research topic of contemporary trends in historical materialism, I ventured this intellectual exploration into a Soviet 'invisible college'. That meeting, followed by talks with individuals at the university, brought to my eyes the sight of enquiring personalities, asking fundamental questions, seeking more freedoms, disenchanted with the hypocrisies that they had to mouth, the repressive phrases, the creaking of the control-fastened, clumsy, unventuresome ponderous economy, and the ugly, sullen indifference in the everyday networks of ordinary dealings; a ruling cadre of frozen-faced pedestrian henchmen, stamped with so many party directives, was almost incapable of uttering a spontaneous word that could spark human feeling. Toward the end of my stay, another such liberal group, more literary in its interests, was uncovered at the university, and expelled. I discontinued my visits to the Seminar in order to avoid endangering any students, and confined myself to the Soviet Academy of Sciences, and its institutes.

I chose to return to the United States several weeks before my allotted time had expired, and would have preferred not to give any farewell lecture. But several of the younger members of the institute urged me to speak on my impressions of Soviet philosophy and sociology. I gathered that they hoped that I might say some of the things they dared not. Both the director and secretary of the institute concurred in the request, misled perhaps by a series of previous American visitors such as C. Wright Mills into expecting a ritualistic tribute to the advancing socialist world. The correspondent for *The New York Times* asked to come along, and I invited him, thinking it wise to have one American witness there to forestall the not unlikely Soviet lies. The ensuing session lasted about three to four hours. The listeners would wander in and out with glasses of tea enjoying the show; as the speaker, however, I had no privilege except talking. I thanked the institute and its members for their courtesies. When later I discussed the limitations of Soviet sociology, I described how I had not been able to find a single sociologist who would study the 'so-called period of the cult of personality', and enquire into such questions as the number, social, and ethnic composition of the victims, the fate of their children, and the character of their accusers. I asked whether a political sociology could arise in the Soviet Union, if such questions, and similar ones concerning their ruling elite, were barred as unproblems. The chairman retained his composure during this portion of my remarks. When I broached the question of anti-Semitism, however, the director lost all decorum, began to scream, shout, gesticulate, while his KGB secretary tried desperately to soothe him to equanimity. I continued telling that in every synagogue I had visited in the Soviet Union I had been told of the existence of a virulent anti-Semitism, but that not a single Soviet sociologist studied the problem; by contrast, I said that in America, where there was much less of the phenomenon, the researches into anti-Semitism, as at Berkeley, were considerable and impressive. Thereupon the director denied the truth of what I said: I observed therefore that in all my months at the institute I had never met a single Jew among their philosophers or sociologists. That produced some consternation; they began to search through the crowded room, and they found one Jew, who had evidently arrived recently from the Ukraine. I did my best not to embarrass him with direct questions, and he, courteous and gentle, avoided any direct contradiction of what I had said.

The fury of the chairman, Konstantinov, however, was unabated; flailing his arms about, he declared how he had served years in the Soviet Army that had fought the Hitlerites; I replied that I too was a veteran which seemed to catch him aback, but he later proceeded to call me a 'warmonger', and to talk of the police dogs that were used against blacks in Alabama, unlike the Soviet society that was one of freedom. Whereupon I asked him pointedly whether he had ever during Stalin's era raised his voice once on behalf of any of those who were sent by the millions to labor camps. The hall was momentarily hushed as he acknowledged with what may have been a touch of sincere sorrow that he had not. Almost all the institute members knew he had served as a false accuser during the Stalinist era. My discussion of the Soviet philosophers aroused less animosity. I remarked that Marxism was becoming a cover-all label, and actually falling apart into several incompatible approaches, bound together only by the fact that each of them culled its own set of appropriate quotations from Marx, Engels, and Lenin. The new existentialists, the scientific realists, the historical determinists, the logical positivists, the pragmatists, the phenomenologists, and linguistic analysts each had their chosen anthology from the Marxist canon. Konstantinov scored a debater's point at the expense of his Marxist impeccability by replying that this showed their scholars were free. I forebore stating how scholars disliked the degrading necessity of citing passages to show they were hewing still to the Marxist line, but did raise the question as to why not a single outstanding work had been written by a Soviet philosopher.

My discussion of Soviet ethics brought a brief interlude of merriment. I remarked how their system of moral education seemed to revolve around the reading of pamphlets concerning the morals of revered figures – thus, Marx was described as an ideal husband and father, his wife as a loyal comrade, the daughters as happily brought-up model children. The fact of the matter was that his favorite child Eleanor was neurotic, victimized by a comrade, and a suicide; the mother was unhappy in their poverty, and there was the shadow of Marx's illegitimate son by their servant woman, the truth of which, I said, was just being made public by Western scholars. For a moment there was silence; nobody quite knew what to say, till one elderly gentleman, evidently a survivor of another era, chirped: "That shows that Marx told the truth when he said [quoting Terence] Homo sum; humani nihil a me alienum puto" (I am a man; nothing human is alien to my thought). Several laughed uproariously; probably they were of that elder generation that still studied Latin. It was the only Marx joke I heard during all my Moscow months at the institute.

Curiously the director asked if I would not appear for another such lecture-discussion. But I declined, which was just as well because I had to prepare for my departure and what proved to be a four-hour interrogation instead in a detention house of the KGB. My wife and daughter had left the Soviet Union by railroad for Finland the same day that I lectured to the institute. In Leningrad, however, Soviet agents stole her notes on her discussions with an elderly, much respected literary scholar, Julian Grigorevich Oksman, who after many years of dungeons, labor camps, and torture had emerged 'rehabilitated' and restored to the staff of the Soviet Literary Encyclopaedia. At the Finnish border they kept the train waiting while they threatened her with dire consequences; our daughter stood by and refused to leave. Arriving at Helsinki, she visited the United States Embassy to express her fears that the KGB intended to arrest me. The day before my leaving, as I was making a farewell courtesy call at the United States Embassy library, I was told that the Ambassador and the Security Officer, neither of whom I had previously met, would appreciate my seeing them. The latter informed me that they had reason to believe on the basis of the information from Helsinki that some kind of retaliation would take place against me, and that they were leaving it completely to my discretion and judgment as to how to handle the situation, but that they hoped I should try to get out of the country by any means I could.

I was naturally puzzled as to what to expect. Had the Soviet secret police somehow learned of my meetings with the Moscow dissident student circle? Could my inquiries into Stalin's past and his alleged expulsion from the Party in Georgia have endangered my Tbilisi friends? Could my conversations with fellow-Jews in several cities have constituted in Soviet eyes an 'ideological espionage'? For in the Soviet Union, no scientific study of 'contemporary trends in historical materialism' or its social tensions and discontents is indeed possible: for any such inquiry, that elsewhere in any civilized country in the world would be regarded as unquestionably legitimate, would in the Soviet order be regarded as a case of 'ideological espionage'. When Alexis de Tocqueville visited Jacksonian America in 1831, he regarded it as absolutely essential to have discussions with such Federalist opponents of the government as John Quincy Adams. The greatness of de Tocqueville's insight into the achievements of democracy and its possible misdirection even to despotism were in large part due to his discussions with those disaffected with the American democracy as well with those who supported it. The Soviet Union, however, is the only industrial society (apart from its satellites) that defines such study as 'ideological espionage'.

When I arrived at the Moscow airport very early in the morning, my customary institute escort suddenly took leave as I prepared to check my baggage. Two Soviet border guards then materialized like gloomy apparitions at a third-rate spiritualist seance, and escorted me into a small house in an outer yard. Three men were waiting to interrogate me, fortifying their questions with the periodic appearances of a burly uniformed officer, a sort of reminder that if here were no abode for Soviet wisdom, there was Soviet power. After some preliminary pseudo-civilities as to how much I had enjoyed and learned from my visit, they produced the documents that had been stolen from my wife mentioning that I had participated in a meeting with Professor Oksman. A long interrogation ensued in which I told them quite truthfully that my conversational Russian was not such as to permit my following a sophisticated discussion of Soviet literary criticism and theory, and the details concerning books and authors I had never read, and authors I had never met. I was prepared to acknowledge only that I had eaten an orange, which I remembered because they were so rare in Moscow. One KGB man insisted that I understood all their own interchanges, but a Foreign Office representative, who seemed to take a sympathetic stand, took my word. In the glare of an intense police lamp, I actually found it difficult to decipher stolen papers, and could never read more than the opening greetings. One was an unfinished letter to Gleb Struve, the son of Lenin's friend and foe, Peter Struve; the KGB agents were much alarmed when I said that Gleb Struve, professor of Russian literature at Berkeley, and I were indeed friends. They were much distraught that in their seeking to discover something sinister that the allegedly sinister name of Struve (whom they knew from their *Short History of the Communist Party*) had obtruded, I may not have helped matters when I humorously remarked that I was always interested in Gleb Struve because he was the only baby that Krupskaya described in the memoirs of her husband, Lenin. Perhaps, too, they feared that my wife or I had left copies of Pasternak's *Dr Zhivago* with Soviet friends, some such work that the Soviet totalitarian system with its gargantuan maw inflated with falsehood, finds indigestible. I added, perhaps not unmischievously, that it was also true that Alexander Kerensky, premier of the provisional government of 1917, lived an hour's drive from me. They looked at each other in some consternation. I explained I had never met Kerensky, but that in America, all sorts of persons could live within the same region freely. As they continued to threaten, however, to detain and arrest me, I demanded they inform the American Embassy. They then proceeded to search my luggage most thoroughly, my stockings, my Moscow-reinforced John L. undewear, my hat and cap. Nothing was there, except some manuscripts on Marx and philosophical and sociological reprints, but as they said, they were interested in neither Marx nor Marxism, but only in what concerned the Soviet Union. They seemed crestfallen that they had not found my diaries and personal notes, and looked at each other in surprise, at the absence of what they had evidently been led to expect. Indeed, for once, I was a scholar without portfolio; I had taken care to divest myself of my diaries and notes, and to arrange for their delivery through diplomatic auspices. At my request, furthermore, the Israeli Embassy had kindly instructed one of their diplomats scheduled to be a fellow-passenger on my flight to Warsaw to verify indeed whether I was seen boarding the plane. The plane had by this time long since departed.

I was pleased, however, to note during the course of the interrogation that the Soviet secret police had no inkling of my meetings with the student 'underground' circle at Moscow; they seemed to have been informed of our meetings with a Moscow Jewish movie critic, whose father had died with Babel in a prison camp, and to have precise details, probably provided by apartment house surveillance, as to the hours of appointment with Oksman. On the other hand, of my later visit to Oksman when he was ill at the academy hospital, and our talk of his executed friend Bukharin, they knew nothing. Finally, the KGB interrogators adjourned to hold a council outside, and probably to call their Ministry for instruction; the Foreign Office man looked at me, and said: "They will let you go, but they will never let you into the Soviet Union again." I remarked that I doubted whether I would ever wish to see the country again. The secret agents returned and told me I could leave by the next plane. As they had told me their names at the outset of the interrogation, I asked them to repeat their names which I wrote down on a piece of paper. The chief interrogator, a youngish man with a long, narrow, pinched face, further pinched by a pince-nez, and with an alternately aggrieved and menacing expression, but speaking English very well, regarded me oddly. They conducted me through a passageway to the airport. The airport clerk said there was a Sabena plane just about to leave for Brussels. I exchanged my ticket, adding my last few dollars, then waving goodbye to the Soviet representatives, said perhaps we would meet again some day. The Belgian plane, almost empty, took off; I asked the stewardess to let me know when we crossed the Soviet border. I could now appreciate my mother's comment when I told her I would visit Russia: "Why do you want to go there? People would do anything just to get out of it." When I reached Brussels, I telegraphed the United States Embassy in Moscow that I had gotten out; years later we heard that when the Israeli Embassy notified them that I had not been seen taking my scheduled flight to Warsaw, they had spent some anxious hours. Some weeks after we left the Soviet Union a Yale professor, friendly and knowledge-seeking, was arrested in Moscow by some Soviet secret police. It took President Kennedy's personal assurance to secure his release. I wondered whether the Soviet agents were angry that I had been allowed home, especially after a long *Times* news article had chronicled my Moscow debate.

At home in the United States, I spoke often to Jewish organizations that wished to be informed as to the lot of Jews in the Soviet Union. I expressed the opinion that if the pressure of world public opinion were kept steadily, and Soviet Jews themselves manifested somehow their wish to emigrate, the Soviet regime would in its own self-interest allow such emigration; thereby it would obviate internal discontent, and acquire some American goodwill that it needed at a cheap price. A few years later a wave of such Jewish emigration was indeed allowed for several years.

My chief achievement in the Soviet Union was to adhere to the principle that a visiting scholar should not connive at the regime's uncivilized practices. He should act as a freeman, and not acquiesce to the Soviet presumption that notwithstanding the fact that its citizens enjoyed the protection of their freedom abroad, no such freedom would be granted to the foreign scholar within its borders. If the Soviet Union chooses to act as a semi-civilized regime, it is not the part of the scholar to join in a chorus that proclaims such behavior as legitimate and civilized. A visiting scholar has made no commitment to de-civilize himself.

In the fall of 1964, a year after my return from the Soviet Union, the Berkeley student uprising began, and was later followed by a whole series of such movements that affected the United States, France, Britain, Italy, and West Germany. I was not unprepared for such a generational movement; since my reading at college in Kropotkin and the back-to-the-people movement, and my own writing in the early 30s for *Revolt* and *The Student Outlook*, and involvement in the first student 'strike' for peace at Harvard in 1934, I had wondered how such movements could be explained, for their motivations stood outside the economic laws that largely governed, for instance, the unrest of workers and their trade union activities. In my first published book, *Psychoanalysis and Ethics*, I had discussed how such factors as generational revolt were involved in revolutionary movements. Then in 1960, when the Japanese student movement led much of their nation into an irrational opposition to the defensive treaty with the United States, a travel grant had enabled me to undertake a journey to Japan. Beside lecturing at the universities of Waseda, Kyoto, and Hiroshima, and meeting as well with working-class groups, I had spent much time talking with the leading members of the Japanese student movement, and had indeed had initiated a major research study, replete with questionnaires and an assistant, that was later published. Since my own writings and activities of dating back to the pre-Second World War years were known to Japanese leftist friends, a common ground for our understanding in discussions existed; we had long arguments over our political disagreements.

My view gradually emerged that student 'idealistic' political movements arise when the motives of generational revolt, converging with those of the intellectual's longing for power, find an occasion for a collective, massive release of re-directed rebellious energy that some large-scale incident of political or military failure provides, and the responsibility for which can be attributed to the older generation. The elder Japanese generations, associated with the leadership in a war that ended in humiliating defeat, became the political target for energies of generational rebellion that otherwise might have expressed themselves in bitter resentments within the strict Japanese families; thus, a revolutionary 'youth movement' was the outcome of the otherwise scattered energies of generational rebellion in family situations being redirected and concentrated within a movement against surrogate university and political authorities. The word 'alienation' was popular among rebellious students precisely because nobody could define it; it expressed those vague nuances of generational resentment whose origins students preferred to leave unanalyzed. I presented a paper, "What is Alienation?", along these lines to the Berkeley Socialist Club; in Japan the irrational potentialities in the use of the concept were made manifest when otherwise intelligent persons veered stridently toward an anti-Americanism, despite the fact that their personal liberties, national independence, democratic constitutional evolution, and growing economic prosperity rested on American support. Indeed, the very fact that the Japanese democratic constitution was conceived under the aegis of the American military authorities made it seem tarnished with the weakness of the parental generation.

To the surprise of its corps of social scientists, the Berkeley student movement rapidly became even more intense and violent than its Asian and Western European counterparts. Almost the very week it began, a conference took place at the Hoover Institution in Stanford to consider the state of Marxism on the hundredth anniversary of the founding of the First International. In my brief contribution, I told the conference that I thought a major new chapter in the history of Marxism might well be making its first overt manifestation in the events taking place at Berkeley. Raymond Aron remarked humorously that if a revival of ideology was going to take place, despite his recent writing about its end, it would give him something again to analyze as its critic. At lunch, he ventured that the Berkeley unrest was a purely local phenomenon, and not to be expected in France. During the next four years, professors at a series of universities – Columbia, Cornell, Harvard, the London School of Economics, the Sorbonne, and Mexico City in turn – were confident that their institutions would be immune to the Berkeley type of irrationalism; then, in each, even more violent generational revolts flared; the issues were often illusory; if, at a large university students complained about the bureaucratic impersonality of the classes, in the more personalized Sarah Lawrence College, the young women fretted at the high degree of personalization in the seminars.

Perhaps the most unusual phenomenon was the collective exhilaration and play-acting that descended upon so many of the university professors. It was as if this new social movement beckoned to all the emotions otherwise strictly controlled within the academic unconscious: one had the sense of learning for the first time what lurked behind footnotes and formulae. At Berkeley, I watched a scholar of Chinese Communist history beating his breast publicly, and confessing his sins of indifference to past classes, and promising that in the new order he would behave differently. Evidently he was re-enacting in Berkeley one of those confessional scenes of penitent intellectuals that he had described in his writings on Maoist China. A youngish philosopher from Oxford engaged with messianic fervor in many speech-acts about the oppressed studentry; later he wrote a book about speech-acts. A historian with chancellorial aspirations who had written about Rosa Luxemberg and the spontaneity of the masses was excited that history was spontaneously uplifting the students and himself. Middle-aged sociologists and librarians-in-exile who hadn't addressed a mass meeting since their Trotskyist youths at the City College in New York felt the exaltation of the pulse-beat of history's sacred flux, and were moved to reaffirm eloquently the radical loyalties of their youth. I became aware how a crowd could recatalyze the unconscious. A pale scholar of the texts of classical political philosophy thrilled to the emergence of 'the political', and was incensed in the pages of a New York literary journal that anyone could think that the psychological phenomena of generational revolt were at all involved. A would-be dean for an experimental college on the old pattern of Alexander Meiklejohn's at Wisconsin more than 30 years previously thrust himself forward, joining his fantasy with the 'movement', and succeeding predictably in wasting years in the lives of an experimental group of students. At Columbia, I heard a well-known professor of drama proclaim that now everything was possible; evidently his studies of Brecht had gone to his head; at least he had the decency to resign. At Harvard a professor of mathematical logic announced he would live in a Maoist-style commune; fortunately Harvard's cultural revolutionaries did not quite emulate the way the Chinese Maoists treated logicians.

The Berkeley experience brought forward sharply the psychological weaknesses of the academic intellectual class. A curious subservience arose, for instance, toward the allegedly higher insight of the young. A distinguished visiting professor from Oxford, eminent for his knowledge of ancient law, said in a public talk that when controversy erupted, he kept to the side of the young, and that that was why he differed with me. The scholar was a Jew; he had seen the rise of the Nazis when the German student movement rallied with immense enthusiasm to Hitler's party and army. Nonetheless this great scholar's political reasoning was as fatuous as that of an uneducated lout. His and my generation had surmounted class ideology; was it now to acquiesce in an equally fallacious generational ideology?

The universities meanwhile had become major centers contributing to the making of foreign and domestic policy because of the highly complex and far-flung character of the problems. Washington and Lincoln would never have dreamed of turning to the universities for advice and knowledge; no president today could consider not doing so. Yet of the instabilities of character in the university intellectuals, and their proneness for making decisions more expressive of their irrationalities than a rational resolve toward problems, we know very little. To what extent is the recurrent 'treason of the intellectuals' (as Julien Benda first called it) a product itself of the psychological formation of intellectuals, their 'intellectualization'? Why, as Winston Churchill asked in 1933, was it that "Our difficulties come from the mood of unwarrantable self-abasement into which we have been cast by a powerful section of our own intellectuals"? At Berkeley, I wrote some papers dealing with this question, and finished a long study 'The Sociology of the Intellectual Class' that circulated in typescript, but that remained unpublished when the volume for which it was intended was abandoned by its academic promoter.

During the Berkeleyan troubles, I attended many student meetings, collected their leaflets and literature of foment and grievance, and kept writing the historical chapters of my book *The Conflict of Generations*. At an oft-described faculty meeting in December, 1964, that enacted a resolution guaranteeing freedom of speech and advocacy, I introduced an amendment that would have maintained such freedom short of an advocacy of the immediate use of force or violence. Several members of the faculty joined to argue for his amendment, but with the huge crowd of 'activists' outside, and the microphones and loud speakers blaring the proceedings to them, a certain non-rational, mood of yielding perhaps possessed genteel academics who were far more at home merely writing in their studies of Milton's *Areopagitica* or Justice Holme's 'clear and present danger' criterion. Our amendment failed as its opponents ridiculed the notion that the Berkeley campus could become a stage for violence. We were charged with lacking faith in the Berkeley studentry. Within two or three years, there followed the unprecedented outbreaks of violence, bombings, arson, and clashes with the National Guard that brought death and blindness as a consequence for individuals; the main room of the library was gutted by fire; the campus indeed did become the staging area for violence as students and non-student auxiliaries (or vice versa) graduated to successive levels of violent opposition during the time of the Vietnamese War. Several years afterwards in Toronto I heard that the Berkeley faculty, chastened by experience, enacted the equivalent of my amendment, though still trying to envelop their earlier lack of judgment or courage with polemics directed against me by name. I was not there to reply. Meanwhile, of course, Ronald Reagan had been elected governor of California by a large majority in an election in which the student violence and the university's capitulation was the chief issue. It was the first election in which the 'treason of the intellectuals' was explicitly in the minds of the voters, and that, of course, had much to do with the shedding of some of the professorial extravagances.

Among the Berkeley faculty who stood out most clearly against the collective irrationality of the intellectuals, a small group composed of such scientists as Hardin Jones, Daniel Arnon, John Lawrence, and Edward Teller, was most outstanding. At Berkeley, there was a veritable anti-Teller cult among the intellectuals. In later years, as I read the documents of the Oppenheimer case, much of that hatred seemed to me to be due to the fact that Teller had been right both in his judgment that a thermonuclear bomb was practicable, and that its construction was essential if the Soviet Union were not, by manufacturing it first, to be encouraged to try to realize its domination in world affairs. As for Teller's view that Oppenheimer's moral and political ideas might well have affected his scientific judgment, was that not after all consonant with much that we know of the sociology and psychology of scientists? The same sociologists of science who documented at length the influence of the scientists' social and political ideas on their scientific work wanted to make Oppenheimer an utter exception to their rule. Einstein, for quite other reasons, bitterly disliked Oppenheimer and, according to Ms Helene Dukas his secretary, once said, tapping his heart: "Where others have a heart, he has nothing." Probably scientific disagreements blended with the moral ones.

If there were friendships that were dissociated during the years of America's academic tubulence, there were others that became more deeply grounded. It was a pleasure to be in the company of Sidney Hook, as he took up the task of clarifying the ethics of the American scholar and scientist, a task momentous because an endemic bitterness against free thought was generating forms of disruption new in America as many students and teachers alike betrayed their calling. Throughout this period, Hook remained always ready to reason, to argue, to point to overlooked considerations, to appeal to the element of rationality that survives in every human being. The society that he improvised, 'University Centers for Rational Alternatives', was the only one in those critical years that functioned effectively to band together their defenders. I had first met Sidney Hook at the Philosophical Circle meetings in Christmas, 1934, when we discussed Stalin's reprisals against the assassination of Kirov. During several years at Philosophical Circle meetings we had argued. Always fair-minded in discussion, always really trying to learn from the other person and not simply to refute him in verbal joust, his philosophy grew in wisdom, became more many-sided, more conscious of the complexities and uncertainties of things. He was known as a feared polemicist, and indeed dogmatists and fellow-travellers avoided encounter with him because he patiently unravelled their unreason; I knew rather the underlying spirit, generous in combat and loyal to the human aspiration.

At this juncture I began to wish very much to leave Berkeley. I wanted to write several books and saw little prospect of doing so in the Berkeley environment. Perhaps the university had lost its energizing stature for me. At Barrows Hall, professors would ask me as they looked about fearfully, whether I wasn't afraid that I would be assaulted on the Plaza. I did for the sake of my daughter and her friends engage in a debate at their high school auditorium with the youthful undergraduate orator who then commanded so much public attention, Mario Savio. It had a humorous outcome as I got him to acknowledge that the accusation against him that he had used violence against a campus guard was true; he had bitten a campus guard in the left thigh, or calf. It was that charge which had elicited the first strike against the alleged 'frame-up' and 'betrayal'. But truth is a minor consideration when the messianic, ideological fever is running high.

The Department of Philosophy, moreover, at Berkeley was a group beset with ugly feuds. When I published a psychoanalytical study of Descartes, Professor X in one faction told me that my method would be highly applicable to Professor Y of the other faction whom he heartily disliked. But Professor Y indicated that my method would find a fruitful subject in the thought and character of Professor X, whom he equally disliked. For once, I agreed with both. Several older men were preparing to retire; a middle-aged group that dominated the department seemed to do little writing, but were exceedingly active in administrative positions; among several able younger men was Thomas R. Kuhn, then holding a joint appointment with History; the middle-aged group wished to oust him from the Philosophy Department, and were finally successful in their cabal. Although only one-third of my work was with the Department of Philosophy, I served frequently on the all-university committees that considered their recommendations for promotion. Thus, it was, for instance, that I was apprised of their wish to rid themselves of the historian of science, Professor Kuhn. On my return from a brief sojourn in Japan, I found myself appointed to a university committee that would consider the curious recommendation of the

Philosophy Department in the case of Kuhn; unanimously, its tenured full professors joined in recommending his promotion provided, however, that he were not allowed to continue in the Philosophy Department but was transferred altogether to the History Department. This recommendation seemed to the committee a bizarre one; for how could the Philosophy Department take part in recommending the promotion of a person in another department who was not be a member of its own? The underlying fact was, however, that Kuhn had incurred the philosophers' displeasure by having had private conferences with the dean as to what could be done to improve their intellectual standing. The university committee unanimously proposed that the chancellor consider overruling the Philosophy Department. I informed Kuhn that if he chose to challenge the Philosophy Department's position, he would find much support. But he was downcast, and preferred to let things take their course. A year or so later he left Berkeley for Princeton.

I also served on the university committee that considered the promotion to a full professorship of the man who later served as chairman during the period of Kuhn's case. This deliberation was unusual because the candidate had never published any independent essay, though he submitted various manuscripts. It transpired that these were mainly works rejected by various journals. The committee accordingly sent in a negative report, but was reversed by the chancellor.

The Department of Philosophy, for its part, took a jaundiced view of an appointment of Karl Popper, precisely because of his famed publications. As a post had become available, I proposed an offer to Karl Popper. One member of the department, said to be learned in ancient philosophy, assailed Popper's writings on Plato, and urged instead the appointment of a friend of his, a legal philosopher; the latter had the advantage of not having published any book, nor as I recall, any major article, and was therefore immune to criticism. Another departmental member, at least as well informed in the classics, noted that outstanding classical scholars agreed with Popper. Nonetheless the opposition to Popper prevented anything but a semester's visiting appointment. He came, was mostly ignored, was never invited to give a public lecture, and was left to his own devices in a strange town. The legal philosopher got the vacant post, and naturally became a militant enthusiast for the redemption of the world by the cohort of students who never studied.

What I observed of the Berkeley Philosophy Department strengthened my long-growing view that significant philosophical work rarely comes from a preoccupation with academic philosophy. Rather some new departure in the sciences provides the experience and motivation in which a fresh philosophical idea is born. Otherwise, the fluctuations of philosophical opinions are cyclical, even predictable phenomena, in accordance with the law of the periodic alternations of ideas, and the principle of wings, which I set forth in my book *Ideology and the Ideologists*. I felt a desire to be free from the academic responsibility of having to spend time keeping up with a literature that had become so cultistic and verbal. Another university committee at that time, whose membership included a Nobel Laureate or two, puzzled over the appointment documents of an English analytic philosopher; they fingered his meagre collection of papers with their prissy, affected titles of word-play, and wondered whether this was meant to be taken seriously.

As it happened, the chance came for me to enjoy a decade devoted more fully to political, historical, and sociological studies. The president of the University of Toronto, Claude Bissell, had written most warmly in a letter to *The New Leader* of an article I had published on the Berkeley events. As most of my uncles, aunts, and cousins still lived in Toronto, and I had warm recollections of my visits with them, the thought occurred to me that perhaps a move there would provide a setting not only helpful for work, but also for the first time, provide an opportunity to enjoy familial associations. President Bissel encouraged my suggestion of such a move; a post was available exclusively in sociology, principally because the salaries paid to philosophers were so low in comparison to the sociologists; also *The New York Times Magazine* had published an article of mine to which, without either my permission or knowledge, they attached the sensationalistic title 'American Philosophy is Dead'. I was quite annoyed at first, but then reasoned that the title was not inapt insofar as American philosophy was dead in the same sense God was. The prospect of giving an honors course in the history of sociological theory, and a seminar in the sociology of ideas was attractive to me. When my wife, too, was offered the chairmanship of what was soon to become the largest Slavic department on the North American continent, I felt that we should move to Canada. A few days prior to our departure from Berkeley, my farewell article, 'The Decline of Freedom in Berkeley' was published in *The Atlantic Monthly*, September, 1966. The managing editor of the magazine told me that it sold more than any other issue in its history, and brought it the largest quantity of mail. The discussion included a letter from Ronald Reagan, the Republican candidate for California's governorship, beside friendly words from Howard Mumford Jones. Most of the Berkeleyans, however, took issue with me, some in rather intemperate ways. Its speech-actor charged me with quoting from an unsigned leaflet, a charge dead wrong, as I pointed out, because the document was signed by an array of names. A professor of American literature pronounced against my style of writing; the editors of a half dozen textbooks and anthologies for English composition, however, differed with him, and reprinted all or parts of my article as an example of masterly English prose. Berkeley's most well-known political sociologist attributed the Berkeley phenomenon to the specifically Californian extremism, evidently rejecting my view that it would spread throughout America, and that a kind of 'Intellectuals' Party' was in formation. Events proved him wrong during the next years, especially when Senator McGovern's followers seized control of the Democratic Party. I had known this sociologist since he was a City College freshman, and can recall nobody who was more consistently wrong in his political predictions, though this consistent record never diminished his dogmatic assurance. I should add that he also cited Aristotle to me on 'the essence of tragedy'; I thought it strange that when university professors are intimidated by neo-barbarian behavior they should be elevated as tragic heroes. What the relative proportions were in the professors' assemblage of rationality, cowardice, confusion, liberalistic messianism, or a sheer indifferent conformity to the idols of the tribe can never be precisely determined, but the record of the debate, and the accounts later published by participants, scarcely would support the belief that the tragic virtues suffused the auditorium, nor that clearly reasoned conviction played much part in the enactment of what became a charter for campus terrorism. In any case, the Berkeley professors sustained only a small penalty for their flawed aberration into revolutionary politics; Governor Ronald Reagan denied them salary increases for a year or two.

My book *The Conflict of Generations* was brought to completion by the end of my first year at Toronto. Delays in publication, however, ensued; meanwhile, the spreading student outbreaks and the fanaticism of the opposition to the Vietnam War intensified the violence on many campuses; it seemed I might always have to add paragraphs about current episodes to preserve timeliness. Finally, the book was published in the spring of 1969. The division of opinion among its critics was foreseeably sharp. There were favorable review from Henry Murray, Otto Klineberg, and the literary critic and historian of anarchism, the genial friend of Orwell, George Woodcock. That by Professor Murray had an especial significance for me. When he undertook to write the review, Murray telephoned me. He wanted to know something about me, where I had come from. He did not know that I had been a faithful auditor of his class on Abnormal Psychology more than thirty years earlier. I reminded him of all sorts of incidents in his class, and sent a letter with the personal details that he requested. Much puzzled by the hostility that the book had evoked even among his colleagues, he wrote in *The American Scholar* of

> the several reviewers that I met foaming with indignation at the author of this book, avowedly for his being a bad theorist but viscerally between the lines, for his not idolizing kids . . . But to have guessed the choleric intensity of the above mentioned critics you would have had to know Professor Feuer's changes of position . . . , and you would also have had to know something about the light brigade of touchy juvenolaters . . . At this date, so far as I know, there is nothing with this coverage, nothing that will bear comparison with the scope and span of Professor Feuer's absorbing history, nothing with its prodigality of relevant, amply documented facts and figures . . . [I]t looks to me as if the possession of this knowledge early last winter would have enabled a person to predict with but few errors the course of the student disruptions that occurred – especially those at Harvard. Not often is it possible to say this much with confidence.

Professor Murray did indeed state his very clear differences with my occasional use of Freud's hypotheses in my attempt to work out a theoretical explanation for the recurrence of generational conflict throughout history. Actually my use of such hypotheses was never logically intrinsic, although consistent, with my generalizations concerning generational conflict as a universal historical theme. And Professor Murray himself might well concede that in the cases of the student leaders and militants there was considerable evidence indicating, for instance, a high frequency of Oedipal problems. Since 1969, newspapers and magazines have from time to time reported on the psychological history of chief figures in the Berkeley student movement; mental collapse, suicide, and other varieties of aberrant behavior have rather lugubriously confirmed the general tenor of my analysis. Neo-leftist critics attacked me for allegedly 'reducing' the explanation of student movements, to one causal factor, namely, that of generational rebellion, and inferentially, to motives stemming from Oedipal complexes. This was a rather obvious psychological mis-reading of my book, because I devoted much time to the fact that no such movement arises unless the elder generation has been de-authoritized, discredited, through the failure of its leadership to guide, for example, a war to victory, or to achieve some promised basic, needed reform. Only when such a situation of de-authorization exists is a political fulcrum provided around which individual resentments of a generational kind can coalesce; otherwise they would probably express themselves in a more sullen, less dramatic way, within their families. Several critics failed to note utterly that a universal theme like generational rebellion takes diverse forms under diverse social conditions. Young Americans in the first half of the nineteenth century often left their father's New England farms to go West; in Eugene O'Neill's play *Beyond the Horizon*, that was indeed the way a young son might resolve his generational conflict, with its Oedipal component, by leaving his father's house. And Alexis de Tocqueville could thus note that adolescence seemed a skipped stage in America. Later, the sons of working men could surpass their fathers and resolve their generational animus by venturing into new industries, and finding better jobs. Within the social setting of universities, generational revolt veers, however, toward more irrational and ideological forms. Historical materialism is stymied when it tries to explain the irrationalities of students, well housed, well fed, well leisured, and enjoying the best in books, culture, and companionship, as the outcome of economic aims or class struggles.

A large batch of reviews of *The Conflict of Generations* culled from leftist journals, many of them briefly burgeoning at that time, was sent to me by the publisher. I read a series of them; some were frenetic, others were vulgar, a few were worse. I recalled Einstein's remark to a friend: "One doesn't have to read the stuff"; it was good advice that I followed. I had begun working on what became another book, and it was futile trying to reply to critics whose expressions showed such transparent resistances against the inroads of psychoanalytical truths concerning themselves. I had evidently tapped a sick nerve beneath all their ideological posturing. Clark Kerr, the president of the University of California, had commented on the risks an inquirer into the psychodynamics of a current movement faces; the unwilling object of such an inquiry, upon finding his rhetoric deflated, tends to flail out at the inquirer. For many years, leftists had revelled in the analytic probing, the 'unmasking' (as they called it) of 'bourgeois' thinkers, conservative, liberal, and democratic leaders, entrepreneurs, magnates, builders, men of the establishment. But to have such a method turned against themselves!

Indeed, the whole outburst of indignation against me for using psychoanalytical, Freudian ideas like the 'Oedipus Complex' was demonstrably bogus. For the same critics were usually enthusiasts at the same time for the writings of Herbert Marcuse, the most popular author among neo-leftist readers, and Marcuse had depicted the motivations of the revolutionary intellectuals and the student 'activists', all making the 'Great Refusal' in terms that were far more Freudian than anything I essayed. What mattered clearly was not whether psychoanalytical ideas were or were not used but that if they were, they should conduct one to the proper revolutionary conclusion and condemnation of bourgeois society and bourgeois science. Since Marcuse and Reich maintained that the repressions in bourgeois society, especially the Oedipal ones, were at a higher level that ever before in world history, and were rendering students so increasingly miserable that a social revolution was absolutely essential, they were applauded by many professors and ideologists as prophetic intellectuals. All ideologists take their support where they can find it; the method employed is of little consequence. The master ideologist was one who endorsed as valid the students' attack on academic freedom and scientific method, and who approved their burnings and bombings of libraries in the 1960s and 1970s as 'creative vandalism'.

Toronto during this period provided a congenial place for study and writing. In the first place, Canadians do not work very hard; the university's academic year was the shortest I had ever enjoyed; classes began in late September, and ended the first week of April and usually met for only twice a week for an hour, so that my weekly teaching was never more than four hours a week; since courses were of year-length, one had to read only one set of final examinations; the students were usually conscientious, though brilliant ones were rare; the library was excellent; and I had more time to pursue my inquiries than I had ever known in my life. Also I had the pleasure denied me since youth of having several Jewish restaurants and bakeries adjacent to the university, where I could imbibe with calm the Kosher soups, sandwiches, rye bread, and even meat dishes, and adhere to my gastronomical Judaism. At *The Bagel* and the *Famous Tel Aviv*, I could meet with colleagues, friends, and students, and have a noodle or vegetable soup that was not defiled by intrusive globules of bacon; discussions of philosophy, politics, and sociology came then so much more easily and I could understand better why the great scientific movement of the seventeenth century found its temples of freedom in their coffee-houses.

'Off-Broadway' and amateur theatre groups abounded in Toronto: moreover, as the bookstores a few streets from our house were open into the late evening, I could stroll down Yonge Street through the snows to enjoy them, or pass an hour or two at the fine, comfortable library at St Michael's College of which I became very fond. We attended our family's weddings and bar-mitzvahs faithfully, and heard enthralling stories of 'olden times' from my grand-uncle Avrahm Moishe and my favorite uncle Alex; once he told me how when he first rode an American train as an immigrant of thirteen, and was eating a strange fruit called 'banana' without peeling it, a black porter approached him and instructed him in that art. My father too travelled from New York for reunions, and discovered that our cleaning woman, Jerre, and her husband had emigrated from the same Polish county.

During my first Toronto years I returned frequently to the United States to give lectures or take part in symposia on subjects ranging from the student unrest, to Soviet society, to the social-intellectual background of Einstein's theory of relativity. Though I avoided, as an American citizen, any personal involvement in Canadian politics, I was the principal speaker at the Young Conservatives Convention in Ottawa, where I enjoyed a talk with Robert Stanfield, the thoughtful leader of the Conservative Party, and another such in Toronto where I shared the platform with the young Minister of Education, Wiliam H. Davis, who later served for many years as the Prime Minister of Ontario. Lectures at Canadian universities took me as far as the fastnesses of Nova Scotia and Newfoundland, where at Memorial University I found a whole company of left intellectuals from New York; they had sought out a retreat whence to excoriate and repudiate the bourgeois civilization (that still cushioned their ease).

Above all, we enjoyed many evenings of discussion with a reporter of the *Toronto Telegram*, Peter J. Worthington, who had just returned in the fall of 1966 from several years as correspondent in the Soviet Union. At considerable peril he had helped his Russian secretary make her escape from the Soviet land, and first met us while he was exploring the ways and means of enrolling her as a graduate student in Russian literature; he telephoned the chairman of the Toronto Slavic Department, my wife, and soon we were comparing our Muscovite notes.

Worthington evinced a tremendously inquiring mind; mention the title of a useful book and you could be sure that by the next time we met, he would have read it. His store-house of experience was the most variegated and interesting of any person I had met; the son of Canada's first tank commander, he himself had fought in Korea with the Canadian Army, been Schweitzer's guest in Equatorial Africa, covered the Biafran war in Nigeria, and befriended the ostracized Soviet attaché, Igor Gouzenko, who had dared reveal the details of Stalin's espionage network in North America. When the *Telegram* ceased publication, Worthington the following day co-founded a new daily newspaper, *The Toronto Sun*, of which he became editor-in-chief. All the left establishment intellectuals predicted its early demise. Instead, it became the most successful newspaper in Canada. Peter was able to test his editorial ideas on my wife and myself; often the subjects pertained to matters of which I was utterly ignorant, but I did my best to place myself with the average Toronto subway rider. The subways, I might note, were truly one of Toronto's prides, clean, swift, efficient, and when we first came, still like the city, with an exceedingly low crime rate. Toronto was justly called Toronto the Good.

To be sure, the Toronto Department of Sociology was ruled, as it had been for many years, by a dictatorial individual who had difficulty controlling his dislike for Jews; such sentiments had long been the norm in Canadian universities. I made allowances for some time, but when his dictatorial behavior at one occasion exceeded even my forebearance, I said in two minutes publicly what I thought of him, whereupon his authority collapsed, and he resigned his chairmanship the next week, after the dean, an unusually fair minded man, proved unsympathetic to his plaint; a hitherto obedient departmental majority withered to nothingness. Many Canadian social scientists at this time began speaking about a distinctive Canadian school; the more mediocre a professor was (including a growing number of anti-American Americans), the more apt he was to imbibe from the nationalistic pick-me-up. When I investigated the matter, it appeared that the tenets of a self-described distinctive Canadian school of thought usually consisted in what its proponents' teachers had learned at Chicago fifty or sixty years before.

Of course, after several years a diffusion of student violence took place from the United States to Canada. Toronto was not spared a collapse of will on the part of many of its officers and professors; all sorts of physical indignities were visited upon the former in the parliamentary chamber in which they convened. Anti-American slogans were rife, as Communist North Vietnam became an adored model, and Che Guevara a 'culture hero' for the leftist psyche. The provincial parliament, pressed by the leftists, enacted a constitution for the university which had the janitors, typists, and cleaning women involved in the process of choosing a university president; students naturally were given a major electoral role; within a couple of years only 11 percent of them were voting, which was just what the leftist sects wanted to ensure a disproportionate influence. As an American, I remained mostly disinvolved with these doings, and since I was not known to be a practitioner of non-violent quiescence, 'neo-leftist' hyper-activists kept away from my classes and paths.

The research and writing of *Einstein and the Generations of Science* was what principally concerned me at this time; it welded together the themes of my previous books as well as affording a chance to develop my own views in the philosophy and sociology of science. While I was a graduate student, and reading as well in the history of socialism, I had been struck by the coincidence that Albert Einstein and Friedrich Adler, the secretary of the Socialist International, had as student friends at Zurich been followers of the same scientific philosophy, that of the positivist Ernst Mach. Then I had listened with much interest to a paper by the physicist-positivist, Philipp Frank, an old friend of Einstein, at the International Philosophical Congress in Paris in 1937. One day in 1938, as I walked along Riverside Drive Park, I saw him sitting on a bench, lonely, staring at the river, a refugee from Hitler's advance into Czechoslovakia. I sat down beside him; he brightened to know that someone recognized him; we talked of many things, and some days later I helped him carry his wife's baggage, when she arrived, to their room near the International House. In time, we became good friends; he was much interested in political and social subjects, and was already preparing to write his biography of Einstein; his wife would recount anecdotes of her days on outings with Franz Kafka ("he laughed at his own stories, and told me not to read them"). During summers at Cambridge, after he was appointed as an Associate in the Department of Physics, I asked him many questions concerning Einstein, Adler, and their milieu. I heard Adler, too, speak with his thunderous voice at a Jewish worker's rally at Carnegie Hall, and one day saw him wandering, worn, in the New York Public Library. It was strange to think that this philosophical man with tired, aged idealistic eyes, a shaggy brow and bent shoulders, had once killed the Prime Minister of Austria, and urged upon young Einstein to join the Machian assault against all metaphysical entities.

When I reviewed Frank's book on Einstein in 1947, I ventured some suggestions as to how intellectual currents in socialist Zurich in the early 1900s might have had a formative influence on Einstein's mode of thought. Frank commented to me that he thought indeed there was a 'connection with 1905', the year in which the revolutionary ferment everywhere was stirred with the revolution against the Russian autocracy. The problem was a crucial one for the sociology of scientific ideas. For whenever discussions had taken place concerning the possible influence of political and psychological factors on the formation of scientific hypotheses, always Einstein's theory of relativity would be adduced as the sockdolager, contrary case; a theory so abstract and removed from social considerations seemed clearly to stand outside the purview of an extra-logical causal line. Nonetheless, it transpired that in Einstein's case, a remarkable drama had been enacted in which a generational circle of youthful rebellious friends, those of the 'Olympia Academy' and such as Adler, had sustained Einstein in transmuting into constructive channels of novel scientific theorizing such feelings as might otherwise have festered in personal discontent. The Machian philosophy that helped him dispense with absolute space and time was at that time highly popular among young intellectual revolutionaries, and 1905 was in several senses a year of revolutionary stirrings. Regrettably Philipp Frank was no longer alive when my first essays on the subject were published in *Annals of Science*.

I then went on to extend this approach to the psycho-social and philosophic backgrounds of Niels Bohr, Werner Heisenberg, and the Prince de Broglie. What impressed me was the truth of the principle that had so upset the Marxists in 1947 – that no particular philosophy or social setting had a privileged place in the formation of scientific hypothesis; Bohr was a kind of Kierkegaardian; Heisenberg was a Platonic, romantic idealist; de Broglie was drawn to the Bergsonian imagery of compenetrating waves. Their socio-political standpoints likewise traversed the breadth of parliamentary chamber, and each likewise had in his youth participated in a generational circle that shared something of a common outlook. The cases of Heisenberg and de Broglie were the most unusual, since they contravened our own American pragmatic, liberal values; Heisenberg's was the most extreme, for as a romantic idealist of the German Youth movement, he had become a follower of the Nazis, while his principle of indeterminacy and mathematical conception of the atom originated in his thinking, as he told, while he stood sentry duty as a schoolboy against the Munich Communist Army. What would Jefferson or John Dewey have made of a scientist who claimed that an anti-democratic animus elicited his highest scientific creativity? De Broglie likewise, a scion of perhaps the most aristocratic family in France that had suffered much during the Revolution in 1789, sought, like de Tocqueville in political theory and Joseph Fourier in physical theory, to sight continuities of superposed waves where others saw only quantum discontinuities, and he stood aloof from the prevalent republican positivism.

Every philosophy, I concluded, was for its individual bearer, associated with certain emotions, images, and longings that, when interacting with the state of knowledge, uncertainties, and problems at the given time, might, in favorable circumstances, incite a heuristic approach for fruitful, novel insights for theoretical generalization or indeed for experimental innovation and finding. In his own lifetime, the scientist's philosophy might change, and lead to still further novel hypotheses; thus, Einstein's philosophy evolved from Hume toward Spinoza, from empiricism to belief in a rational universe governed by simple, mathematical laws; this latter, relatively more mystical frame of mind helped him conceive his general theory of relativity and cosmological conception of a static, finite universe.

Ultimately, the character of the scientist determines the kind of law of nature that he will spontaneously seek to discover. The concept of 'scientific character' is one that I should hope to explore more deeply than I have. Laws of conservation, for instance, have curiously been discovered by men with conservative feelings – Joule, Mayer, Helmholtz. Laws that employed the idea of equilibrium have been pre-eminently preferred by such men as Pareto, Duhem, and Willard Gibbs who shared an aversion to all socially disequilibrating situations. Probabilities and deviations found their logical pioneers in the nineteenth century in two men who were personally in extreme rebellion against social norms: Charles S. Peirce and Karl Pearson. Deterministic explanations, on the other hand, have been much preferred by such Jewish scientists as Einstein, Jacques Loeb, and Freud. Why? Largely, as I have suggested, because the determinist mode of analysis was a weapon of intellectual retaliation on the part of members of a defenceless minority against the cruelties of a majority; to explain the Nazi movement deterministically was a way of exposing the lowly, unflattering causes of their high-flown pretensions. When one's resentment against cruelty, injustice, and physical oppression cannot express itself in direct aggression against the oppressor, it issues in the indirect aggression of a determinist explanation of the oppressor's acts. Nothing so much irritates a tyrant and his henchmen as a deterministic explanation of their behavior. The phenomenalist mode of scientific laws, we might similarly observe, was especially cultivated by French scientists in the early nineteenth century; having experienced most acutely the arbitrary cruelties, even executions, of their family members by the Revolution's Reign of Terror, they were tired of 'ideologies', especially of the materialistic variety, and preferred phenomenalist accounts that no new turn of fashion could upset; such were Fourier, Ampère, Gay-Lussac, Carnot. In 1976, I set forth these views in a long paper to the Conference on the History of Physical Science held at Baltimore in 1976, and I returned to this subject in a paper 'Teleological Principles in Science', written within a year after we have moved to the University of Virginia in the summer of 1976. It might be said that no scientist ever discovered a law, or even conceived a hypothesis, that he didn't like; the hypothesis, in a psychological sense, is a teleological principle; its use projects the desire that the world will prove to be of a kind that somehow conforms to the scientist's deepest unconscious longing. Where the scientist has experienced a conflict of dual feelings with respect to his hypothesis, as Charles Darwin did with respect to his theory of natural selection, his unconscious may be the stage for a long battle within himself to surmount certain resistances, that derive from his early education in the family circle and religion. Darwin's long, melancholy delay in publishing *The Origin of Species* was mainly due, in my opinion, to the struggle within himself between his own agnostic pessimism on the one hand and the religious optimist philosophy and acquiescence to the Church of England with which he had begun his voyage on the H.M.S. *Beagle* on the other. And if a great book remains unwritten, it may well be, as with John Stuart Mill's unwritten *Ethology* and Lord Acton's *History of Liberty*, because the facts that they collected never quite seemed to fit the teleological principles that they longed to see fulfilled – a law of progress and the advance of liberty.

The popularity that the idea of 'scientific revolution' acquired, however, seemed to me sociologically the outcome in large measure of a kind of 'religion of revolution' which, during the thirties especially, became the badge of the intellectuals. Even in the New York Philosophical Circle where the notion of a 'dialectical method' in logic was almost altogether rejected, still it was said that the historical evolution of scientific ideas might well exhibit a dialectical structure – that is, each system of dominant scientific ideas would engender in the course of time its own theoretical or experimental 'contradiction' that would be eliminated only by formulating a 'qualitatively novel' set of concepts and laws. Once I had a long discussion with Samuel Wolfenstein, later professor of mathematics at Paris, as to whether this pattern, exemplified in the case of Einstein's theory of relativity, was generally applicable to the history of science, or only partially so. By the 1960s, however, I had come to the conclusion that the idea of 'scientific revolution' was beginning to be used with deleterious consequences, and that Boltzmann had enunciated an important, sometime forgotten truth when he denied that there were experimental revolutions; for those observations recorded from a given base of scientific instruments still remain observations that are valid relative to those experimental conditions; the theoretical law that has explained those experimental findings is not "negated" by the more general theoretical law that supersedes it; rather it survives as a special, limiting case of the more general law under specified conditions. Nothing of this kind is true in social or economic revolutions; the capitalist system of pre-revolutionary Moscow did not become a special case of the Soviet socialist economic system, nor were feudal privileges and immobilities, such as that of the serfs, special cases of the economic laws of the free market, although all revolutions as well as social systems are themselves governed by sociological laws of a higher generality. Moreover, the novel scientific ideas as of the quantum, relativity, and natural selectionist theories have usually not found themselves barred from publication; nor have the papers presenting them been denied a hearing at scientific meetings. Those sociological traits that characterize revolution as a social process have been notably lacking in the evolution of modern scientific ideas. Though some of the psychological characteristics of the revolutionist as an individual will be found in the scientist who fashions a novel mode of thought, those character-vectors have been channelized within the constitutional process of scientific verification, into the constructive procedures of the growth of science rather than toward an obliterating of the accumulated previous findings. Social revolutionists of 1794 and 1917 used the language of unfettered power and the suspending of all judicial safeguards by a dictatorship of the Jacobin Club or the Politburo of the Bolshevik Party. The language of 'scientific revolution' has been correspondingly a metaphor much prized by fixated generational rebels, and persons touched by the ideologies of fellow-travellerism and Marxism among the intellectual class. Far more congenial to the American pragmatic spirit have been the metaphors rather of the frontiersman, the pioneer, the explorer. By contrast, the linking by Leninists of the notion of 'scientific revolution' to their social one evoked the demand for a new, revolutionary, 'class science', a 'proletarian science' that would overthrow 'bourgeois science'; the persecutor of genetics, Lysenko, invoked the ideology of scientific revolution to justify himself, as have the mob-like followers of Velikovsky at a session of the American Association for the Advancement of Science.

To explore further the traits that differentiate ideology from science, I wrote a short book *Ideology and the Ideologists*. It differed from other works on the subject insofar as it traced a common pattern in all ideologies; all were variants of the Mosaic narrative telling a story of the rescue of some oppressed group, class, race or people by a youthful elite. All ideologies, furthermore, derive their 'truth' by a fallacious deduction from the scientific or philosophic ideas that are prevalent at the time, and all of them evolve rapidly into forms of anti-science as they insist on denying every crucial experiment that contravenes their claims. The ideologist is perforce a fixated generational rebel; whereas Marx and Engels believed that ideology is propelled by unconscious economic motives, it is much rather the case that ideological thought is propelled by unconscious aims of generational rebellion, that of the sons against the fathers, what Antonio Gramsci in his prison's silence when he put Marxism far behind him, called in appreciation of Freud, "the eternal conflict between father and son" (Gramsci, 1979). Again, as was to be foreseen, ideologist-critics railed against my book, though their indignation seemed inversely proportional to the weight of their contrary evidence. On the other hand, Professor Maurice Cranston, of the London School of Economics, whose work I much respected, welcomed my book in the *Lugano Review* (1975/76) as "arguably the most important sociological study of ideology to have yet appeared". One critic raised the point that since Zionism too was an ideology, my sympathies for the state of Israel rested therefore on a tenuous foundation. What this argument overlooked, however, is how little the emergence of the state of Israel had to do with Zionist ideology. During the 1920s in the

United States, the number of members in the Zionist Organization had dwindled to about 11,000. Then the advent of Hitler and his National Socialist movement ignited the most virulent anti-Semitism in the world's history; European peoples and governments proved incapable, if not unwilling, to save their Jewish citizens, or rescue foreign Jews. Israel became a last alternative for those who, if God's Chosen People, were also the Rejected of the Nations. Israel was more the consequence of the Holocaust's flames than the corollary of a doctrine that had found comparatively so few followers.

In studying the relation of ideology to the retrogression of societies, I was much taken with the analogy between the role of Marxist ideology in the contemporary world and that of Christian religion in the ancient world. Engels had been fond of that comparison, but it never seems to have occurred to him that just as Christianity made for the decline of Roman civilization and impeded the possible development of a scientific movement, so likewise, in the twentieth century, Marxism was making for the decline of Western civilization, rationalism, and science. Something that Marx always feared – a world of a Bakuninist 'barracks communism', in which people's lives and thoughts were controlled and directed to the maximum possible by a dictatorial committee of anti-intellectual, anti-scientific, party men – might well come into existence. In part, this work led me to study the theories that sociologists had advanced to account for the decline of Roman civilization; Weber's economic theory, that emphasized the growing shortage in the available supply of slaves, foundered on such facts as the widespread manumission of slaves that kept taking place; Pareto's theory of a failure in the circulation of the elites was contravened by the extraordinary social mobility that made emperors, generals, and ministers out of barbarians, freedmen and their sons; the theory advanced by Marxists that Rome was a kind of Nazi state overlooked the fact that the Roman Empire brought peace and prosperous cities throughout the Mediterranean shores for the first time in the world's history, and that such minority peoples as the Jews largely flourished in the growth of trade that was made possible. On the other hand, the Roman economy gradually became one centrally directed by the state; the population declined in numbers as voluntary childlessness and homosexuality became norms among the upper classes; the rate of taxation rose as the tax base declined, and people fled the responsibilities of governmental duties; the state's economic needs never diminished for the burden of the

Roman legions defending the imperial borders was a fixed one, and the doles of food, meat, grain, oil, and entertainment for the permanently non-working Roman proletariat were a virtual constant. The coming of Christianity with its cult of celibacy, its diatribe against the family and the Roman military, and its masochist exaltation of the barbarians destroyed whatever chances the Roman Empire might have had for a revival along new scientific, industrial directions. Even so today do new leftists, Third Worldists, pacifists, deviants, anti-scientists, communitarians, avail themselves of various forms of Neo-Marxist ideology to assail the system of liberal democracy founded primarily on free enterprise.

An 'explanatory crisis', in my view, has virtually put an end to Marxist theory in its classical scientific sense. Not only was there an observable tendency of all communist economies to evolve toward a 'barracks communism', but imperialism as well as war had turned out to be far more intrinsic to communist existence than they had been under capitalism. The Russian communists warred against the Chinese; the Chinese embattled the Vietnamese communists; the latter massacred Cambodian communists; the Cambodian communists conducted genocide against their own people; the Yugoslavs charged the Russians with 'great power chauvinism', while the Cuban communists said the same about the Chinese, who in turn charged the Russians with 'global hegemonism'. Meanwhile, communism has become the ideology for the explicit rule of the 'new class', mainly composed of semi-intellectuals; what then would be the fate of the world's culture and science if the competing presence of the United States and the Western European liberal democracies were gone, and all the planning of cultural and scientific life were, as in the Soviet Union, concentrated in the dictatorship of the semi-intellectuals? All the fears that Mill had had for the decline of liberty under the rule of the envious and the mediocre would probably be realized. The exponential pattern of scientific and technological progress that was realized under the stimulus of capitalism is already encountering the restraint of a near-exponential increase in the cost of scientific research; a world communist order would probably regard the investigations of pure science as expendable much as Bakunin, Stalin, and Mao had.

Apart from my theoretical writings, I undertook several historical inquiries to correct the worst of the inaccurate hagiographies in ossified Marxist scholarship. For several years, in lectures to my classes and discussions with my friends in New York, especially Irving Kristol and Gertrude Himmelfarb, I had expressed my doubts as to the authenticity of the famous 'Darwin–Marx' correspondence; I even wondered whether they were forgeries by the notoriously dishonest Edward Aveling. But as I read and re-read the phrase in Darwin's main letter referring to his correspondent's association with the Freethought movement it seemed to me that Aveling could scarcely have been its contriver, for he would certainly have known of Marx's hostility to both that movement and its leader, Charles Bradlaugh; it then struck me that the letter rather than having been forged by the unscrupulous Aveling might well have been addressed by Darwin to Aveling himself, who, as I remembered, was not only active in the Freethought leadership but had also published a *Student's Darwin* at that time in 1881. All the references in Darwin's letter thereby fell into place; the request for permission to dedicate a book to Darwin was for no work of Marx's (which nobody had been able to identify), but rather for Aveling's own handbook for students. I wrote the essay slowly and carefully, and used my intermittent journeys to New York to examine the articles by Aveling that were unavailable in Toronto, and then wrote an essay, published in *Annals of Science* that reconstructed the hypothetical letter from Aveling to Darwin, and its probable contents; I called upon the Curator of the Darwin Archives to search among its holdings for this communication to which the letter of Darwin, mistakenly regarded as having been sent to Marx, had been indeed a reply. A young graduate student who was working in England on the Archives shortly after they were opened notified me that he had indeed found such a letter. Meanwhile, however, I had come across a letter by Engels to Eduard Bernstein, which scholars had overlooked, that confirmed the authenticity of an earlier letter in which Darwin had thanked Marx for sending him a copy of *Das Kapital*, whose pages however were still almost all uncut.

The news of my article 'Is the "Darwin–Marx Correspondence" Authentic?' and its verification spread among Soviet scientists; after their first consternation, its validity was acknowledged in an article in a Soviet scientific publication. I further undertook to explain why the Soviet communist journal in 1931 had first spread the unverified story of Darwin's long letter to Marx; I associated it with the trial of the Mensheviks at that time, and the outster of Ryazanov, a distinguished scholar and a long-standing opponent of Stalin who was director of the Marx-Engels Institute; evidently the publication with world-wide publicity of the purported 'Darwin–Marx letter' with its talk of a dedication to Darwin was tied to Stalin's demand for a more aggressive Bolshevik line in Soviet Marxist scholarship, a more aggressive 'Darwinism,' combined with an attack on 'bourgeois agnosticism'. My hypothesis was sustained a few years later when the Soviet emigrant-dissident, Anton Antonov-Ovseyenko, son of the commander of the seizure of the Winter Palace, disclosed that Stalin had charged Ryazanov in the case of the Mensheviks

> with the crime among other things, of stubbornly refusing to hand over a letter Darwin had written Marx . . . This struck Stalin as so odd and suspicious that he decided to banish the obstreperous scholar . . . And a few years later he had Ryazanov done away with altogether. (1981)

Thus Stalin was the chief agent of the Marx-Darwin dedication legend that Western Marxists parroted for forty years.

Scholars, moreover, scarcely noted that after 1859 Marx gradually became less of an Hegelian, and more and more of a Darwinian evolutionist in the English style. Indeed, his outlook converged with what today is called the 'sociobiological' standpoint, with the 'dialectic' method being regarded as a terminology that was equivalent to the evolutionary method. Both Marx and Engels, who consigned many peoples to their list for 'extermination' as unfit for participating in technological progress, would have made Social Darwinists and Social Spencerians look like very genteel folk. An essay 'Marx and Engels as Sociobiologists' presented the supporting evidence for the change in Marx's method after he read Darwin's *The Origin of Species*.

Marx's own vulnerability in youth to leftist Hegelian ideology, moreover, was not unrelated to the religious emptiness in his family's home where the father was compelled against his wish to convert to Christianity, and have his children baptized. An essay by Marx's father, that was almost forgotten (only one scholar had mentioned it), showed the deep loyalty that Marx's father felt to the Jewish people and their religious rituals; movingly he defended the nobility of their character in face of persecution. Yet his appeal to the Prussian authorities that he be allowed to pursue his legal career while remaining a Jew was ignored. The burden of this capitulation must have been a heavy one for the religion-seeking young Karl Marx; the Promethean path he chose was consonant with his own rebellion against the Jew and his pseudo-Christianity.

As I had long felt that Marx in his last years had probably developed basic doubts concerning the validity of his revolutionary approach and dialectical method, I looked for some personal evidences concerning such a last intellectual episode. I found them in the friendship that Marx developed in his last two years for young English thinkers who rejected the basic premises of Marx's outlook, especially with Edwin Ray Lankester and Charles Waldstein. The first, Lankester, a great zoologist and the valued associate of Darwin and Hurley, published a celebrated address, *Degeneration*; though its evolutionary emphasis on retrogression scarcely fitted the 'dialectical' model of progression, Marx forebore from criticism. Toward Waldstein, a young lecturer on art and archaeology, Marx felt a genuine intimacy despite the fact that the former rejected both the materialist conception of history and the socialist ideology alike. And Marx indeed was likewise enthusiastic for the poetic translations of James Thomson, well known as the poet of the pessimistic landmark, *The City of Dreadful Night*. If the choice of comrades, as Ignazio Silone said, is the deepest sign of the incipient choice of ideas, Karl Marx toward his life's end had grown restive with his own formal philosophy. No wonder that his last published article appealed to French socialists to shelve their sectarian differences, and to contribute to social reforms within the structure of the French Republic by joining together in a common statistical study of the conditions of the French working class.

Still another scholarly investigation of mine was directed against the growing mythology concerning the Frankfurt School, as it was called, the Institute of Social Research, of which Herbert Marcuse became the most well-known member. Its leading historian repeated naively a story as to how they were welcomed to Columbia University. The Columbia archives, however, to which I was granted access, told another tale; the negotiations had been conducted on the institute's behalf by a quondam operative for more than a decade of the Communist International, and assisted by its Columbia lobbyist who misrepresented the institute as an 'association of liberal scholars'. According to the testimony of the agent's former wife, the Frankfurt Institute was regarded as a 'cover' for agents of the Communist International. Moreover, as attested in the autobiography of the president of Brandeis University, Herbert Marcuse had been the 'intellectual mentor' of the biggest Soviet operation in espionage (later exposed) in wartime Washington. Such was the pragmatic significance of Critical Theory.

Naturally, such investigations did not endear me to the array of curious crackpots, indurated leftists, and embittered envious who constitute the mainstay of the neo-Marxist journals. Articles were published that made claims of priority that were demonstrably false; one stormy critic allowed herself to be bamboozled by the Soviet Institute of Marxism–Leninism into believing that there were mysterious, informative notes on the Lankester letters of which I possessed copies; finally, the Amsterdam Institute of Soviet History rebuked her, and sent me copies of their correspondence. The chief generalization that I have seen confirmed through many years' observation is the law of diminishing integrity in the Marxist intellectual world.

During my Berkeley and Toronto years, I, from time to time, published studies in what H. G. Wells once called 'The Psychoanalysis of Philosophy'. Especially did I find this method illuminating in my researches on Descartes and Kant. The writing of each essay required an extensive reading of the given philosopher's works as well as the biographies concerning him, but the reward came in discovering the probable emotional basis for tenets that the philosopher accepted as self-plausible even though their logical grounds were slender or unconvincing. Descartes's mother died when he was an infant; his father was mainly absent, and then feared; the traumas, doubts and guilt that Descartes concealed as if behind a mask, as he said, were, in all likelihood, the psychological sources for such notions as that animals lacked feelings and that God was conceivably deceiving him with the purported perceptions of an external world, and above all, for the three dreams as well in which, according to Descartes, his philosophy was essentially revealed to him on the night of November 10, 1619. The ontological argument I generally tried to show in my essay *God, Guilt, and Logic* was associated with such philosophers from Anselm to Royce in whom guilt-experience induced a response of what we might call a 'logical masochism' of the intellect, an anxiety-induced inability to doubt the existence of God; the more optimistic philosophers of the eighteenth century – Locke, Berkeley, Voltaire – found the ontological argument a bit of unconvincing word-play. Similarly, Kant's utter mistrust of our sensations, our perceptions of an external world, was less a matter of argument or evidence than the outcome of an intense revulsion against virtually every sense of the human body and all its physiological functioning. The senses of hearing, sight, smell, the thirst that he might have, the garters that might slip and lay bare his legs, the secretions of his body's heat, were all ominous for Kant with occasions for traumatic experience. The *a priori* forms of intuition, and the categories that the understanding 'legislated' were, in Kant's view, like the decrees of a judge imposing order on otherwise 'lawless' data that indeed sometimes disturbed his sexual equanimity.

During the course of these studies and further reflection, I came to have certain reservations concerning the psychoanalytical approach. I once heard Anna Freud state that the psychoanalytical method was unique insofar as its method of research was also its method of therapy. I still think that it is a powerful tool of research, but I doubt whether its hopes in therapy have been fulfilled. Even as a continuous use of X-rays damages human tissue, thus likewise the continuous exposure of the unconscious to psychoanalytical probing seems to me to have frequently damaging effects on the person's character. The unconscious, too, we might say, has a teleological function that should be respected, and Freud was perhaps impairing a human resource for coping with traumatic memories and fears when he made central the dredging of the unconscious to full consciousness. Secondly, the intuitionist philosophy, from Descartes onward, has not been altogether refuted by the psychogenetic analyses of Locke, Mill, Freud, and Piaget. Locke, for instance, tried to show that 'innate ideas' (or intuitions) as, for instance, of God's existence, always originated causally from social psychological influences; to which Leibniz rebutted, successfully in my opinion, that after all allowance has been made for the effect of such social circumpressures, there still remains as irreducible intuitive predisposition to believe that God exists. A methodological indeterminacy, however, attaches itself to the status of such irreducible intuitive predispositions. Let us suppose that we have an intuitive propensity toward beliefs in the existence of God and the holiness in the human being; nevertheless it would not be clear to us whether this propensity has been implanted in us by God, or whether it is an evolutionary-psychological variation that occurred early among hominoids, and proved 'adaptive', perhaps accumulating and intensifying during the glacial ages, or whether, indeed, the genetic mutations on which such variations were based could be altogether physico-chemically explained, or were themselves propelled in part by a divine agency, intervening within zones of indeterminacy in ways compatible with the existence of probabilistic causal laws.

In short, the psychoanalytical critique of ethical concepts, which, in my view, still has much to offer, is valid at that level where the lines of social psychological causation are ascertainable; it becomes more questionable, however, where biogenetic or possibly, meta-genetic factors, are consistent with the evidence, and not themselves explicable on psychogenetic grounds.

In the mid-seventies, I began to write again on some of the metaphysical themes that had engaged me in my youth. In part, this was the natural outcome of my writing of *Einstein and the Generations of Scientists*, for Einstein's thought posed once again the status of the belief in God, as the creator of a universe governed by simple mathematical laws, as a heuristic scientific principle. I also received invitations to take part in commemorative symposia in 1977 on Spinoza's thought at the Jewish Theological Seminary, the Catholic University of America, and the Hebrew Union College which provided formal occasions for discussing such topics. I wrote two essays during those years that set down ideas that had gradually evolved – 'Teleological Principles in Science' and 'Noumenalism and Einstein's Argument for the Existence of God'. The word 'noumenalism' (which perhaps I have been the first to use) was used in contrast to phenomenalism. Thus, while Einstein in his youth was largely guided by a phenomenalist standpoint – that is, the aim to describe the laws of observed phenomena in the most economical fashion we can with variables that are all observationally defined – in his later years that standpoint was replaced by the conviction that the universe was governed by rational, beautiful, simple mathematical laws, that on purely *a priori* grounds would have been infinitely improbable compared to the infinite possible number of irrational chaotic universes. From the noumenalist standpoint, the existence of such simple mathematical laws, logical in their form, is inexplicable unless they are taken as the expressions of a rational God; our human minds have therefore in scientific thinking managed at least partially to re-trace God's thoughts.

A theism such as Einstein's, with its corresponding noumenalist conception of scientific law, has seemed to stand on strong grounds insofar as the judgment of many mathematical physicists has been concerned. No sociologist, however, has taken a similar delight in the workings of a purportedly divine sociological intelligence. Such diverse social theorists as Mill, Marx, Pareto, Durkheim, and Weber all knew too keenly the cruelties, pains, and ensnarements of social existence, while Freud, James, Darwin, and Huxley were appalled by the spectacle of human suffering against the pervasive background of biological misery.

Mill and James experimented with the idea of a finite God, an idea that contributed to the hypothesis of emergent evolution much favored by writers such as Bernard Shaw and H. G. Wells, and elaborated by philosophers and biologists such as C. Lloyd Morgan, Joseph Needham, Samuel Alexander, Whitehead, H. C. Jennings, and Lovejoy. But the notion that there are emergentist laws for life, consciousness, and thought does violence to the central principle in causal explanation that Emile Meyerson, the epistemological scholar most admired by Einstein, had emphasized: the requirement of a conservational identity between the antecedent and the consequent in a causal law. Emergent evolutionists did indeed respond by calling upon us to accept the occurrence of emergences with 'natural Piety', and to set aside any misgiving that emergentist laws contravened the requirement of identity; was this not, however, tantamount to a call to join in collectively repressing our natural longing for a causal explanation? Charles Peirce once noted that for an explanation to be satisfactory it had to resolve the feeling of discomfort that psychologically located the existence of a problem. An extra-natural divine agent for emergences might answer the need for a conservational source, but the pain and suffering in our temporal world would still mar the notion that this was a world of beautiful sociological laws. Perhaps the equations of existence house a solution in which time's traits are largely altered; in such a world of a kind that Kurt Gödel's imagination might conceive, one conjugate with ours in which somehow sociological disfigurements vanish, their values diminishing to zero, would a theism, such as Einstein's be as consonant with the laws of persons as with those of physical things.

My 'noumenalism' essay evoked a generous response from the foremost figure in American metaphysical thought, Brand Blanshard:

> Your splendid paper reached me yesterday. I have read it carefully through with sustained admiration and assent . . . I did not realize how much my gropings after a rationalistic view of things had in common with Einstein's. . . . I wish all members of the APA could read it.

And when Sidney Hook – whose critical acumen has kept American pragmatic naturalism an ever live and growing philosophy that has not rigidified into an orthodoxy – spoke warmly of the new arguments and considerations I had advanced (though criticizing me for introducing my own views through a discussion of Einstein's), I felt that with two such readers my essay was amply rewarded.

Throughout my life I have always retained a strong interest in the history of sociology of the Jewish people. Jewish customs, habits, and underlying feelings were part of me; though the East Side had often seemed to me ugly and cruel with its quarrelsome disrespect for fellow human beings and its cluttered litter, my aversion to un-Kosher foods was invariant, and for years I wore the cap of the East Sider. I was an officer of Young Judaea at my Talmud Torah, I helped found a Jewish club at high school, joined the Menorah Association at college, worked as a summer counsellor in a YMHA camp, and saw a great deal of our Chaplain, Morris Chapman, in New Caledonia. At Vermont during the summer of 1953, I had many evenings' discussions with the theologian Will Herberg, who was completing his journey from 'right-deviationist' Marxism to existentialist Judaism. Our disagreements sometimes aroused Herberg to a thundering declamation; only once did I see Herberg at a loss for words. That was when my six-year-old daughter Robin thrust herself into the fray to protect me; gathering herself bravely against Will's torrential words, she said: "My daddy knows how to talk. You don't have to teach my daddy how to talk." Will was subdued.

As I turned sixty years of age, I began to write more frequently on Jewish subjects; for one, the deaths of Wolfson, Schapiro, Kallen, and the publication of a biography of Morris R. Cohen, recalled to me, as I wrote memorial essays, the settings of their work and, above all, the sheer obstruction that the first American Jewish university thinkers had faced as they knocked at the doors at American intellectual institutions. I wrote essays on the legend of the socialist East Side, the history of Jewish professors in the United States, and the political bearings of Jewish tradition that were far more frequently in a conservative rather than a radical direction; I investigated the causal role of the Jewish marital ethic and its rejection of celibacy in maintaining the genetic stock of Jewish intelligence while Christian Europe was decimating its most intelligent by designating them for monasteries and churches. I even unravelled from Virginian archives the full truth of what befell on its grounds to the first Jewish professor at an American university, the immense genius and wonderfully courageous James Joseph Sylvester, destined to become Britain's foremost mathematician, yet on a March day in 1842, forced to flee for his life from Thomas Jefferson's 'academical village'. My chief aim in these Jewish studies was to tell a story of hope and accomplishment, not of ideology nor of prophetic criticism, but of work done and a position achieved through the respect that the love of work and its realization have always inspired. To me the story of the first Jewish professors and their researches in the face of adversity was more thrilling than the sagas of left-wing parades, rallies, and demonstrations.

The Holocaust had, during the post-war years, a profound effect on Jewish theological speculations. In my view, however, Holocaust theology was, from a logical standpoint, ill-conceived; the Holocaust was, though unprecedented as the genocide of one advanced people by another, not unique, not a member of a class necessarily with only one member; it could recur, as the very emphasis on remembering it so that it should not happen again, indicated. Its places were not the locations of God's presence, for the experience of its survivors showed that it had mostly reduced rather than advanced their belief in God's existence. Only the reality of a non-temporal universe in which the evils of the temporal one were expunged as drosses from the human ore; only such a noumenalist conception of God could salvage human faith and values from the shattering impact of the Holocaust.

If there were a book I should hope to write, it would be on *The Seven Pillars of Wisdom*, to recapture the meaning of wisdom before it was perverted with Utopian projects, speculative dogmas, and alleged ultimates in knowledge. For the ancient Greek sages were men whose reflections were the fruit of their own practical experience as statesmen in human affairs; they were men very much as Benjamin Franklin was in the eighteenth century. They were characterized by what they called the 'principle of harmony', which we would describe today as an equilibrium between our consciousness and our unconscious. Wise men recognized the 'principle of impotence', that, for instance, Utopian all-satisfying societies are as impossible as perpetual motion machines; the one violates the laws of human nature as the other does those of physical existence. They have accepted the 'principle of geneticism' – namely, that all philosophies are bounded by the psychogenetic handicaps and limitations of their conceivers and that our judgment of them should be tempered by the recognition that all are beset by patent inadequacies not glossed over by our human devices; that work, the principle of 'jobbism' as Justice Holmes called it, provides our energies with their most absorbing and long-run satisfying activity and, that the 'principle of science', of verified hypotheses and ascertainment of facts, has been a mainstay to the life of wisdom from Thales to Franklin to Einstein. If our human goals and knowledge are limited, partial, never total, always edged with uncertainty, we can still cherish the 'principle of religion', the postulate, perhaps intuitively based, that the human episode is not altogether a transient one but reaches somehow into the abiding nature of the universe; and lastly, the 'principle of humor', that joys are available to every human life, and that those who would maim their experience by superimposing a burden of primary guilt on every human existent are moved both by a self-enmity and an enmity toward those they would profess to save.

REFERENCES

Antonov-Ovseyenko, A. 1981. *The Time of Stalin: Portrait of a Tyranny*, transl. George Saunders. New York, p.129.

Gramsci, Antonio. 1979. *Letters from Prison*, transl. Lynne Lawner. London, p. 163.

Irwin, Elizabeth A. and Marks, Louis A. 1924. *Fitting the School to the Child: An Experiment in Public Education*, New York, pp. 90–91, 219–249.

Jowett, B. (transl). 1875. *The Dialogues of Plato*, Vol. II. Oxford, p. 361.

Numberg, Herman and Federn, Ernst (eds). 1967. *Minutes of the Vienna Psychoanalytical Society*, transl. M. Nunberg, Vol. II: 1908–1910. New York, pp. 495–496.

LIST OF PUBLISHED WRITINGS OF LEWIS S. FEUER

1. 1932. 'The Strange Career of Walter Lippmann', *Revolt* **I**, No. 2 (Dec.), 9, 22.
2. 1933. 'Economy in Education', *The Student Outlook* **I**, No. 5 (May), 16.
3. 1933. 'Conditions for Revolution', Review: Curzio Malaporte, The Technique of Revolution, *The Student Outlook* **II**, No. 1 (Nov.), 18–19.
4. 1933. Letter, 'Socialism and War', *The Student Outlook* **II**, No. 1 (Nov.), 23.
5. 1934. 'Philosophy Follows the Flag', *The Student Outlook* **II**, No. 3 (Feb.), 5, 6, 19, 20, 21.
6. 1934. Democritus, 'Harvard' in 'Students Strike against the War', *The Student Outlook* **II**, No. 5 (May), 12–13.
7. 1934. 'Upton Sinclair, Socialist?', *The Student Outlook* **III**, No. 1 (Oct.), 7–8.
8. 1934. 'Hillquit's Service', Review: Morris Hillquit, Loose Leaves from a Busy Life, *The Student Outlook* **III**, No. 1, (Oct.), 18–19.
9. 1934. Reply on Upton Sinclair's Candidacy, in 'Free for All', *The Student Outlook* **III**, Nos. 2–3 (Nov.–Dec.), 39–40.
10. 1934. 'On the Use of "Universe"', *Mind* **XLII**, 346–348.
11. 1934. Review: Morris R. Cohen and Ernest Nagel, An Introduction to Logic and Scientific Method, *The City College Alumnus* **30** (Sept.), No. 7, 106.
12. 1936. Review: Collected Papers of Charles Sanders Peirce, Vol. VI, Scientific Metaphysics, *Isis* **XXVI**, 203–208.
13. 1937. Summary of Harvard Theses: 'The Philosophical Analysis of Space and Time', *Summaries of Theses, 1935*. Cambridge, pp. 328–331.
14. 1939. Review Maurice Mandelbaum, The Problem of Historical Knowledge, *Science and Society* **III**, 417–419.
15. 1940. 'The Economic Factor in History', *Science and Society* **IV**, 168–192.
16. 1941. 'The Development of Logical Empiricism', *Science and Society* **V**, 222–233.
17. 1941. 'Dialectic and Economic Laws', *Science and Society* **V**, 336–361.
18. 1941. 'Towards a National Front', *Threshold* **I**, No. 2 (Dec.), 23–26.
19. 1942. Reviews: Max Lerner, Ideas for the Ice Age; Henry M. Magid, English Political Pluralism, *Science and Society* **VI**, 180–184.
20. 1942. 'Ethical Theories and Historical Materialism', *Science and Society* **VI**, 242–272.
21. 1943. 'Ethics and Marxism', *Science and Society* **VII**, 260–267.
22. 1943, 'Ethics and Historical Materialism', *Science and Society*, **VII** 357–360.
23. 1944. Review: Paul A. Schilpp (ed.), The Philosophy of G. E. Moore, *Science and Society* **VIII**, 189–192.
24. 1945. 'Metaphysics and Social Science', *Science and Society* **IX**, 255–260.
25. 1946. 'Cartel Control in New Caledonia', *Far Eastern Survey* **XV**, (June 19), 184–187.
26. 1946. 'End of Coolie Labor in New Caledonia', *Far Eastern Survey* **XV**, (Aug. 24), 264–267.

S. Hook, W.L. O'Neill and R. O'Toole (eds.), Philosophy, History and Social Action, 87–101.

27. 1946. With Paul M. Sweezy, 'Has Colonialism a Future?', *The New Republic* **155**, (Nov. 25), 687, 691–693.
28. 1947. 'Analysis and Scientific Practice', *Analysis* **8**, 28–30.
29. 1947. 'Philosophy and the Theory of Relativity' (review of Philipp Frank, Einstein: His Life and Times), *Science and Society* **XI**, 259–269.
30. 1948. 'Materialism, Idealism and Science', Review of Maurice Cornforth, Science and Idealism, *Philosophy of Science* **XV**, 71–75.
31. 1948. 'Indeterminacy and Economic Development', *Philosophy of Science* **XV**, 225–241.
32. 1948. Reviews: Pierre Naville, Psychologie, Marxisme, Materialisme; George Politizer, Le Bergsonisme: Une Mystification Philosophique, *Philosophy and Phenomenological Research* **8**, 467–472.
33. 1948. 'Russia and the Liberals', *The New Republic* **119**, (Nov. 8), 14–16.
34. 1949. 'Mechanism, Physicalism, and the Unity of Science', *Philosophy and Phenomenological Research* **IX**, 627–643.
35. 1949. 'Restoration of Coolie Labor', letter to *New York Herald Tribune* (Jan. 24, or Feb.).
36. 1949. 'Changes in Marxist Doctrine', *The New Republic* **120**, (May 16), Section Two, 'How Strong is Soviet Russia?', 26.
37. 1949. 'John Stuart Mill and Marxian Socialism', *Journal of the History of Ideas* **X**, 297–303.
38. 1949. 'Talmud to Mill', reviews of Morris R. Cohen, A Dreamer's Journey; Morris R. Cohen, Studies in Philosophy and Science, *The New Republic* **120** (March 28), 21–23.
39. 1949. 'The Future of International Socialism', review, Paul M. Sweezy, Socialism, *The New Republic* **121**, (Oct. 31), 17–18.
40. 1949. 'Dialectical Materialism and Soviet Science', *Philosophy of Science* **16**, 105–124.
41. 1950. 'The Philosophy of Morris R. Cohen: Its Social Bearings', *Philosophy and Phenomenological Research* **X**, 471–485.
42. 1950. Review: Henri Lefebvre, A La Lumière du Matérialisme Dialectique, *Philosophy and Phenomenological Research* **X**, 591–593.
43. 1950. 'The Quality of Life in Israel's Collectives', *Commentary* **9**, 497–507.
44. 1950. 'Israel After Two Years', *The New Republic* **123**, (July 31), 14–16.
45. 1950. 'Leadership and Democracy in the Collective Settlements of Israel' in Alvin W. Gouldner (ed.), *Studies in Leadership: Leadership and Democratic Action*. New York: Harper & Brothers, 363–385.
46. 1951. Review: Morris R. Cohen, Reflections of a Wondering Jew, *Philosophy and Phenomenological Research* **XI**, 437–438.
47. 1951. 'The Paradox of Verifiability', *Philosophy and Phenomenological Research* **XII**, 24–41.
48. 1952. 'What is Philosophy of History?', *The Journal of Philosophy* **XLIX**, (May 8), 329–340.
49. 1951. Review: 'F. A. Hayek, John Stuart Mill and Harriet Taylor: Their Correspondence and Subsequent Marriage', The University of Chicago Press. *Philosophy and Phenomenological Research* **XIII**, (Dec. 1952), 246–248.
50. 1953. 'Sociological Aspects of the Relation between Language and Philosophy', *Philosophy of Science* **20**, 85–100.

51. 1953. 'The Social Motivation of Spinoza's Thought', *Proceedings of the XIth International Congress of Philosophy* **XIII**, 36–42.
52. 1953. 'Thorstein Veblen: The Metaphysics of the Interned Immigrant', *American Quarterly* **V**, 99–112.
53. 1954. Review: 'John Wisdom, Philosophy and Psychoanalysis; Roger Bastide, Sociologie et Psychoanalyse; Ernst Kris, Psychoanalytic Explorations in Art', *Philosophy and Phenomenological Research* **XIV**, 407–410.
54. 1954. Review: 'Gertrude Himmelfarb, Lord Action: A Study in Conscience and Politics', *The Standard* **XL**, 104–106.
55. 1954. 'The Brighter Side of Coexistence', *The New Republic* **131**, (Nov. 1), 10–12.
56. 1954. 'Causality in the Social Sciences', *The Journal of Philosophy* **LI**, 681–695.
57. 1954. 'Morris Cohen's Last Book', Review: Morris R. Cohen, American Thought: A Critical Sketch', *The New Leader*, **XXXVII** (Dec. 6), 23–24.
58. 1954. Review: 'John Oulton Wisdom, The Unconscious Origin of Berkeley's Philosophy', *Philosophy and Phenomenological Research*, **XV**, 259–260.
59. 1955. Letter on Oppenheimer Case, *New Leader*, **XXXVIII** (March 7), 28.
60. 1955. 'Political Myths and Metaphysics', *Philosophy and Phenomenological Research* **XV**, 332–350.
61. 1955. Review: 'Jean Paul Sartre, Existential Psychoanalysis', *Philosophy and Phenomenological Research* **XV**, 427–429.
62. 1955. 'Bertrand Russell: The Pilgrimage of Scientific Philosophy', *University of Toronto Quarterly* **XXIV**, 217–233.
63. 1955. *Psychoanalysis and Ethics*. Springfield, Illinois: Charles C. Thomas, p. 130.
64. 1955. 'A Theory of Meaning', review of Ernest Nagel, Sovereign Reason, *Commentary* **20** (Oct.), 389–392.
65. 1956. Review of 'Edmond Cahn, The Moral Decision', *New York University Law Review* **31** (March), 627–630.
66. 1956. 'The Aims of a Philosophy of Education', *Harvard Educational Review* **XXVI**, 112–113.
67. 1957. 'Dialectic', 'Dialectical Materialism', *The Encyclopedia Americana*, Vol. IX, 58–59.
68. 1957. 'Clarence Darrow: 1857–1957', *The Ethical Outlook* **XLII**, 112–115.
69. 1956. Japanese translation, *Psychoanalysis and Ethics*, transl. Kazuko Tsurumi. Tokyo: Iwanami Gendai Sosho.
70. 1957. 'The Principle of Simplicity', *Philosophy of Science* **24**, 109–122.
71. 1957. 'Letters of H. A. P. Torrey to William T. Harris', *Vermont History* **XXV**, 215–219.
72. 1957. 'The Dream of Benedict de Spinoza', *The American Imago* **14**, 225–242.
73. 1958. 'The New York State University Fight', *The New Leader* **XLI** (Feb. 24), 18–19.
74. 1958. 'H. A. P. Torrey and John Dewey: Teacher and Pupil', *American Quarterly* **X** (Spring), 34–54.
75. 1958. 'James Marsh and the Conservative Transcendentalist Philosophy', *The New England Quarterly* **XXXI** (March), 3–31.
76. 1958. *Spinoza and the Rise of Liberalism*, Boston: Beacon Press; 308 pp, with Index, 323 pp.
77. 1958. 'John Dewey's Reading at College', *Journal of the History of Ideas* **XIX**, 415–421.

78. 1958. 'The Politics of the Death Wish', *The New Leader* **XLI** (Sept. 1), 17–19.
79. 1958. 'The Rosenberg Case: Five Years After', *The Ethical Outlook* **XLII** (Sept.–Oct.), 163–166.
80. 1958. Letter to *the New Leader*, 'Death Wish', **XLI** (Sept. 29), 30–31.
81. 1958. 'The Sociology of Philosophic Ideas', *The Pacific Sociological Review* **I**, 77–80.
82. 1959. Letter on Rosenberg Case in Reply to Prof. Lewis D. Wormuth, *The Ethical Outlook* **XLV**, 15–16.
83. 1959. 'A Historic Opportunity', *The New Leader* **XLII** (Jan. 26), 11–14.
84. 1959. 'A Tribute to J. Salwyn Schapiro', *The New Leader* **XLII** (Mar. 2), 10.
85. 1959. 'Rejoinder on the Principle of Simplicity', *Philosophy of Science* **26** (Jan.), 43–45.
86. 1959. 'The Bearing of Psychoanalysis upon Philosophy', *Philosophy and Phenomenological Research* **XIX**, 323–340.
87. 1959. 'Coexistence', *The New Leader* **XLII** (Aug. 27), 29–30.
88. 1959. Anthology: Marx and Engels: *Basic Writings on Politics and Philosophy*. New York: Doubleday & Co., Anchor Books; 'Introduction', pp. xi–xxi.
89. 1959. 'John Dewey and the Back to the People Movement in American Thought', *Journal of the History of Ideas* **XX**, 545–568.
90. 1959. 'The Presidency and the Church', *The New Leader* **XLII** (Nov. 9), 11–14.
91. 1960. 'Soviet Students on an American Campus', *The New Leader* **XLIII** (Feb. 29), 8–11.
92. 1960. Review: C. Wright Mills, The Sociological Imagination, 'A Neo-Marxist Conception of Social Science', *Ethics* **LXX** (April), 237–240. Reprinted in *Berkeley Journal of Sociology* **V** (Fall 1959), 119–126.
93. 1960. Review: Sidney Hook (ed.), Psychoanalysis, Philosophy and Sciencific Method, *Philosophy and Phenomenological Research* **XX**, 550–552.
94. 1960. 'Ralph Waldo Emerson's Reference to Karl Marx', *The New England Quarterly* **XXXIII**, 378–379.
95. 1960. 'The Standpoints of Dewey and Freud: A Contrast and Analysis', *Journal of Individual Psychology* **16** (Nov.), 121–136.
96. 1961. 'Youth in the '60's', *The New Leader* **XLIV** (March 6), 18–22.
97. 1961. 'A Talk with the Zengakuren', *The New Leader* **XLIV** (May 1), 16–20. Reprinted in *Japan Times* (May 24, 1961).
98. 1961. Rejoinder on 'The Role of Sexuality in the Formation of Ideas', *The Journal of Individual Psychology* **17**, 110–111.
99. 1961. 'Impressions of Japanese Radicalism', *Bulletin of the International House of Japan*, No. 7, pp. 2–13. Japanese translations of this essay were published in Kaiban Kokusai (Jan. 1961), No. 6, 2–15. *Ekomisuto* **39**, No. 10, 6–15.
100. 1961. Reprint: 'Youth in 60's' in Spanish translation, 'La juventud estudiantil norteamericana', *Cuadernos*, Numero 49 (Junio), 85–90. *Current* (May 1961), No. 13, 15–18.
101. 1961. Japanese translation, 'John Dewey and the Back to the People Movement in American Thought', *Shiso*, No. 2, 72–82; No. 7, 106–114.
102. 1961. Contribution to Symposium, 'Intellectuals: East and West', *Shiso no Kogaku, Science of Thought*, No. 25, 8–20.
103. 1961. 'Cult of the Rebel', Review of Paul Goodman, Growing up Absurd, *The New Leader* **XLIV** (June 19), 22–24.

104. 1961. 'Reply to I. I. Morris on Zengakuren', *The New Leader* **XLIV** (July 3–14), 29–30.
105. 1961. 'Natural Rights', *Atti del XII Congresso Internazionale di Filosofia*, Volume Ottavo, Firenze, pp. 97–104. Reprinted in *Handbook on Human Rights*, No. 9, 'Fundamentals of Human Rights', Academy of Human Rights, Switzerland, Zurich, 1961, pp. 1–4.
106. 1961. Review: Adam B. Ulam, The Unfinished Revolution: An Essay on the Sources of Influence of Marxism and Communism, *American Sociological Review* **26**, 802–803.
107. 1962. 'Introduction: A Critical Evaluation, Currents in Japanese Socialist Thought', *New Politics* **I**, No. 2 (Winter), 116–123.
108. 1962. Review: Walter Kaufmann: The Faith of a Heretic, *The New Leader* **XLV** (March 5), 25–27.
109. 1962. 'Marxism as History', Review: George Lichtheim, Marxism: an Historical and Critical Study, *Survey*, No. 41 (April), 176–185.
110. 1962. Review: Max Black (ed.), The Soviet Theories of Talcott Parsons, *The Journal of Philosophy* **LIX** (March 29), pp. 182–193. Reprinted in *Berkeley Journal of Sociology* **VII** (1962), 129–136.
111. 1962. 'Youth in the 60's', reprinted in Albert T. Anderson and Bernice P. Biggs (eds), *Focus on Rebellion*. San Francisco: Chandler Publishing Co., pp. 142–151.
112. 1962. 'What is Alienation? The Career of a Concept', *New Politics* **I**, No. 3 (Spring) pp. 116–134. Reprinted in Maurice Stein and Arthur J. Vidich (eds), *Sociology on Trial*. Englewood Cliffs: Prentice-Hall (1963), pp. 127–147.
113. 1962. Review: Robert C. Tucker, Philosophy and Myth in Karl Marx, *The New Leader* **XLV** (Jan. 11), 26–28.
114. 1962. 'American Travelers to the Soviet Union 1917–32: The Formation of a Component of New Deal Ideology', *American Quarterly* **XIV**, No. 2 (Summer), 119–149.
115. 1962. 'Shirley's Can-Can', *San Francisco Chronicle* (June 28), 3.
116. 1962. Prefatory Note to John Wright Buckham, 'A Vermont Boyhood', *Vermont History* **XXX** (July), 201–203.
117. 1962. Interview: *Tokyo Afternoon Shimbun* (Aug. 19).
118. 1962. 'Marxian Tragedians', *Encounter* **XIX**, 23–32. Reprinted in Hebrew translation, *Molad* (Dec. 1962), 608–616. Reprinted, 'Socialism and Love in the Family of Karl Marx' in G. R. Urban (ed.), *Talking to Eastern Europe*. London: Eyre and Spottiswoode, 1964, pp. 56–73. Reprinted in *Scaling the Wall: Talking to Eastern Europe*, George R. Urban (ed.). Detroit: Wayne State University Press (1964), pp. 56–73.
119. 1963. 'Eleanor Marx' (Reply to Eugene Kamenka), *Encounter* **XX** (April), 106, 108.
120. 1963. 'The North American Origin of Marx's Socialism', *The Western Political Quarterly* **XVI** (March), 53–67.
121. 1963. 'The Dreams of Descartes', *The American Imago*, **20** (Spring), 3–26.
122. 1963. 'The Philosophical Method of Arthur O. Lovejoy: Critical Realism and Psychoanalytical Realism', *Philosophy and Phenomenological Research* **XXIII**, No. 4, 493–510.
123. 1963. 'Dilemma of the Soviet Intellectual', *The New York Times Magazine* (Aug. 18), pp. 13, 80–81. Reprinted, Western edition (Aug. 20, 1963), p. 9. Reprinted,

Current (Oct. 1963), No. 42, 21–24.

124. 1963. 'Marx and the Intellectuals', *Survey*, No. 49, (Oct.), pp. 102–112. Translated into Japanese, *The Ushio* (March 1964), pp. 96–104. Also, published as 'Marxism and the Hegemony of the Intellectual Class', *Transactions of the Fifth World Congress of Sociology*, Vol. IV. Louvain (1964), pp. 83–96.
125. 1963. 'Jews in the Soviet Union', *The New Republic*, **149** (Nov. 30), 10–12.
126. 1963. 'Anxiety and Philosophy: The Case of Descartes', *The American Imago*, **20**, 411–449.

126a. 1963. *The Scientific Intellectual*. New York: Basic Books.

127. 1964. Symposium: The Meaning of the Life and Death of John F. Kennedy; 'The Pervasiveness of Violence', 'The Reaction of his Peers', *Current* (Jan.), No. 45, 18, 28.
128. 1964 'The Soviet Jews: Resistance to Planned Culturocide', *Judaism* **13**, No. 1 (Winter), 90–100.
129. 1964. 'Discussion of Anti-Semitism: Soviet Institute of Philosophy Meeting', *Jews in Eastern Europe* **II**, No. 4 (Feb.), 5–11.
130. 1964. 'Meeting the Philosophers', *Survey*, No. 51 (April), pp. 10–23. Reprinted in *Dissent* **XII**, No. 1 (Winter 1965), 60–74.
131. 1964. 'Problems and Unproblems in Soviet Social Theory', *Slavic Review* **XXIII** (March), 117–124.
132. 1964. 'Marxisms – How Many?', *Problems of Communism*, **XIII**, No. 2 (March–April), pp. 48–57. Spanish translation, 'Marxismos¿ Cuantos?', *Problemas del Communismo* **XI**, No. 2 (Marzo–Abril, 1964), 52–62.
133. 1964. 'Lenin: the Idol and the Man', Reviews: Angelica Balabanoff, Impressions of Lenin; Louis Fischer, The Life of Lenin; Robert Payne, The Life and Death of Lenin; Stefan T. Possony, Lenin: The Compulsive Revolutionary; *The New Leader,* **XLVII** (Aug. 3), 18–21.
134. 1964. Review: Milton Halsey Thomas, 'John Dewey: A Centennial Bibliography', *American Sociological Review,* **29**, 631.
135. 1964. Reprint: 'Dialectical Materialism and Soviet Science', in Leonard M. Marsak (ed.), *The Rise of Science in Relation to Society*. New York: Macmillan-Collier, pp. 108–121.
136. 1964. Letter on Lenin, (reply to Stefan Possony), *The New Leader* **XLVII** No. 20 (Sept. 28), 25.
137. 1964. 'Passage Beyond Marxism', *The New Leader* **XLVII** (Oct. 26), 9–11.
138. 1964. Letter, *The New Leader* **47**, p.
139. 1964. 'Rebellion at Berkeley: The New Multiversity: Ideology and Reality', *The New Leader* **XLVII** (Dec. 21), 3–12.
140. 1964. 'Anti-Semitism: Soviet Style', the complete version of what was partially published in *The New Republic* (Nov. 30, 1963), *The Burning Bush*, No. 2 (Sept. 1964), Hillel Foundation, San Francisco State College, pp. 14–20.
141. 1965. 'A Reply', *The New Leader* **XLVIII** (Jan. 4), 15–17.
142. 1965. 'Inevitability and Institutes', *The New Leader* **XLVIII** (April 12), 9–10.
143. 1965. 'Pornopolitics and the University', *The New Leader* **XLVIII** (April 12), 14–19.
144. 1965. Review: John Dewey and Arthur F. Bentley, 'A Philosophical Correspondence, 1932–1951, Sidney Ratner and Jules Altman (eds), *The Annals of the American Academy of Political and Social Science* **360** (July), 222–223.
145. 1965. Reprint, 'Causality in the Social Sciences' in Daniel Lerner (ed.), *Cause and Effect*. New York: Free Press, pp. 191–208.

146. 1965. Review: Raymond Aron, Main Currents in *Sociological Thought*, Vol I. New York: Basic Books. *The American Journal of Sociology* **LXXXI** (1965), 331–332.
147. 1966. 'Intellectuals as a Class of Men', Review, Lewis A. Coser, Men of Ideas, *The New York Times Book Review* (Jan. 16), 4, 5, 44.
148. 1966. 'Alienation: the Marxism of Contemporary Student Movements' in *Marxist Ideology in the Contemporary World – Its Appeals and Paradoxes*, Milorad M. Drachkovitch(ed.), New York: Hoover Institution, Frederick A. Praeger, pp. 37–59.
149. 1966. 'American Philosophy is Dead', *The New York Times Magazine* (April 24), 31, 122, 124.
150. 1966. 'Ideology and Religious Dualism', *Reconstructionist* **XXXII** (May 27), 7–15.
151. 1966. 'The Decline of Freedom of Berkeley', *The Atlantic Monthly* **218**, No. 3 (Sept.), 77–87.
152. 1966. 'The Risk is "Juvenocracy"', *The New York Times Magazine* (Sept. 18), 56, 57, 59, 60, 64.
153. 1966. 'The Influence of the American Communist Colonies on Engels and Marx', *The Western Political Quarterly* **XIX** (Sept.), 456–474.
154. 1966. 'Professor Feuer Replies', *The Atlantic* **218**, No. 5 (Nov.), 45–47.
155. 1966. Reprint, 'Meeting the Soviet Philosophers' in Alex Simirenko (ed.), *Soviet Sociology: Historical Antecedents and Current Appraisals*. Chicago: Quadrangle Books, pp. 260–274.
156. 1966. Translation of 'American Philosophy is Dead' in *Literaturnaya Gazeta*, Moscow, (Aug. 25).
157. 1966. With Mervyn W. Perrine, 'Religion in a Northern Vermont Town: A Cross-Century Comparative Study', *Journal for the Scientific Study of Religion* **5**, No. 3 (Fall), 367–382.
158. 1967. 'University of Toronto is entering on an era of greatness', *The Varsity* **86**, No. 37 (Wednesday, Jan. 4), 8.
159. 1967. 'The Student Left in the U.S.A.', *Survey*, No. 62 (Jan.), 90–103.
160. 1967. 'The Elite of the Alienated', *The New York Times Magazine* (March 26), 22–23, 74–77.
161. 1967. Translation of 'What is Alienation?' by Yasuhide Sudo, *Chuo-University Review* **19**, No. 1, 126–148.
162. 1967. 'Eric Hoffer vs. the Intellectuals', *Books* Vol. 4, No. 9, (Oct.), 1.
163. 1967. 'On Civil Disobedience', *The New York Times Magazine* (Nov. 26), 29, 122.
164. 1968. Reprint, 'The Risk is Juvenocracy' in Edward Quinn and Paul J. Dolan (eds), *The Sense of the Sixties*. New York: The Free Press, pp. 30–39.
165. 1968. Reprint, 'Decline of Freedom in Berkeley' in J. Ashmead, Rinaldo C. Simonini, Jr, and L. R. Sharp, *English 12: Composition and Language*. Boston: Ginn and Co., p. 83ff.
166. 1968. Reprint, 'The Psychological Revolution: the Emotional Source of the Scientific Movement' in George Basalla (ed.), *The Rise of Modern Science*. Lexington, Mass.: D. C. Heath, pp. 69–75.
167. 1968. Reprint, 'Sociological Aspects of the Relation between Language and Culture' in Robert A. Manners and David Kaplan (eds), *Theory in Anthropology*, Chicago: Aldine Publishing Co., pp. 411–420.
168. 1968. Review, ' Kurt H. Wolff and Barrington Moore, Jr (eds), The Critical Spirit:

Essays in Honor of Herbert Marcuse', Boston: Beacon Press, 1967, in *American Sociological Review* **33** (June 1968), 465–467.

169. 1968. 'From Ideology to Philosophy: Sidney Hook's Writings on Marxism' in Paul Kurtz (ed.), *Sidney Hook and the Contemporary World: Essays on the Pragmatic Intelligence*. New York: The John Day Company, pp. 35–53.
170. 1968. Review: Chushichi Tsuzuki, The Life of Eleanor Marx, 1855–1898: A Socialist Tragedy, Oxford 1967; *International Journal* **XXIII**, No. 4 (Autumn, 1968), 627–628.
171. 1968. 'The New Marxism of the Intellectuals', *The New Leader* **LI**, No. 21 (Nov. 4), 7–11.
172. 1968. 'The Socio-Psychological Transformations of Soviet Society' in Milorad M. Drachkovitch (ed.), *Fifty Years of Communism in Russia*. University Park: Pennsylvania State University Press, pp. 99–137, 299–303.
173. 1968. 'Karl Marx and the Promethean Complex', *Encounter* **XXXI**, No. 6 (Dec.), 15–32.
174. 1968. 'God, Guilt, and Logic: the Psychological Basis of the Ontological Argument', *Inquiry*, **II**, 257–28l.
175. 1968. 'Patterns of Irrationality', in 'Students and Revolution: A Discussion', *Survey*, No. 69 (Oct.), 43–5l.
176. 1968. Reprint, 'Beyond Ideology' from Psychoanalysis and Ethics, in Chaim I. Waxman (ed.), *The End of Ideology Debate*. New York: Funk & Wagnalls, pp. 64–68.
177. 1969. 'Conflict of Generations' (edited excerpts), *Saturday Review* **LII**, No. 3 (Jan. 18), 53–55, 66–68.
178. 1969. 'Berkeley and Beyond', *Change in Higher Education* **I**, No. 1 (Jan.–Feb.), 47–51.
179. 1969. Review: 'Herbert Marcuse: An Essay on Liberation', *Book World* **III**, No. 8 (Feb. 23), 14.
180. 1969. Reprint: 'We Should Distinguish between Disobedience and Resistance' in Hugo Adams Bedeau (ed.), *Civil Disobedience: Theory and Practice*, New York: Pegasus, pp. 204–205.
181. 1969. *The Conflict of Generations: The Character and Significance of Student Movements*. New York; Basic Books, pp. ix, 543.
182. 1969. British edition, with new introduction, *Marx and Engels, Basic Writings and Philosophy*. London: Fontana Books, new introduction, pp. 11–41.
183. 1969. Reprint: 'American Philosophy is Dead' in Christopher J. Lucas, *What is Philosophy of Education?*, New York: Macmillan, pp. 39–41.
184. 1969. Reprint: 'We Should Distinguish between Disobedience and Resistance', in *New Dimensions of Freedom in America*, Frederick M. Wirt and Willis D. Hawley (eds). San Francisco: Chandler Publishing Co., pp. 101–103.
185. 1969. *Anarchy, Authority and Human Freedom*. Boulder: Institute for the Comparative Study of Political Systems and Ideologies, University of Colorado, pp. 123.
186. 1969. *Ibid.*, reprinted in *Liberty under Law: Anarchy, Totalitarianism: This is the Choice*, American Bar Association, The Standing Committee on Education about Communism and to Contrast with Liberty under the Law (no peace), (July), pp. 55–72.
187. 1969. 'Advice to the Reillusioned': Reviews: "Bertram D. Wolfe, An Ideology in

Power: Reflections on the Russian Revolution; Zhores A. Medvedev, The Rise and Fall of T. D. Lysenko', *The New Leader* **LII**, No. 19 (Oct. 13), 16–19.

188. 1969. 'John Dewey's Sojourn in Japan', *The Record*, Teachers College, Columbia University, **71**, No. 1 (Sept.), 123–145.

189. 1969. Reprint: 'What is Alienation?: The Career of a Concept' in David J. Borrows and Frederick R. Lapides (eds), *Alienation: A Casebook*. New York: Crowell, pp. 87–98.

190. 1969. 'The Favorite of the Party', Review: " George Katkov, The Trial of Bukharin', *The New Leader* **LII**, No. 23 (Dec. 8), 24–26.

191. 1969. *Marx and the Intellectuals: A Set of Post-Ideological Essays*. New York: Anchor Books, Doubleday & Co., pp. 301.

192. 1969. 'The Autonomy of the Sociology of Ideas: A Rejoinder', *Inquiry* **12** (Winter), 434–445.

193. 1969–70. Review 'Milovan Djilas, The Unperfect Society: Beyond the New Class', *International Journal* **XXV**, No. 1, 217–218.

194. 1970. 'The USSR on Trial', Reviews: 'Pavel Litvinov, The Demonstraion in Pushkin Square; Anatoly Marchenko, My Testimony', *The New Leader* **LIII**, No. 4 (Feb. 16), 22–24.

195. 1970. Italian Translation: *L'Intellecttuale Scientifico*, traduzione di Edevo Suriani. Bologna: Nicola Zanichelli, 405 pp.

196. 1970. Reprint: 'Jewish Students as Bearers of Generational Conflict'. *The Jewish Digest* **XV**, No. 5 (Feb.), 1–9.

197. 1970. Review: Robert C. Tucker, The Marxian Revolutionary Idea, *The Slavic Review* **29**, No. 2 (June), 330–331.

198. 1970. 'Ernst Mach: The Unconscious Motives of an Empiricist', *American Imago* **27**, No. 1 (Spring), 12–40.

199. 1970. Reprint from *Judaism*, 1964. 'Soviet Marranos', *The Jewish Digest* **XV**, No. 10, 54–64.

200. 1970. Reprint from *The Scientific Intellectual*, 'The Scientific Intellectual', Jack D. Douglas (ed.), *Freedom and Tyranny: Social Problems in a Technological Society*. New York: Alfred A. Knopf, 85–95.

201. 1970. 'The Strife of Generations' in *Students in the University and in Society*, American College Personnel Association, Student Personnel Series 13, John M. Whitely (ed.), pp. 46–55.

202. 1970. 'Lenin's Fantasy: The Interpretation of a Russian Revolutionary Dream', *Encounter* **XXXV**, No. 6 (Dec.), 23–35.

203. 1970. 'The Intelligentsia in Opposition', *Problems of Communism*, **XIX**, (Nov.–Dec.), pp. 1–16. *Ibid.*, Spanish translation, 'La Intellectualidad en Oposicion', *Problemas del Comunismo* (Nov.–Dec.), 34–48.

204. 1970. 'Lawless Sensations and Categorial Defences: The Unconscious Sources of Kant's Philosophy' in Charles Hanly and Morris Lazerowitz (eds), *Psychoanalysis and Philosophy*. New York: International Universities Press, pp. 76–125.

205. 1971. Reprint: 'Patterns of Irrationality' in Paul D. Knott (ed.), *Student Activism*. Dubuque: Wm. C. Brown & Co., pp. 26–34.

206. 1971. 'The Political Linguistics of 'Intellectual' 1898–1918', *Survey*, No. 1 (78) (Winter), 156–183.

207. 1971. Letter, 'Lenin's Fantasy', *Encounter* **XXXVI**, No. 6 (June), 92–93.
208. 1971. *USSR: The Intelligentsia in Opposition*, pamphlet, reprinted from *Problems of Communism*, Washington, D.C., pp. 1–16.
209. 1971. 'The Social Roots of Einstein's Theory of Relativity, Part I', *Annals of Science* **27**, No. 3 (Sept.), 277–298.
210. 1971. 'Is the Cold War Really Over?' *The New Leader* **LIV**, No. 20 (Oct. 18), 7–12.
211. 1971. Reprint: 'The Risk is Juvenocracy' in Helen Willis, *Basic Usage, Vocabulary, and Composition*. New York: Holt, Rinehart, and Winston, pp. 415–421.
212. 1971. Spanish translations of 'Is the Cold War Really Over?', '¿Termino realmente la guerra fria?', *La Prensa de Buenos Aires*, 8 de diciembre de 1971; 'Objectivos morales en la politica internacional de Estados Unidos', 12 de diciembre de 1971; Estados Unidos, Rusia y China', 14 de diciembre de 1971; "La moral, en la politica internacional de Tuo Sam', *El Tiempo de Bogota*, 2 de diciembre de 1971.
213. 1971. 'The Social Roots of Einstein's Theory of Relativity, Part II', *Annals of Science* **27**, No. 4 (Dec.), 313–344.
214. 1971. Reprint: 'Between Fantasy and Reality: Lenin as a Philosopher and Social Scientist' in *Lenin and Leninism: State, Law, and Society*, Bernard W. Eissenstat (ed.), Lexington: D. C. Heath and Co., pp. 59–79.
215. 1972. Spanish translation: *Los Movimientos Estudiantiles. Las revoluciones nacionales y sociales en Europa y el Tercer Mundo*. Buenos Aires: Paidos, transl. Advebo Adolfo Negrotto, pp. 1 – 436.
216. 1972. Hebrew translation: 'Is the Cold War Really Over?' in 'Ha-Milchamah Ha kirah', *Hadoar* (New York) **LI**, No. 14 (Feb. 11).
217. 1972. 'A Lesson from Hobbes', Review: 'John P. Diggins, Mussolini and Fascism: The View from America', *The New Leader* **LV**, No. 18 (Sept. 18), 21–22.
218. 1972. 'Generations and the Theory of Revolution', *Survey* **18**, No. 3 (84), (Summer), 161–188.
219. 1972. Reprint: 'Einstein and Counter–Culture', *Jewish Spectator* **XXVII**, No. 7 (Sept.), 8–14.
220. 1972. 'Democrats versus Democracy', *National Review* **XXVII**, No. 42 (Oct. 27), 1178–1180.
220a. 1972. Selection from 'The Decline of Freedom in Berkeley', in Steward La Casce and Terry Belangery, *The Art of Persuasion: How to Write Effectively about Almost Anything*, New York: Charles Scribners Sons, pp. 47–48.
221. 1972. 'Fascism', Letter in reply to John P. Diggins, *The New Leader* **LV**, No. 22 (Nov. 13), 27.
222. 1972. 'Student Unrest in the United States', *The Annals of the American Academy of Political and Social Science* **404** (Nov.), Marvin Bressler (ed.), *American Higher Education: Proposed Choices*, pp. 170–182.
223. 1972. 'The Conversion of Karl Marx's Father', *The Jewish Journal of Sociology* **XIV** (Dec.), 149–166.
224. 1973. 'Ideology and No End: Some Personal History', *Encounter* **XL**, No. 4 (April), 84–87.
225. 1973. 'Scientific Creativity and Decentralized Societies: The Swiss Cantons and the Rise of the Social Sciences', *The Kansas Journal of Sociology* **IX**, 47–53.
226. 1973. Reprint: 'The Intelligentsia in Opposition' in Samuel Hendel (ed.), *The Soviet*

Crucible: The Soviet System in Theory and Practice. North Scituate, Mass. (March), pp. 215–221.

227. 1973. Reprint: *Psychoanalysis and Ethics*. Conn.: Greenwood Press, Westport.
228. 1973. Statement in Sidney Hertzberge, 'Kennedy and the Intellectuals', *The Washington Post* (Nov. 18).
229. 1973. Review: 'David Caute, The Fellow Travellers: A Postscript to the Enlightenment', *The Russian Review* **33**, No. 2 (April), 231–232.
230. 1974. Letter: 'The "New York School of Sociology"', *Encounter* **XLII**, No. 1 (Jan.), 95.
231. 1974. 'Dwelling on the Superficial', Review of George Dykhuizen, The Life and Mind of John Dewey, *The New Leader* **LVII**, No. 2 (Jan. 21), 18–20.
232. 1974. 'The Achievement of Max Nomad, Impossibilist, 1881–1973', *Survey* **20**, No. 1 (90) (Winter), 159–163.
233. 1974. 'The New Imperialism', *National Review* **XXVI**, No. 13 (March 29), 369–372, 386, 387, 389. Reprinted (and attacked) in *The Renaissance*, Nigeria, No. 546 (April 11, 1974), 13–14.
234. 1974. 'The Fellow-Travellers', *Survey* **20**, No. 2/3 (91/92), (Spring-Summer), 206–210.
235. 1974. Spanish translation, Lewis S. Feuer, *La Nocion Marxista de alienacion y los movimientos estudiantiles*, Una Polemica: el estado capitalista, Nicos Poulantzas, Ralph Milibard, Quito: Editorial Universitaria, 82 pp. Coleccion Cuadernos Culturales, no. 7.
236. 1974. Review: Jügen Habermas, Theory and Practice, *Contemporary Sociology* **3**, No. 4 (July), 318–319.
237. 1974. 'The Preconceptions of Critical Theory', Review Article of Martin Jay, The Dialectical Imagination, *The Jewish Journal of Sociology* **XVI** (June), 75–84.
238. 1974. 'The Illusions in the Search for East-West Accommodation', *International Perspectives* (July-Aug.), 12–15.
239. 1974. *Einstein and the Generations of Science*, New York: Basic Books (May), pp. 374.
240. 1974. 'Karl Marx', *The New Encyclopedia Britannica*, 15th edn. Vol. II. Chicago, pp. 549–553.
241. 1974. 'Karl Popper: Humanist Philosopher', *The Humanist* **XXXIV**, No. 6 (Nov.-Dec.), 37–41.
242. 1974. Reprint article in Japanese translation from *The Conflict of Generations*, in Gendai No Esprit, No. 86. *l'Esprit d'Aujourd'hui*,
243. 1975. *Ideology and the Ideologists*. Oxford: Basil Blackwell (Feb.), pp. 220.
244. 1975. Selection from 'The Decline of Freedom at Berkeley' in John Ashmead, Ann Christine Heintz, Elaine Schuster, *Synthesizing: An Elective Course in Composition*. Lexington, Mass: Ginn and Company, pp. 83–84.
245. 1975. Letter to *Contemporary Sociology* **4**, No. 1 (Jan.), 8–9.
246. 1975. 'Is the "Darwin-Marx Correspondence" Authentic?', *Annals of Science* **32**, No. 1 (Jan.), 1–12.
247. 1975. 'Why not a Commentary on Sevareid?', *National Review* **XXVII** (Aug. 15), 874–876.
248. 1975. 'The Pragmatic Wisdom of Sidney Hook', *Encounter* **XLV** (Oct.), 37–45.

249. 1975. 'The Fission of the Intellectuals', *National Review* **XXVII** (Dec. 19), 1475–1476.
250. 1975. 'Response' in Jacob Katz (ed.), *The Role of Religion in Modern Jewish History*. New York: Ktab Publishing Co., pp. 19–22.
251. 1976. 'Is History in Kissinger's Way?', *The New York Times* (May 27), 35; reprinted, "Will fate of U.S. be that of Ilium?', *Montreal Star* (June 10, 1976); also, *Windsor Star* (June 8, 1976).
252. 1976. 'Recollections of Harry Austryn Wolfson', *American Jewish Archives* **XXVIII**, No. 1 (April), 25–50.
253. 1976. 'John Stuart Mill as a Sociologist: The Unwritten Ethology' in John M. Robson and Michael Laine (eds), *James and John Stuart Mill: Papers of the Centenary Conference*. University of Toronto Press, pp. 86–110.
254. 1976. 'The 'Darwin-Marx correspondence': a correction and revision', *Annals of Science* **33**, 383–384.
255. 1976. With P. Thomas Carroll (my part was almost nominal), 'Further evidence that Karl Marx was not the recipient of Charles Darwin's letter dated 13 October 1880', *Annals of Science* **33**, 385–387.
256. 1976. 'Method in the Sociology of Science: Rejoinder to Professor Agassi', *Philosophy of the Social Sciences* **6** (Sept.), 249–253.
257. 1976. Reprint: 'Generational Struggle in Plato and Aristotle' in Stephen Bertman (ed.), *The Conflict of Generations in Ancient Greece and Rome*. Amsterdam: B. R. Gruner, pp. 123–128.
258. 1976. 'What is an Intellectual?', in Alexander Gella (ed.), *The Intelligentsia and the Intellectuals: Theory, Method and Case Study*. Beverly Hills, California: Sage Publications, pp. 47–58.
259. 1976. 'America's Medicine Man', Review of Niel Coughlan, Young John Dewey: An Essay in American Intellectual History, *Times Literary Supplement* (Dec.3).
260. 1976. 'From the Class of' 3l', *The New York Times* (Dec. 27), p. A23.
261. 1977. 'Mendicant Friar of the Revolution', Review: Michael Harrington, The Twilight of Capitalism, *The Humanist* **XXXVII** (Jan.–Feb.), 46–48.
262. 1977. 'A teacher from the tenements', Review: David A Hollinger, Morris R. Cohen and the Scientific Ideal, *The Times Literary Supplement*, No. 3, 920 (29 April), 521–522.
263. 1977. 'Arthur O. Lovejoy', *The American Scholar* **46** (Summer), 358–366.
264. 1977. 'The Achievement of Will Herberg: Herberg as Political Philosopher', *National Review* **XXIX** (Aug. 5), 882–883.
265. 1977. 'Bertram David Wolfe, 1896–1977', *The Russian Review* **36**, No. 4 (Oct.), 533–535.
266. 1977. 'Historical Method in the Sociology of Science: the Pitfalls of a Polemicist', *Philosophy of the Social Sciences* **7**, 255–26l.
267. 1977. 'Dissenso', *Enciclopedia Italiana* **II**, 154–169.
268. 1978. 'The Legend of the Socialist East Side', *Midstream* **XXIV**, No. 2 (Feb.), 23–35.
269. 1978. 'The Professionalization of Philosophy', Review: Bruce Kuklick, The Rise of American Philosophy, *The Chronicle of Higher Education* **XVI**, No. 2 (March 6), 21–22
270. 1978. 'Our Immigrant's Lot', Letter to *The New York Times* (April 23, 1978), (Sunday), p. 18, Section Four.

271. 1977–78. 'Bertram David Wolfe 1896–1977', *Survey* **23**, No. 1 (102) (Winter), 194–201.
272. 1978. 'The Case of the 'Darwin-Marx' Letter: A Study in Socio-Literary Detection', *Encounter* **LI** (Oct.), 62, 64, 66, 70, 72, 74, 76, 78.
273. 1978. 'University Marxism', Reviews: Perry Anderson, Considerations on Western Marxism; Mark Poster, Existential Marxim in Postwar France: from Sartre to Althusser; Michael Löwy, Pour une sociologie des intellectuels; l'evolution politique de Lukacs, 1909–1929, *Problems of Communism* **XXVII** (July-Aug.), 65–72.
274. 1978. 'University Marxism', reprinted without my permission in *The Rising Tide,* **VII**, No. 12 (Nov. 12), 6, 3.
275. 1978. German translation, Ideology and the Ideologists, *Die authoritäre Vermessenheit: ideologen im Dilemma*, Verlag Styria, Graz Wien Koln, transl. Walter Sauer.
276. 1978. 'Teleological Principles in Science', *Inquiry* **21**, No. 4 (Winter), 377–407.
277. 1978. French translation, *Einstein et le Conflit des Générations,* préface de Serge Moscovici, Editions, Complexe, distribution Presses universitaires de France, 376 pp., 99F. (Reviewed by Emmanuel Todd, 'D' ou viennent les Révolutions Scientifiques? A partir d'Einstein, une approche psychoanalytique de l'histoire de la physique', *Le Monde* (Nov. 17, 1978), pp. 19, 26.
278. 1977–78 'Marx and Engels as Sociobiologists', *Survey* **23**, No. 4 (105), (Autumn), 109–136.
279. 1979. Japanese translation, *Einstein and the Generations of Science*. Tokyo: Bunka Hoso Development Center Co.and Charles C. Tuttle Co.
280. 1979. 'Cultural Detente at Scientific Congresses: The Secret Report of the Czech Sociologists', *Orbis: A Journal of World Affairs* **23**, 115–127.
281. 1979. 'American Activitists and Soviet Power: Supporting Liberalization in the Soviet Union' in Raymond D. Gastil (ed.), *Freedom in the World: Political Rights and Civil Liberties 1979*. Freedom House, pp.161–172.
282. 1979. 'The Friendship of Edwin Ray Lankaster and Karl Marx: The Last Episode in Marx's Intellectual Evolution', *Journal of the History of Ideas* **XL** (Oct.–Dec.), 633–648.
283. 1979. 'Science and the Ethic of Protestant Asceticism: A Reply to Professor Robert K. Merton' in *Research in Sociology of Knowledge, Sciences and Art: A Research Annual*, Robert Alun Jones and Henrika Kuklick (eds), Vol. 2. Greenwood, Connecticut: JAI Press, Inc., pp. 1–23.
284. 1979. 'Some Irrational Sources of Opposition to the Market System', in *Capitalism: Sources of Hostility*, (ed) Ernest van den Haag. New Rochelle, New York: The Heritage Foundation, Epoch Books, pp. 103–152.
285. 1980. 'Spinoza's Thought and Modern Perplexities: Its American Career' in *Spinoza: A Tercentenary Perspective*, Barry S. Kogan (ed.). Cincinnati: Hebrew Union College-Jewish Institute of Religion, pp. 36–79.
286. 1979. 'Arthur O. Lovejoy' in *International Encyclopedia of the Social Sciences: Biographical Supplement*, Vol. 18. New York: The Free Press, pp. 464–469.
287. 1980. 'Spinoza's Political Philosophy: The Lessons and Problems of a Conservative Democrat' in *The Philosophy of Baruch Spinoza*, Richard Kennington (ed.). *Studies in Philosophy and the History of Philosophy*. Vol. 7. Washington, D. C., pp. 133–153.
288. 1980. 'Time to Drop Obsolete Party Slogans', Letter to *The New York Times* (Aug. 13), p. A22.

289. 1979. Hebrew Translation, *Aynshtayn u-bnai doro*, Am Oved Publisher, Sifriat Afikim, Tel Aviv, Israel, transl. Gad Levi, pp. 430.
290. 1980. 'On Will Herberg' in 'Reasoning Together' in *The Canadian Review of American Studies* **11**, No. 2 (Fall), 262–264.
291. 1981. 'Arthur O. Lovejoy' in Joseph Epstein, *Masters: Portraits of Great Teachers*. New York: Basic Books, pp. 119–139.
292. 1981. 'Introduction: Bertram David Wolfe 1896–1977' in Bertram D. Wolfe, *Revolution and Reality: Essays on the Origin and Fate of the Soviet System*. Chapel Hill: The University of North Carolina Press, pp. IX-XXI (reprint).
293. 1981. 'Captive Beliefs', *The Toronto Sun* (June 22), p.11.
294. 1981. 'The Frankfurt Marxists and the Columbia Liberals', *Survey* **25**, No. 3 (112) (Summer 1980), 156–176. (Actually revived Sept. 8, 1981.)
295. 1981. 'Generations of Terrorists', *The Toronto Sun* (Thursday, Nov. 5), p. 61.
296. 1981. 'Machiavelli as Unperson', *The Toronto Sun* (Thursday, Dec. 17), p. 11.
297. 1982. 'Reds through Tinted Glasses', *The Toronto Sun* (Friday, Jan. 22), p. 12.
298. 1982. 'Why the Party Likes Reds', *The Toronto Sun* (Monday, Jan. 25), p. 11.
299. 1982. 'John Reed and Red Scare', *The Toronto Sun* (Wednesday, Jan. 27), p. 12.
300. 1982. 'Disillusion of John Reed,' *The Toronto Sun* (Thursday, Jan. 28), p. 11.
301. 1982. 'Arkansas' Monkey Trial,' *The Toronto Sun* (Tuesday, Mar. 11), p. 11.
302. 1982. 'Prologue', *Einstein and the Generations of Science*, Second Edition, New Brunswick, New Jersey: Transaction Books, Rutgers, The State University, pp. ix–xlv.
303. 1982. 'Einstein and the Prague Circle', *Midstream* **XXVIII**, No. 6 (June-July), 36–39.
304. 1982. 'The Stages in the Social History of Jewish Professors in American Colleges and Universities,' *American Jewish History* **LXXI**, No. 4 (June), 432–465.
305. 1983. 'The 'Explanatory Crisis' of Marxism,' *Midstream* **XXIX**, No. 1 (Jan.), 22–31.
306. 1983. 'The Social Role of the Frankfurt Marxists; A Rejoinder,' *Survey* **26**, No. 2 (115) (Spring 1982), 150–170. (Actually published in Winter, 1983).
307. 1983. 'The Genetic Fallacy Re-examined,' in Paul Kurtz (ed.), *Sidney Hook: Philosopher of Democracy and Humanism*. Buffalo: Prometheus Books, pp. 227–246.
308. 1983. 'The Sociobiological Theory of Jewish Intellectual Achievement: A Sociological Critique' in *Ethnicity, Identity, and History: Essays in Memory of Werner J. Cahnman*, Joseph B. Maier and Chaim I. Waxman, (eds.), New Brunswick, New Jersey: Transaction Books, pp. 93–123.
309. 1983. *The Case of the Revolutionist's Daughter: Sherlock Holmes Meets Karl Marx*, Buffalo, New York: Prometheus Books, pp. 1–159.
310. 1983. 'The Disease of Idealism': Review, Sidney Hook, Marxism and Beyond, *The American Scholar* **52** (Autumn), pp. 558, 560, 562–563.
311. 1983. 'Noumenalism and Einstein's Argument for the Existence of God,' *Inquiry* (University of Oslo) **26**, No. 3 (Sept.), 251–285.
312. 1983. 'South Pacific Memoir', *The New Leader* **LXVI**, No. 12 (June 13), 9–10.
313. 1983. 'Sidney Hook on Marxism': Review, Sidney Hook, Marxism and Beyond, *Midstream* **XXIX**, No. 9 (Nov.), 51–54.
314. 1983/84. 'Anti-Science: The Irrationalist Vogue of the 1970's,' *Free Inquiry* **4**, No. 1 (Winter), 36–46.
315. 1984. 'Grenada Coverage,' *The New Leader* **LXVIII**, No. 1 (Jan. 9), p. 22.

316. 1984. 'Horace M. Kallen on War and Peace,' *Modern Judaism* **4**, No. 2 (May) 201–213.
317. 1984. 'Meyer Liben Remembered', *Midstream* **XXX**, No. 6 (June-July), 42–46.
318. 1984. Letter, 'From the Kremlin, A Hint of Neo-Stalinism,' *The New York Times* (July 17), A20.
319. 1984. Letter, 'Historical Fiction,' *Midstream* **XXX**, No. 7, (Aug.-Sept.), 63–64.
320. 1984. 'What Orwell Means For Our Time,' *Survey* **28**, No. 2 (121) (Summer), 155–164.
321. 1984. Marx and Engels, *Basic Writings on Politics and Philosophy*. Fontana Book Reader, Aylesbury: Fontana-Collins, with several new pages, pp. 39–42, which is an amended introduction.
322. 1984. 'Charisma and Leadership', unedited remarks, recorded, with opening passages unrecorded, in *Leadership*, R. S. Khare and David Little (eds.). Boston: University Press of America, pp. 49–52.
323. 1985. Lewis S. Feuer, in United States Department of State, *Contemporary Soviet Propaganda and Disinformation: A Conference Report*. Airlie, Virginia: Department of State Publication 9536, (June 25–27), pp. 1, 49–51.
324. 1985. 'America's First Jewish Professor: James Joseph Sylvester at the University of Virginia,' *American Jewish Archives* **XXXVI**, No. 2 (Nov. 1984 actually Feb. 1985), 152–201.
325. 1986. 'The Reasoning of Holocaust Theology,' *Judaism* **35**, No. 2 (Spring) 198–210.
326. 1986. *Imperialism and the Anti-Imperialist Mind*. Buffallo, New York: Prometheus Books, pp. i–vi, 1–265.
327. 1986. Review: A. P. French and P. J. Kennedy (eds.), Niels Bohr: A Centenary Volume, *American Journal of Physics* **54**, No. 8 (Aug.), 762–764.
328. 1986. 'Jefferson wouldn't tolerate secret societies,' *The Cavalier Daily* **97**, No. 12, Charlottesville, Virginia: University of Virginia, (Sept. 18, 19), p. 2.
329. 1987. 'Sylvester in Virginia', *The Mathematical Intelligencer* **9**, No. 2, New York: Springer-Verlag. Shortened version of No. 323.
330. 1987. Letter to the Editor, 'Soviet Poet Scheduled to speak at the University,' *The Cavalier Daily* **97**, No. 130, Charlottesville, Virginia: University of Virginia, (April 23) p. 2.
331. 1987. Review: Alan J. Friedman and Carol C. Donley, *Einstein as Myth and Muse*, Cambridge University Press, 1985, *Annals of Science* **44**, (Jan.), 85–87.
332. 1987. 'The East Side Philosophers: William James and Thomas Davidson,' *American Jewish History* **LXXXVI**, No. 3 (Mar.), 287–310.
333. 1987. 'Is the Jewish Tradition Radical or Conservative?' *Midstream* **XXXIII**, No. 6 (June-July), 14–21.
334. 1987. 'Freud's Ideas in the Soviet Setting: A Meeting with Aleksandr Luria', *Slavic Review: American Quarterly of Soviet and East European Studies* **46**, No. 1, (Spring), 106–112.
335. 1987. 'Introduction: John Stuart Mill as a Sociologist: The Unwritten Ethology', in John Stuart Mill, *On Socialism*. Buffalo, New York: Prometheus Books, pp. 9–50. Reprint of No. 253.
336. 1987. *Spinoza and the Rise of Liberalism*, Introduction to the Transaction edition, reprint of No. 76, pp. XIII-XXX. New Brunswick, U.S.A. and Oxford U.K.): Transaction Books, (Sept.).

WILLIAM J. BAUMOL

ON THE REALITY OF ECONOMIC ILLUSION

Economists have long taken the reasonable position that a consumer's satisfaction cannot be evaluated correctly by counting his tangible possessions; rather, his welfare must in some way reflect the utility he derives from different quantities and combinations of such possessions, given that person's psychological make-up. Yet this way of looking at choice and welfare has not kept us from what can be considered an analogous fallacy in which consumer's preferences are taken to be free from 'irrationalities' and illusions, terms whose connotations will become clearer as we proceed. On this basis, at least some economists have advocated measures which are widely disliked, presumably upon the grounds that, despite their unpopularity, they really serve the public's true welfare. An example is the view that income tax is to be preferred to an excise, both because the latter is generally more regressive and because it may lead to a poorer allocation of resources (though lately the literature on the 'second best' has cast serious doubt on this second contention). I will suggest that the contrary is likely to be true: that even considering its distributive effects the excise is often likely to be, shall we say, *nearly* Pareto preferable: it may well leave nearly everyone better off.

1. ILLUSION AND CONSUMER SATISFACTION

Illusions beset our consumption patterns and our economic activities in a variety of ways. A simple example should bring out the point: on a voyage by ship from England to Denmark I was distressed to discover that passengers were expected to purchase their meals and to pay just for what they ordered. Unlike a transatlantic ship voyage, food was not included in the price of the passage. Of course, I was well aware that a calculation based on cost per mile, including the cost of the meals, would show that this trip was much more reasonably priced than the transatlantic trip. Despite my consequent conviction that I was being irrational, I was not *really* reassured, and suffered a very real loss of

S. Hook, W. L. O'Neill and R. O'Toole (eds.), Philosophy, History and Social Action, 103–117.

utility because the meals on the Scandinavian ship were not 'free'.

A psychologist friend once told me, somewhat uneasily, that despite a period of many months spent demonstrating and explaining a set of optical illusions, when he left the platform and joined the audience, they still fooled him every time. Economic illusions have a similar property: those of us who have been subject to such illusions are left with the uncomfortable feeling that we should not succumb to them, yet somehow we cannot help it.[1]

We can all cite many more examples. Some of my colleagues make certain that substantial income tax deductions are made from their salary checks so they will have little or nothing left to pay at the end of the year. Obviously, they would gain financially if they were to hold back until the last minute every tax dollar they legally can, earning interest on the money in the meantime. But, to those persons, avoidance of the pain of writing out a large check on April 14 is apparently worth the interest foregone. It is then surely not their decision that is irrational but whatever disquiet they feel in sticking to it.

Illusion may also play a role in the pattern of use of time at conferences and over the telephone. It is hard to avoid the impression that people are generally much briefer in long-distance calls than they are in face-to-face conferences involving comparable distances. This may in part be a response to the rate structure that is rational in the conventional sense: the longer one talks on the telephone, the more the call. costs. But most of the cost of a trip is a fixed charge so that a conferee may, perhaps subconsciously, suffer conscience pangs unless his meeting lasts a sufficient amount of time to have given his money's worth to whoever is paying the fare.

Commercial pricing policies sometimes take economic illusions into account. Prices set at $4.99 and $29.99 are an obvious response to the phenomenon. Consumers are probably not fooled – they are only made to feel better about their outlay. Some public utilities permit prepayment on anticipated monthly charges and make periodic adjustment payments to balance off the tally. It is said that their experience has led these firms to a policy of systematic *overestimation* of future payment levels on the ground that goodwill is engendered when the customer receives a rebate rather than a bill.

Economic illusions are also to be found in high places. In its subbasement vaults where foreign gold is stored, the Federal Reserve Bank of New York caters to them. In response to its demand, each nation's gold is kept in a separate and identifiable pile, presumably because of the reassurance and satisfaction provided to a visiting representative of a central bank when he is shown the gold that is beyond any question *his*.[2]

2. PRICES AND UTILITY

Before investigating how illusions such as those which have just been mentioned affect standard analysis it is necessary to be a bit more specific about some of the ways in which they may arise. It seems quite unlikely that there is any shortlist of sources which give rise to such phenomena. Illusions, by their very nature, seem likely to be unpredictable in character, and there may be no limit to the ability of the human mind to come up with surprising varieties.

Still, it is easy enough to identify some of the important sources of illusions, meaning by this variables which, in the economist's sense, do not correspond to 'real' phenomena, but which nevertheless enter many individuals' utility functions. Prices and monetary payments are among the prime candidates for this role.

In this respect the phenomenon usually denoted by the term 'money illusion' is suggestive. Economists are undoubtely right when they point out that consumers will consider prices to be undesirably high if they have recently risen substantially even if consumers' money incomes simultaneously rise proportionately. However, we can easily draw an untenable conclusion from this observation. It is our instinct to argue that people somehow 'ought to know better'. Rather, the proper lesson may be simply that a rise in prices is likely to be very painful. The rise in costs which is involved may indeed be illusory, but the resulting loss in utility is very real and should be taken into account in our calculations.

It is important for economists to understand this because there are so many problems which we propose to resolve by a rise in prices. We often recommend as the means to reduce pollution, a tax upon emissions. Such a tax is designed, in part, to raise the prices of outputs whose production generates pollution, these price rises serving to discourage their consumption. Similarly, if a crop fails, we generally argue against price controls, and defend price increases as an effective means to ration the products and to insure that they are consumed slowly enough to prevent the shortage from becoming truly critical before the next crop becomes available. I do not mean to dispute the validity of these agruments, of which I remain thoroughly convinced; but I do believe that they are incomplete. For, as they stand, they do not take account of the fact that the recommended medicines are painful, even if their pain does rest on illusion. One thing that follows is that in cases where the next best policy measure would otherwise be 'almost as good', the inherent disutility of price increases may in fact make the 'second best' solutions preferable to the pricing approach. Second, it follows that even where the pricing approach is nevertheless to be preferred, it may be optimal to swallow a dose of this medicine milder than would otherwise be the case. Slightly lower effluent charges and slightly higher rates of emission than those which appear optimal on the usual calculation may in fact be called for. The third, and most important, implication is that the widespread opposition which such proposals often encounter among politicians and members of the public is not to be ascribed merely to ignorance and inflexibility. A more sympathetic and understanding view of the public's reaction to such measures may well facilitate communication and perhaps even make it easier to explain their benefits to people outside our profession.

There are at least two ways in which money and prices may significantly affect an individual's utility: (i) As we have just seen, the absolute amount of money paid out by the person may enter his utility function. The larger the payment that is required at a particular juncture, the lower the utility the individual will enjoy, even if the values of all real variables are fixed. (ii) The disutility produced by a given payment may be affected by the *form* taken by the payment – whether it is laid out in a lump sum, takes the form of a fixed charge per use, or assumes some other shape. The previous section offered some examples of this phenomenon. I shall now discuss a few more examples.

3. ALTERNATIVE PRICE STRUCTURES AND THEIR UTILITY IMPLICATIONS

I have already observed that many people derive pleasure from the prepayment of expenses, and enjoy the illusion that their subsequent consumption is 'free'. Many people prefer to arrange for vacations with 'all expenses paid', Apparently, many persons would prefer to buy their telephones rather than paying a monthly rental for the indefinite future, even if the payment stream is equivalent in expected present value to the cost of buying the instrument outright. It is not difficult to find other examples.

However, one must be careful in interpreting these conclusions. There may be reasons other than the inherent attractiveness of a low (or zero) price per use which may induce people to prefer prepayment. For one thing, prepayment may be a way of reducing uncertainty. When one does not know what expenses may arise (as on a vacation trip) or how long payments will continue (as in a telephone subscription), prepayment may be the buyer's easiest way of shifting risk to the supplier. Second, prepayment is a way of reducing the marginal cost of additional consumption. Thus the very feature of 'free' supply of resources that is anathema to economists – its inducement of wasteful consumption – may be a prime source of attraction to buyers. People with large appetites are apt to like prepaid (buffet) meals, and sickly persons are likely to prefer health care programs with fixed payments.[3]

Moreover, preferences are not uniform in this respect and may well vary from person to person and from case to case. I have already alluded to the apparently widespread support of excise taxes – a case of many small payments, whose total amount varies with the number of purchases a person chooses to make – as against income tax – which comes closer to a single fixed payment. I have no evidence whether there are many people who simultaneously exhibit both of these apparently conflicting preferences, nor is it my goal to attempt any explanation or reconciliation. Rather, I want only to emphasize the complexity of the phenomena with which we are dealing and the consequent inevitability of oversimplification in the discussion in the next few sections.

Despite the reservations and qualifications I have expressed on the subject I shall proceed on the illustrative assumption that there are a number of goods and services for which many people prefer to pay under an arrangement that involves a relatively large fixed charge and a relatively low user charge. That is, all other things being equal, they would prefer to make their payment in the form of a fixed charge rather than paying a (slightly) smaller amount in total, the bulk of which take the form of user charges.

4. DIGRESSION ON NON-LINEAR PRICING SCHEMES

The choice among pricing arrangements naturally brings up the sort of pricing schemes to which recent discussions refer as 'non-linear pricing arrangements'. An important example of research in this area is the recent work of Faulhaber and Panzar in which they propose an extremely clever alternative, an intermediate step between a system of fixed (parametric) prices, and the perfect price discrimination which one can never achieve in practice. They consider a *variable* two-part tariff for any commodity which cannot easily be repackaged and resold by one customer group to another. One can think of the two prices as a *license price* and a *unit price*. The first of these is the cost of a license which entitles its holder to buy any desired quantity of the good, at the associated (fixed) unit price. However, unlike the usual two-part tariff, under this arrangement the consumer is offered his choice of a variety of such price packages. Some packages combine a high license fee with a small unit price while others do the reverse. These packages can range all the way from the case where the unit cost is zero (or is, at least, equal to marginal cost) and the license fee is very high, to the opposite end of the spectrum, where the license fee is zero and the unit price is very high. This last package is, of course, nothing other than the ordinary fixed-price arrangement. Similarly, one can construct more complicated price arrangements with n parameters (an n-part tariff). If non-linear pricing each customer is always permitted to select whichever of the available pricing arrangements best serves that person's interests.

Robert Willig has proved that under a non-linear tariff it is always possible to attain a state of welfare as good as, or superior to, that which can be achieved under normal (parametric) pricing arrangements. That is, there always exists an appropriately chosen set of non-linear arrangements which is Pareto equivalent or superior to any fixed price arrangement, benefiting the supplier as well as every consumer.

As is often true of such a clever idea, after it has been discovered, the reason it works is not very difficult to see. From the point of view of consumers the variable multi-part tariff is usually Pareto superior (and certainly never inferior) to a system of parametric prices because one of the packages that can be included in the range of options is the polar case in which the license fee is zero, i.e. the parametric price option. Since any consumer is free to select whatever pricing option that person finds most advantageous, another option will be chosen only if it yields a utility greater than the fixed price option. Indeed, it can be shown that *all* consumers can be enabled to gain by the broadening of the range of choices offered by the variable two-part tariff, as long as consumer tastes are not identical. For if it is possible to select a price arrangement that benefits some consumers and which harms no one else, then continuity must permit a slight modification of these prices which brings some benefits to those who would otherwise neither have gained nor lost.

It is not quite so easy to see why there must be a net gain to suppliers – the only other party involved (assuming the absence of externalities). This is so because no consumer need pay more under the arrangements we are discussing than he would with a single fixed payment; and since some consumers will pay less, we can expect some increase in demand to follow. The trick, then, is to select a set of prices for each of the options such that marginal revenue is at least equal to marginal cost. The expansion in demand can then produce an increase in profit.

One can, in fact, use a Pareto optimality approach to the determination of both the license fees and the unit prices of each of the package arrangements. Since this calculation has more degrees of freedom than the ordinary Pareto optimality analysis (more prices to select) and since one possible solution to the extended problem is simply to discard all packages except the one selling at the optimal price of the standard variety (all other prices, in effect, being set at infinity) it follows that the extended optimality solution must always be at least as desirable as the traditional solution.

The variable two-part tariff is, incidentally, generally *not* capable of yielding a 'first-best' optimum where there are economies or diseconomies of scale. The explanation is rather complicated. But it is suggestive to note that the new pricing approach can improve matters because it gives the decision-maker more decision variables, whose values he can select in a way that increases the value of the welfare maximum. For this reason, a three-part tariff can do at least as well as one composed of only two parts, and so on up the line — the greater the number of parameters that compose the price, the better it may enable the economy to perform in terms of allocative efficiency. However, as long as the number of these variables is finite this will not, in general, enable him to construct a curvilinear price–response function, which alone can guarantee to bring the community to an optimal point if returns to scale at that point happen not to be constant.

We may note, finally, a significant and attractive feature of the variable multi-part tariff proposal – the self-selection of customers, who each determine for themselves which of the available price arrangements will apply to them. The bane of any ordinary scheme of price discrimination is the determination of which customers should be charged which price, a decision that calls for information about each customer's price elasticity of demand. Under the non-linear pricing arrangements (such as the variable multi-part tariff), however, each customer can choose whichever package is best for him. The large-volume buyer will tend to select a package with a high license fee but a low unit charge and the reverse will be true of the small buyer. In this way there will be self-matching between the customer and the price package best adapted to his needs.

Non-linear pricing analysis is relevant for our purposes here because it illustrates some of the different pricing arrangements which are available for the pricing of any particular commodity. In standard non-linear pricing analysis, as in standard resource allocation theory, a particular pricing arrangement is useful only as a means to an end. It is never desirable or undesirable in itself – it never enters the utility function of the consumer. But non-linear pricing analysis does enable us to see how the structure of a price arrangement can be tailored to the consumers' *pricing* preferences.

However, on our illustrative hypothesis that, because of price illusion user charges have an inherent disutility, we can also imagine the consequences for the optimal level of a user charge in a non-linear pricing arrangement. We can also imagine the consequences for allocation of resources. User charges are, of course, the payment whose purpose is to discourage 'excessive' use of resources. We tend to blame water shortages, mounting flows of solid wastes and other misuses of resources on failure to set a user charge sufficiently high upon the resources employed in these activities. We usually conclude on this sort of basis that the appropriate user charge is the marginal opportunity cost of such resources.

Though I have not proved the results formally, the following conjectures at least seem plausible:

1. Given the appropriate concavity–convexity conditions, the optimal two-part tariff under the postulated form of price illusion will involve a lower user charge and a higher entry fee than would otherwise be optimal.

2. The result is likely to be a smaller number of users of the affected commodities, though this need not be true if the utility gain resulting from reduced user charges is sufficiently large.

3. Neglecting the income effect, each consumer of a given commodity will use more of it than he would have otherwise.

4. The net effect is likely to be larger output quantities and greater resources use than would have occurred otherwise.

5. APPLICATION: PRICE DISCRIMINATION AND THE BURDEN ON CONSUMERS WITH INELASTIC DEMANDS

A case can be made for the view that a price increase affects in some sense most severely the buyers whose demands are inelastic – the consumers whom the market is likely to offer no good substitute products, who have 'nowhere else to go.'

It is noteworthy that, in practice, price discrimination is likely to give rise to such problems in aggravated form. For we know that a discriminating monopolist who 'charges what the traffic will bear' generally obtains from customers with inelastic demands a price higher than that imposed on other buyers if the marginal cost is the same for all units supplied. We also recall that the same is likely to be true under Ramsey second-best pricing, at least where scale economies are present, since then price will tend to vary inversely with demand elasticity.

Yet, as we have noted, many who have thought about the matter seem to judge that there is something inequitable about price differences that discriminate against consumers with inelastic demands. If, as an illustration, we consider the price mechanism as a means to preserve scarce resources, e.g. the elimination of price ceilings on scarce raw materials, something more can in fact be said for this viewpoint. Recognizing the psychic pain caused consumers by any rise in prices, *efficiency* considerations require us to ask what pattern of discriminatory prices can achieve any given level of conservation with minimum (weighted average) rise in prices to different consumers. Suppose we divide consumers into different categories with different price elasticities and ask what discriminatory prices will achieve this objective. We find that, defined suitably, minimization of the price changes required to achieve a given restriction in use of scarce resources calls for price discrimination in which *higher* prices are charged to consumers with *more elastic* demands, just the reverse of the pattern with a seller who charges what the traffic will bear. Thus, we shall show now

PROPOSITION 1. *If prices are used to restrict consumption of some resource to some target quantity and it is desired to do so in a way that minimizes consumer expenditure on the resource then* (a) *no consumer group will purchase a quantity at which marginal revenue is positive*; (b) *if all cross elasticities are zero, no consumer group will purchase a quantity at which demand is elastic, i.e. any group whose demand is elastic throughout will, in effect, be faced by an 'infinite price', and its consumption driven to zero*; (c) *with zero cross elasticities, among consumer groups that do purchase non-zero quantities of the resource, the more elastic the demand of a group at its equilibrium point, the higher the price it will be charged.*

Proof: Let y_i be the quantity of the resource purchased by group i and let p_i be the price charged to the group. Then the problem is to

minimize $\Sigma p_i y_i$

subject to $\Sigma y_i = k$ (the target level of resource consumption)

whose Lagrangian is

$L = \Sigma p_i y_i + \lambda(\Sigma y_i - k)$, with $\lambda \geqslant 0$, $y_i \geqslant 0$, and whose

Kuhn–Tucker conditions include

$$\mathrm{mr}_j + \lambda \equiv \partial \Sigma \frac{p_i y_i}{\partial y_j} + \lambda \geqslant 0 \qquad y_j(\mathrm{mr}_j + \lambda) = 0.$$

Here mr_j represents the marginal revenue derived from group j.

Part (a) of Proposition 1 now follows from the second of our Kuhn–Tucker conditions, for if $y_j > 0$ we must have

$$\mathrm{mr}_j = -\lambda \leqslant 0.$$

Part (b) also follows at once from the standard relation for the case with zero cross elasticities:

$$\mathrm{mr}_j = p_j\left(1 - \frac{1}{E_j}\right) \leqslant 0 \text{ if } E_j \leqslant 1,$$

where $E_j \geqslant 0$ is the price elasticity of demand of group j. Finally, we come to part (c) of the proposition, which follows from the result that for y_i, $y_j > 0$, with zero cross elasticities

$$\text{mr}_x = p_x\left(1 - \frac{1}{E_x}\right) = -\lambda, \qquad x = i, j,$$

so that

$$\frac{p_i}{p_j} = \left(\frac{1}{E_j} - 1\right) / \left(\frac{1}{E_i} - 1\right).$$

Thus, if $1 \geqslant E_j > E_i > 0$, we have $p_i/p_j < 1$, which is our result.[4]

How can one evaluate this conclusion? It would appear to say that in this case there are efficiency as well as equity grounds that favor lower prices to customers with inelastic demands. Consider an extreme possibility – that the price rise occurs through a tax whose proceeds are not spent by the government so that, through readjustment of prices, taxpayers suffer no loss in general purchasing power as a result. What then is the purpose of minimization of the payments required to achieve the desired reduction in resources use? The answer is that if payment of taxes is painful (as it surely is) the fact that the loss in taxpayer's income is illusory is once more beside the point. Since it is desirable to achieve our conservation goal with as little pain and suffering as possible, there is an efficiency argument that favors the favoring of consumers with inelastic demands.

But there is another side to the matter, both in terms of efficiency and equity. It is to be noted that the discriminatory solution we are now discussing *cannot* be Pareto optimal since it must leave unrealized gains from trade. If the relative prices faced by different consumer groups are not the same, in equilibrium their marginal rates of substitution will be unequal and this cannot be optimal. Thus the result is inefficient. While Pareto optimality in the presence of price illusion may well call for *somewhat* lower prices to consumers whose demands are inelastic than would otherwise be the case, we should not expect it to lead all the way to minimization of the total rise in payments. Rather, one should expect the optimal solution to involve a trade-off between the ordinary efficiency requirements and the disutility costs of the expenditure increase.

From the point of view of equity one may also question the desirability of the expenditure-minimizing solution if the price of only one commodity is involved. Those who are favored by it – the consumers whose demands are most inelastic – are an arbitrarily chosen group of persons who may not be particularly poor or particularly deserving on any other grounds.

Only if a considerable number of items is involved and if different persons' demands are likely to exhibit inelasticity of demand for the different goods, so that everyone is likely to share the benefits, is such a proposal apt to pass a test of equity. Otherwise, such a discriminatory policy, though it may indeed offer benefits to those who are affected by price illusion of the sort postulated, may well be far from ideal either in terms of efficiency or equity.

6. APPLICATION PAYMENTS ILLUSION: INCOME VS EXCISE TAXES

As was already noted, in the case of tax payments people may characteristically be subject to an illusion which is opposite in direction from the sort of price illusion posited in earlier sections of this paper. There we proceeded on the hypothesis that people often prefer their payments to take the form of fixed charges rather than a user charge which extracts an additional payment with every increase in use. In taxation there is some evidence that many people prefer a multitude of small payments which are based on quantities of goods purchased (and which they are perhaps able to conceal from themselves) rather than large periodic payments whose magnitude is independent of quantities purchased.

Illusion (or perhaps delusion) is surely at the heart of the excise tax issue. In the course of recent attempts to introduce a broad based tax in a number of states some observers seem to have been taken aback by the evidence of opposition to a progressive income tax among lower income earners, many of whom seem to prefer a sales levy. After all, the main objective of those who advocated an income tax was to avoid the shifting of a disproportionate share of the burden to those who are less affluent. But apparently the fact of the matter is that most people would rather pay their taxes in small unnoticeable amounts – that the annual extraction of a relatively large lump sum or the direct reduction in the size of the paycheck through a withholding arrangement is far more painful. This is just another manifestation of the early prepayment syndrome which characterizes even some economists. Are we, then, really entitled to say that advocacy of a sales tax by a fairly impecunious individual can only stem from ignorance or irrationality? True, he may not know just how large a price he is paying for the privilege of making piecemeal payments but it is possible that he does not *want* to know, and even this may be a rational attitude if he believes he is happier not knowing.

Despite these observations, it must be admitted that there are those who would rather live with a tax on income rather than a tax on sales. Paradoxically, there is some evidence suggesting that the group which is most opposed to an excise tax is composed of financially more comfortable and better educated individuals whose pangs of conscience, when they do not bear their proportionate share of the tax burden, motivate them to join the good fight for the 'fairer' income tax. They want such a tax program for the lower incomes earners' 'own good' even though it may be anathema to the objects of concern of the advocate of the 'fairer' tax. But which of them is then being irrational?

7. ILLUSION AND SELF-SELECTION IN NON-LINEAR PRICING

Finally, we may draw a more general conclusion from our discussion, the conclusion that wherever it is feasible, non-linear pricing rather than any modification in a fixed pricing arrangement is the ideal way of dealing with the allocative issues raised by illusion.

Once asserted, this point is self-evidently true. The pertinent feature of non-linear pricing is the variety of pricing options it offers the consumer among which he is free to choose. He is given the opportunity to select a prepaid vacation package, or a pay-as-you-go program. She can rent cars at a fixed charge per mile, or a high rental charge with no payment for distance, or something in between.

For us, the main feature of these arrangements, one which has already been emphasized, is their self-selection, each consumer deciding for herself which pricing plan best suits her purposes. Now, this choice can, of course, be based simply on the usual pricing considerations; for example, the car rental customer who has far to drive can be expected to select the rental plan with a high fixed fee and little or no charge based on distance.[5] But the customer subject to price illusion can (and very likely will) also take this source of preference into account. Those who enjoy the illusion of 'free meals' will select the prepaid vacation while those who are disproportionately distressed by a large payment may avoid it assiduously.

The most attractive feature of this way of dealing with the problem is its avoidance of the need for information which may often be all but unattainable, but which may be needed for more direct adaptation to price illusion. The policy-maker need not know whether people prefer prepayment or the avoidance of large outlays or have some other form of preference among pricing patterns. With non-linear pricing consumers are left to determine this for themselves and to act accordingly.

Indeed, if, as seems likely, the form as well as the strength of price illusion varies substantially from person to person there is no way to deal with the issue. Any price arrangements which are well suited to one person's set of pricing preferences will be ill suited to another's. But with the range of choice offered under non-linear pricing there is room for satisfaction of all pricing tastes.

Looked at another way, pricing illusion means that consumers may benefit from differentiation of products in terms of pricing arrangements, as well as in terms of the physical or other attributes of the goods which are usually cited as examples of product differentiation. And, as usual, the market does in fact sometimes provide such choices where they stimulate enough consumer demand to make such options profitable. But as in the standard analysis of monopolistic competition and product differentiation there seems little reason for confidence that the optimal range of options will be offered. The theory of monopolistic competition suggests the possibility that it will extend too far, the range of alternatives offered perhaps exceeding what can be justified by the benefits *vis-à-vis* the administrative and other costs imposed by a great proliferation of options. But the presence of a variety of price illusions surely calls for the provision of a larger range of options than it would otherwise be possible to justify.

Princeton and New York Universities

NOTES

[1] Another example will underscore the point. A friend of mine told me recently that though she recently moved to New York City she keeps her car and pays the enormous insurance and garage costs so that she can go on the weekend trips she enjoys so much. Though she is a very capable economist and understands the price relationships, she does this because of her awareness that she will never be able to bring herself to pay the 'high' price of car rental for weekend jaunts although over the year this would save her a great deal. She knows perfectly well that an illusion is involved, but that does not make it vanish. In light of these facts, surely her decision is entirely rational!

[2] Yet sometimes a view that is apparently illusory turns out, on closer inspection, to have some basis in fact or in logic. Thus, consider the following proposition (Baumol's tautology): A majority of the vehicles travelling a two-lane road will be found in the more crowded lane. Does this not help to explain the widespread impression of drivers that they usually end up in the 'wrong lane'?

[3] Such incentive effects are tricky matters, however. Exactly the same phenomenon can lead someone with a weight problem to avoid prepayment of meals. The same may be true of people with little self-control and weak digestions. At a recent visit to a chuckwagon restaurant with no limit on the number of helpings, I saw a number of customers who looked more than a little uncomfortable as they were leaving. Another example is provided by our earlier illustration of the way in which a face-to-face conference of people from scattered geographic locations is likely to encourage garrulousness, while a conference telephone call is likely to encourage brevity because the visible cost of the one is largely independent of the amount of time consumed while the apparent cost of the other varies directly with the time consumed in the conversation. Consequently, employers who wish to minimize waste of time by their staff may on this account be led to favor conference calls over conferences with their relatively large prepaid costs.

[4] We obtain a result that is more striking if we attempt to minimize not $\Sigma y_i p_i$, but $\Sigma y_i^* p_i$, where y_i^* is the fixed 'initial' consumption of the resource by group i. Then the Lagrangian becomes $\Sigma y_i^* p_i + \lambda(y_i - k)$ which yields, for $y_i > 0$, and zero cross elasticities, $y_i^* \mathrm{d}p_i/\mathrm{d}y_i + \lambda \equiv p_i(y_i^*/p_i)(\mathrm{d}p_i/\mathrm{d}y_i) + \lambda = 0$, or $p_i/p_j = E_i^*/E_j^*$ where E_i^* is the bastard elasticity $E_i^* = -(p_i/y_i^*)\,\mathrm{d}y_i/\mathrm{d}p_i$. In other words, in this case the optimal prices will vary strictly proportionately with the bastard elasticity of demand.

[5] This example was suggested to me by Gerald Faulhaber.

WILLIAM BREIT

INSTITUTIONAL ECONOMICS AS AN IDEOLOGICAL MOVEMENT

In the 1920s, when it was in its ascendancy as a movement that seriously challenged the hegemony of neo-classical economics, institutional economics could boast of having some of the most distinguished economists in America in its ranks. Thorstein Veblen was the most famous, if not the most respectable, member of the group. His students and disciples held posts at some of the country's most distinguished universities: men like Wesley C. Mitchell, J. M. Clark, Walton Hamilton, and Morris Copeland. All of these were highly acclaimed economists whose publications appeared frequently in the leading professional journals. Indeed, in 1928, when Paul Homan published his well-known book summarizing the main positions of contemporary economists, he included five major figures to represent the leading viewpoints of the time. Two of them (Veblen and Mitchell) were American institutionalists, and one (J. A. Hobson) was an English cousin of institutionalism. Only J. B. Clark and Alfred Marshall could be counted among the representatives of orthodoxy and by 1928 they had already passed from the scene, (Homan, 1928). In Homan's opinion, Wesley Mitchell, Veblen's favorite student, was the man of destiny: "If Veblen is the Messiah, Mitchell is at least the high-priest of what has come to be called 'institutional economics', the devotees of which are coming to include a considerable proportion of the more capable younger economists" (Homan, 1928, p. 414). As yet, Homan declared, there is no systematic statement of their philosophy. But "they entertain a . . . hope that out of their labors may ultimately arise some sort of synthesis that will constitute a generally acceptable body of economic theory" (Homan, 1928, p. 414). And he concluded, "The view of economics of which Mitchell is the outstanding figure represents the most marked trend of the present time in the United States" (Homan, 1928, p. 414).

Notwithstanding Homan's belief, in 1928, that institutionalism "represents the most marked trend of the present time in the United States", within four years Homan could declare at the annual meeting of the American Economic Association, that institutionalism "has had its day, and done its work, and may be consigned to the lower regions without

S. Hook, W.L. O'Neill and R. O'Toole (eds.), Philosophy, History and Social Action, 119–136.

sorrow" (Homan, 1932). Although Homan was somewhat premature in pronouncing the last rites over the movement, the school of economic thought that had so recently dominated American economics clearly was moving into disrepute, and by the end of the Second World War lingered on only in splinter groups, having adherents mainly at the University of Texas and at the University of Wisconsin.

Why did institutionalism not succeed in maintaining its status and achieving its ambitious goal of replacing the neo-classical school by becoming the mainstream paradigm of American economics? A rehearsal of the treatment of institutionalism in the interpretive literature of economics will clarify the career of the institutionalist movement as an ideology and help provide an answer to this question.

The history of the institutionalist movement reveals remarkable similarities with ideological movements chronicled by Lewis Feuer (see Feuer, 1975). As in the case of other ideologies, in its beginnings institutionalism attracted young intellectuals in generational revolt against the dominant view of the economics profession – that of neo-classical economics which reached its apogee in the writings of Alfred Marshall at Cambridge. Indeed, in the very life of the founding figure of the movement, Thorstein Veblen, the younger generation of economists saw the working out of what Professor Feuer calls the Mosaic myth (Feuer, 1975, pp. 1–16). Veblen was the counterpart of the young revolutionary intellectual depicted in Lewis Feuer's drama. He took up the cause of those exploited by the adherents of neo-classical economics.

To Veblen, neo-classical price theory rationalized the *status quo*. Its models extolled the businessman as carrying the forces of progress in his hands and, under the guise of marginal productivity doctrine as enunciated by John Bates Clark, rationalized the unequal distribution of income as being 'just'. Since everyone in a free market system would be paid in accordance with his 'marginal productivity' – that is, in accordance with what each person contributed to the value of society's output – who could complain? In taking up the cause of the exploited 'common man' Veblen suffered exile – shunted from university to university – never rising above the rank of associate professor. Hounded out of Chicago, Standford, and the University of Missouri, he joined other refugee intellectuals on the first faculty of the New School for Social Research. But in developing an alternative economic theory to that of neo-classicism he led his followers to their historic victory, the rise of institutional economics which, as Paul Homan noted in 1928, had, at least for one brief moment, replaced neo-classical orthodoxy as the dominant approach to economic analysis in America. Veblen lived to see the rise of some of his best students to positions of power and prominence within the intellectual community. At the end of his life he was within sight of the promised land. Virtually on the eve of the Great Crash and the onset of the Great Depression (which was supposed to relegate orthodox economics to the trash bin of history once and for all) he died. But he was vouchsafed a distant view of the new society and lived on in the memory of his followers.

Thorstein Veblen's life precisely follows Feuer's description of the Mosaic myth. The career of institutionalism reveals other close parallels with ideological systems analyzed by Feuer, including its issuance of a manifesto. But it was not institutionalism's father-figure who was to provide the charter for the movement. As Feuer has shown, young intellectuals, in their yearning for ideology, require a charter by which to define themselves and separate their ideas from those of the older generation. The call for the dethronement of the elders was provided by Veblen's student, Walton Hamilton. Walton Hamilton's charter not only defined for the younger generation of Veblenians their historic mission, but provided them with a name for their movement. It was Hamilton who in 1918 coined the term 'institutionalist' to be used as the rubric for the generation of young economists in revolt against orthodoxy. He did so in his article called, 'The Institutional Approach to Economic Theory' (Hamilton, 1919).

In this charter, Walton Hamilton wanted to stake a claim to the whole of economic theory, to deny the use of the term 'theory' to others. He did not attack the importance of the work of other approaches in economic thought, only their claim to being 'economic theory'. The hallmark of the 'institutional approach' was to be its theoretical character.

Economics, he said, had developed two approaches to economic theory. The first is represented by the older generation's treatises concerned with relative prices – value theory. Boehm-Bawerk's *Positive Theory of Capital* and J. B. Clark's marginal productivity theory were two of the chief exemplars of this approach. The other side was represented by Edwin Cannan's account of the function of property in economic organization, and Thorstein Veblen's discussion of "the cultural incidence of the machine process". So there are two kinds of economics: the older generation's value economics, concerned with relative price determination; and the younger generation's institutional economics, concerned with the complex relations between human welfare and industry. Unlike value economics, institutional economics is a generalized description of the economic order.

According to Hamilton, both approaches try to explain the extent of order that exists among economic phenomena. But order cannot be explained through a theory of price, "Its quest must go beyond sale and purchase to the peculiarities of the economic system which allow these things to take place upon particular terms and not upon others." Institutional economics must be a study of the conventions, customs, habits of thinking (what Veblen called 'institutions') and modes of doing which make up the arrangements we call the 'economic order'.

And so Hamilton sounded the clarion call for the younger generation. To be 'economic theory' the new approach must meet five crucial tests:

(1) *Economic theory must unify economic science*. To Hamilton, value economics was divisive rather than unifying. The various fields within its corpus ('money', 'taxation', 'corporate finance') each had its own method and point of attack. Only institutional economics could unify. How? By showing the parts each institution plays in the larger whole of society.

(2) *Economic theory should be relevant to the modern problem of control. Laissez-faire* has left a bad taste with most scholars and institutional economics is relevant to problems of control because it deals with and understands that institutional arrangements are social arrangements subject to change.

(3) *The proper subject-matter of economic theory is institutions*. By this he meant a knowledge of the price system, the level of prices, the place of credit and the relationship between prices and the maintenance of the economic order. To understand any aspect of economic life requires an understanding of the institutions of that economy. The development of industrial society needs control, and control requires knowledge of that which is controlled.

(4) *Economic theory is concerned with matters of process*. Value theory deals with its phenomena as if they were unchangeable substances, complete and independent. To Hamilton institutions are always in a process of change. A law, a court decision, a change in tastes: all affect property rights. Economics should deal with the unified whole in a process of development. This, he takes care to emphasize, is not a call for historical analysis or a search for origins. But it is a recognition of the fact that the modern

economy is a complicated affair that has emerged and is still emerging and changing. It should be treated as a process.

(5) *Economic theory must be based upon an acceptable theory of human behavior.* Hamilton attacked the extreme rationality assumptions and utilitarianism of neo-classical economics. To Hamilton a theory of human motivation must be developed in harmony with social psychology. Instinct, impulse and institutions determine behavior. Hamilton suggested that the sources for building a true theory of economic order would come from left-wing intellectuals: Marx, LaSalle, the Webbs, Tawney, Hobson, Thorstein Veblen and Graham Wallas. It would take a decade at least to work through a process of trial and error that would produce the consistent body of thought, not as clear-cut, definite and articulate as neo-classical analysis, but superior to it in other more important ways: It would never be a formal system of laws and principles, by its very nature, because it is concerned with a changing economic life. Relevancy and truth would replace formal precision in statement.

Hamilton's article is a remarkable document. Its few pages express almost every principle that has guided left-wing institutionalists from that day to this: the rejection of price theory as *the* theory of the economy; the stress on 'realism' versus 'abstraction'; the strong emphasis on understanding institutions in a descriptive way; the attack on *laissez-faire* in favor of 'control'; the emphasis on the importance of technology in creating wealth and giving welfare; the notion of process versus statics; the emphasis on the evolutionary nature of the economic system; the belief that assumptions of human nature must be 'realistic' and based on modern social psychology. Little is left out that we can identify with later institutionalists. Although the stress may be different in different interpreters, Hamilton set forth the charter which would in a decade or more, he hoped, bring about a full-fledged unified field theory of economics, integrating all of these elements into a theory of economic order and process. Many hoped Hamilton himself would bring it off. But, unluckily for the institutionalists, he joined the faculty of Yale Law School, entered the New Deal, and ended up a member of the law firm of Arnold and Porter. To others fell the task of developing the definitive statement of Walton Hamilton's institutionalism.

The first major attempt to clarify the role of institutionalism came at the American Economic Association's annual meetings in 1930. The topic of a panel discussion was 'Institutionalism: What it is and What is Hopes to Become'. Two papers were presented, one by Morris Copeland on 'Economic Theory and the Natural Science Point of View' (Copeland, 1931); the other by Eveline Burns on 'Does Institutionalism Complement or Compete with Orthodox Economics?' (Burns, 1931). It was Copeland's position that neo-classical economics does not put its propositions into testable form and does not engage in enough quantification and statistical work. He attacked marginal utility theory as tautological and metaphysical and urged that it be replaced by socio-historical theories. What is more, the perfectly competitive and perfectly monopolistic market theory should be replaced by an analysis of 'competing monopolies'. Competing monopolies are more and more the general rather than the exceptional case. The paper is typical of much institutionalist literature in attacking marginal productivity theory as being unrealistic and confused. In addition, Copeland presented an attack on the quantity theory of money and a defense of some aspects of mercantilism.

Eveline M. Burns' paper is evidence that institutionalism as a movement had begun to lose its self-confidence as early as 1930. She attempted to analyze the reasons for institutionalism's lack of success in replacing orthodox approaches. Her argument was that the notion of institutionalism is vague. It is identified with quantitative economics, with welfare economics, and with a social psychology which undermines hedonism as the fundamental economic assumption. But, Burns said, these are erroneous identifications. Institutionalism, "is an approach which stresses the interactions between social institutions and economic relationships and aspects of behavior, aims to present an orderly arrangement of phenomena, in which institutions are elevated from the status of the exception and the footnote, and integrated with the main body of economics." She explicitly identified Veblen, J. M. Clark, Commons, Berle, Means, and Walton Hamilton as being in the movement.

In his discussion of these papers J. J. Spengler made the point that institutionalism is concerned primarily with the influence of institutions upon economic behavior (Spengler, 1931). But he did not believe that this approach needed to wait upon the development of a new theoretical framework. Any theory must be unrealistic and institutional theory would be no different in that regard from value theory. Since all economists of the future would be concerned increasingly with social engineering and control, all economists will have to consider the factors emphasized by the institutionalists.

Another panelist, O. H. Taylor, argued that the older political economy was institutionalist in the sense that it was intended to be a science of what society's legal and public policies must be if the economic system is to function effectively in achieving larger output and greater wealth (Taylor, 1931). Individuals can remold their institutions. Taylor interprets the newer institutionalists as believing that institutions must be studied from the point of view of their evolution with a strong emphasis on historical determinism. Institutions make the men. To Taylor both views are healthy correctives to each other.

William Jaffé's contribution to the panel was to define an institutionalist as an author who uses institutions not as background but as *dramatis personae* in the economic play of forces, and only when he uses these *dramatis personae* as capable of profoundly affecting economic relationships (Jaffé, 1931). The institutionalist approach is irritating and even infuriating to orthodox price theorists, but the two modes of analysis can exist side by side.

The next year the American Economic Association tried again to differentiate the institutionalist product, this time with a major paper by Paul T. Homan, who had become one of the severest critics of the movement (Homan, 1928). It was Homan's belief that institutionalism must be defined in terms of an orientation taken from Veblen, namely, as denying the possibility of describing the economy in quasi-mechanistic terms imbedded in 'systematic economic theory', in particular eschewing the notion of equilibrium. Beyond this, there is no common point in their works. The distinguishing mark of people who call themselves institutionalists is that they emphasize change, reject orthodox price theory as being irrelevant to the problems of control, advocate descriptive work, extol the virtues of modern psychology as offering the best model of human nature, and hope that all of these elements will be brought together in an integrated restatement of economic theory. Had Veblen not lived, the language of the institutionalists would have been different. Institutionalists are dependent upon the existence of an evolutionary orientation. But this is not unique to institutionalism. William Stanley Jevons' study of the coal industry and Alfred Marshall's studies of industrial combination are examples. Do they then become institutional economists? "If so, then years of theoretical controversy have been pointless" (Homan, 1928, p. 15). Homan then threw out a challenge that was to become familiar to institutionalists from its repetition over the years: " . . . when one wishes to analyze the probable effects of changing the tariff, levying a tax, raising the discount rate, subsidizing the farmer, repealing the Sherman anti-trust act, raising railway rates, or inflating the currency, does one bother about evolutionary hypotheses?" The answer is that mechanistic orthodoxy provides the clearest way to think about such matters. Institutional economics is an intellectual fiction, substantially devoid of content. "But it has had its impact. The structure of economic theory has been subject to searching scrutiny and has been reduced in importance to its minimum terms of intellectual and analytical usefulness. One would be ignorant indeed to deny the importance of the work of men called institutionalists. He would be misinformed if he supposed that their faith was not in some measure responsible for directing them toward the tasks they have taken up." But how can you distinguish J. M. Clark's studies in costs, Wesley Mitchell's studies of the business cycle, Walton Hamilton's studies of the coal industry, John R. Commons' study of legal institutions (all called institutionalist) from Jacob Viner's study of dumping, W. Z. Ripley's of railroads, Frank Taussig's of tariffs (all non-institutionalist)? Institutionalism has had its day. "In the meantime it has inspired some good work, muddled the minds of graduate students and crippled their training."

The attempt to find a coherent statement of institutionalism was taken up in the professional literature. Two are worthy of note. A. L. Harris in 'Types of Institutionalism' distinguished three types of institutional economics: the quantitative statistical represented by Mitchell; the critical-genetic represented by Veblen; and the class-struggle represented by Marx. What ties them together is a rejection of the value and distribution theory of neo-classical economics and a departure from the concept of equilibrium (Harris, 1932).

One year later Karl Anderson (1933) attempted to delineate institutionalism by contrasting Veblen's approach with that of Alfred Marshall. What Anderson called the 'unity' of Veblen's theoretical system was its emphasis on cause and effect; Veblen tried to apply an evolutionary approach to the study of economic phenomena. The controversy between institutionalism and orthodoxy comes down not to a difference over the usefulness of statistics, nor with the adequacy of equilibrium, nor with merits of descriptive studies. The question is whether economics is to be enlarged into a theory of social evolution, a theory of the evolution of those institutions which affect economic phenomena, like distribution and production.

Although periodic attempts were made following these papers to defend and attack institutionalism, there were few discussions that added much to what had already been stated. Meanwhile the institutionalists themselves were at work. Joseph Dorfman's biography of Veblen (Dorfman, 1934) stimulated interest in the topic of institutional economics, and in Great Britain, John Hobson (1936) introduced British readers to the father of institutionalism in his book *Veblen*. It may be said that an entire Veblen industry grew up, having only the most tenuous connection with institutionalism *per se*. Veblen was interesting in and of himself[1] but work outside the Veblenian Canon went on during the thirties. In 1934, John R. Commons' book *Institutional Economics: Its Place in Political Economy* developed a rather idiosyncratic approach to economics in which the notion of the 'transaction' is scrutinized as the basic building block of the economic order. Also of interest is the Clarence Ayres–Frank Knight discussion in the *International Journal of Ethics*[2] in which Clarence Ayres admitted that institutional economics lacked a theoretical formulation and conceded the point that it had not lived up to its potential. The differences between the two approaches were sharpened further by this debate.[3] Ayres continued to attempt to restate the economic problem from an institutionalist point of view in articles and in his book *The Problem of Economic Order* (Ayres, 1938), but not until *The Theory of Economic Progress* (Ayres, 1944) does institutionalism receive a theoretical formulation that incorporates the Veblenian evolutionary approach into Walton Hamilton's original charter. This work contains the first complete statement of institutional economic theory in a single volume.

The volume said much that had not been stated in the debates up to that time. What Ayres did was to provide a system of economic order which clarified the institutional approach, separating it out from all others. To Ayres the characteristic tool of analysis is the dichotomy between 'technological' and 'ceremonial' behavior. Indeed, Ayres makes this instrument the linchpin of institutionalism, making it possible for generations of students to make sense of Veblen, seeing him as much more than a brilliant satirist. This insight was important. Indeed, so crucial is the distinction between technological and ceremonial activities in Ayresian analysis that the dichotomy might with more justice be called the 'Ayresian dichotomy'. No one before Ayres had singled it out as the distinguishing trademark of the movement. As Warren Samuels has shown,[4] the dichotomy is a major technique employed by the contributors to the *Journal of Economic Issues*, the main outlet for contemporary institutionalist research. Modern-day institutionalists are probably more Ayresian than Veblenian in their orientation. Incorporated into *The Theory of Economic Progress*, and later made accessible to undergraduates in the textbook *The Industrial Economy: Its Technological Basis and Institutional Destiny*, is a system of analysis that shows how the dichotomy can be used to understand the industrial revolution and its occurrence in the Western world, the problem of unemployment, and economic development in general (Ayres, 1952). Combining this dichotomy with John Dewey's theory of valuation, Ayres set forth the clearest and most impressive statement of institutionalist principles that had yet been made. None of this had been clear in the writings of Walton Hamilton, J. M. Clark, Morris Copeland, or any other of the institutionalists before Ayres.

Clarence E. Ayres had been trained as a philosopher under pragmatists at the University of Chicago. He was an ardent follower of John Dewey. His other great hero was Thomas H. Huxley, whose biography he had written. He was able to provide for the institutionalists something that was lacking in Walton Hamilton's manifesto: a new vision written in the language of philosophy and science.

Close on the heels of Ayres' reconstruction of institutional economic theory came two works of analysis attempting to codify the institutionalist canon. The first, by John S. Gambs (1946), was a small volume called *Beyond Supply and Demand: A Reappraisal of Institutional Economics*. The second, in 1947, was a husky tome by Allan G. Gruchy called *Modern Economic Thought: The American Contribution*.

Gambs' work was an attempt to examine the usefulness of institutionalism against the background of then current problems; the problems being, as Gambs put it, "a world of atomic bombings, uneasy peacemaking, threatened revolutions", and (oddly enough for an institutionalist) "dangerously advanced technological developments". The book was inspired by John Maynard Keynes' *General Theory* and its aftermath. The younger generation of economists had by now turned from Veblen and institutionalism to Keynes. As Gambs put it: "[Keynes] promises something that cannot be resisted: full employment and high levels of consumption without serious dislocations of our institutions.

Veblen promises the possibility of an even brighter future: a utopian industrial republic – but only alternatively and possibly at great cost" (Gambs, 1946, p. 3). It was for the future to decide whether the plans of Keynes would win out over the prophecies of Veblen. Gambs makes the point that the term 'institutional' would not have been chosen by Veblen to describe his way of thinking. He would have preferred the term 'evolutionary' economics. Gambs then goes on to describe the characteristics of institutionalism.

What holds all institutionalists together, says Gambs, is an "unrecognized premise". That premise, the identifying badge of institutionalism, is "coercion in human affairs" (Gambs, 1946, p. 11). Coercion is as pervasive as the air we breathe. Coercion and aggression, not competition, are the dominant themes of institutional economics.

The second fundamental principle is the "the doctrine or organic unity". Here Gambs drew on Gestalt psychology. In doing so, he used Veblen's method of going below the surface to look at preconceptions, only this time the preconceptions of institutionalism itself were examined. Gambs concluded by noting that institutionalism had disappointed hopes held out for it and "is today in a rather seedy state" (Gambs, 1946, p. 86). He attacked Veblen for his social satire and difficult style. Veblen attracted non-conformists, reformers, dissenters, and inconsequential rebels, but not many men of genuine stature. Keynes deflected to himself the forces of attraction that might have been directed toward Veblen. To Gambs the accidents of personality and events explain why "institutional economics is run down at the heels" (Gambs, 1946, p. 86). He did not hold out much hope for its future.

The larger work by Allan Gruchy which followed Gambs by one year claimed victory almost at the same moment that Gambs announced defeat. Gruchy's work is optimistic in tone, claiming that institutionalism has made and will continue to make great contributions to our understanding. The book is an attempt at interpretation and is a notable achievement of summarization and synthesis. According to Gruchy, institutionalism, or what he preferred to call 'holistic' economics, is a distinct new school of economic thought. It has brought about "a genuine reconstruction which will turn out to be of lasting significance". Gruchy attempted to prove this assertion by giving detailed surveys of the ideas of Thorstein Veblen, John R. Commons, Wesley C. Mitchell, John Maurice Clark, Rexford Guy Tugwell and Gardiner C. Means. These writers represent a coherent movement because all of them see the economic system as an "evolving, unified whole . . . in the light of which the system's parts take on their full meaning" (Gruchy, 1947, p. viii). In this volume Gruchy provided the most useful survey of the ideas of the chief institutionalists that had yet appeared.

Another detailed treatment of institutionalist economists was to be made in the post-war period by Joseph Dorfman (1959) in the concluding volumes of *The Economic Mind in American Civilization*. Volumes IV and V covered the period 1918–1933. The major institutionalist writers, among other American economists, are summarized in these books. On the other hand, the standard histories of economic thought published in this period treated institutionalism, when they treated it at all, in chapters devoted to Thorstein Veblen. On occasion Wesley Mitchell received chapter-length treatment, but, on the whole, textbooks relegated the topic to footnotes.[5]

At the annual meeting of the American Economic Association in 1950 the subject of institutionalism was again on the agenda after a hiatus of almost two decades. On the program was a paper by Clarence Ayres. It was called 'The Co-ordinates of Institutionalism'.[6] In it, Ayres tried o show that institutionalism had three distinguishing features which gave it a single intellectual massif:

(1) A rejection of price theory as being based on defective psychology.
(2) A replacement of the orthodox economist's conception of man with the anthropologist's concept of the culturally determined being whose likes and dislikes are not antecedent to the economic system. To Ayres, human beings are different from all other animals because they are social animals with a continuous culture.
(3) An acceptance of John Dewey's instrumentalism as one of the foundation principles of institutionalism. As Dewey had shown, technological progress provides a valid criterion for human welfare and therefore contains within it the locus of value.

The paper was a formalization of ideas Ayres had expounded in his earlier books, but they are stated in a more succinct fashion than ever before.

Six years later the Association invited one of the leading mainstream economists, Kenneth Boulding, to provide 'A New Look at Institutionalism' (see Boulding, 1957). Clarence Ayres, who chaired the session, took advantage of his position to make some closing remarks.

It is some measure of the continuing importance of institutionalism that Boulding considered his remarks to be conciliatory, having been a critic in the past. Boulding called institutionalism an historical interlude, but an interlude that made real contributions to the mainstream of economic thought. But for all practical purposes the movement had ended in the thirties. However, Boulding admitted that this judgment is only true in a narrow sense. In the broader sense institutionalism is part of a much larger movement of dissent from orthodox economics. Also it is a dissent from ethically undesirable institutions. He saw the institutionalist contribution as an attempt to make economics dynamic, integrated with other social sciences, and with empirical feedback. Wih this goal, Boulding was in complete sympathy.

Needless to say, the institutionalists present were not pleased. Gruchy pointed out that, far from being an historical interlude, institutionalism is the ideology of the progressive wing of the Democratic Party. In what became a widely quoted phrase, Gruchy said: "While institutionalism is not at present prominently situated in academic halls, it does appear to have permanent riparian rights on the banks of the Potomac River."[7] Ayres took the liberty as chairman to enter a few remarks into the record. Two points were made: First, the market is not the allocator of resources; the institutions allocate. The market only gives effect to those institutions. Second, the state of the industrial arts determines the relative scarcity or plentifulness of resources – a theme he had developed in his books and articles for two decades.

Another major work of reinterpretation of institutionalism occurred in 1963 with the publication of *Institutional Economics: Veblen, Commons, and Mitchell Reconsidered* (see Dorfman *et al.*, 1963). In this book, Joseph Dorfman sketched the background against which institutional economics developed; and Robert Aaron Gordon, in tracing institutional elements in contemporary economic, tried to show that Joseph Schumpeter was an institutionalist by most of the criteria which one would use to brand someone with that label. Simon Kuznets wrote a paean of praise for Wesley Mitchell, arguing that institutionalism's attack on accepted economic theory lessened the dogmatism with which it was preached. Neil Chamberlin argued that Commons was an abstract economist who tried to build generalizations with predictive value. He argued that Commons provided two 'naked insights': (1) the insight that economics can never be an exact science since there are always elements of disorder and discontinuity which must be resolved by human wills; (2) the insight that collective activity is pervasive in the economy which implies that automatic individual adjustment is unlikely. An integral part of all of economics must be a study of the process of bargaining and agreement; that is, conflict resolution. C. E. Ayres reappraised Veblen's 'legacy'. To Ayres this legacy was not simply a different economics, but a different conception of the economy itself in which industrial technology is the real substance. It was Veblen's recognition of the two aspects of the economy – the Veblenian dichotomy – which is most significant. Veblen's conception of the economic life process has been validated by such developments as the high importance given to empirical studies (Ayres, 1963).

The past two decades have seen fewer round-table discussions of institutionalism at the annual meetings of the American Economic Association, probably because during this period the institutionalists formed their own society and held annual meetings at which such topics are commonplace. *The Journal of Economic Issues*, the organ of the Association for Evolutionary Economics, has been published since 1967 and is the chief outlet for institutionalist writers. Moreover, the 1970s saw the publication of more books devoted to institutionalism, the most ambitious of which was Allan Gruchy's long-awaited sequel to his *Modern Economic Thought*. The book, *Contemporary Economic Thought: The Contribution of Neo-Institutional Economics* appeared in 1972. In separate chapters Gruchy treated the ideas of Clarence E. Ayres, John Kenneth Galbraith, Gunnar Myrdal, and Gerhard Colm as being neo-institutionalists. However, the book was not as impressive as Gruchy's earlier volume since he did not succeed in his efforts to present these four writers as being part of the same coherent movement.

In looking back over the sweep of institutionalist literature and its interpretation by those inside and outside the movement, a sense of paradox emerges. Institutionalism started as a left-wing movement of dissent from orthodoxy, gained ascendancy in the 1920s, was seen as 'the trend of economics' by friends and critics alike, and then seemed to lose its self-confidence and its future in the 1930s. From that time onward institutionalism was subjected to some scathing criticisms. Since it was a rather inchoate and amorphous movement, the attack of people like Paul Homan, O. H. Taylor, J. J. Spengler, among others, dealt it blows from which, in a sense, institutionalism never recovered. After 1930 it did not gather enough strength to launch a counter-attack. Ayres' restatement in 1944 came too late.

A number of reasons have been given for this inability to regain the initiative. Gambs gave the credit to Keynes who attracted away some of the best younger economists who might have otherwise opted for Veblen, Commons and Mitchell-type economics. But that cannot be the whole story.

Orthodox price theory also went through some important restatements in the 1930s. The outcome was to make it more realistic and appealing. The Chamberlin–Robinson 'revolution' must be given much of the credit for rehabilitating price theory. By abandoning the perfectly competitive model as the norm, it provided more realistic assumptions and contained implications that seemed to square with observations made by the man on the street. Product differentiation, excess capacity, advertising: these manifestations of Veblenian 'sabotage' were incorporated into the framework of static price theory. Moreover, the developments of welfare economics at the hands of A. C. Pigou, A. P. Lerner, Abram Bergson, and others indicated that price theory techniques could come out with results that did not involve *laissez-faire* policy implications. The tools of price theory increasingly lost their identification with a conservative ideology. But many institutionalists (the great exception is J. M. Clark)[8] did not recognize these changes as significant.

In the 1960s orthodox economics became increasingly imperialistic, moving into areas of research outside the traditional bounds of mainstream economics. This, too, obviated the need for a broad-gauged economics of the kind offered by institutionalists for those disenchanted with the narrow concerns of neo-classical economics. The public choice and property rights theorists began doing work in areas that institutionalists once thought to be their domain. But these approaches were totally unacceptable to institutionalists of the Veblen–Hamilton–Ayres stripe. These interlopers found the tools of orthodox price theory indispensable for an understanding of institutions and, to make matters worse, they espoused a conservative *laissez-faire* ideology. Thus Harold Demsetz used the 'Chicago school' right-wing theory of property rights to explain how private land ownership emerged among the Indians of the Labrador Peninsula but not among the Indians of the Southwestern plains (Demsetz, 1967). James M. Buchanan, another Chicago-trained economist of the right, used public choice theory to cast light on the emergence of legal systems, contracts, and even constitutions (Buchanan, 1975). Still another Chicago product, Roland N. McKean, attempted to examine the origins and functions of Veblenesque rules of etiquette utilizing the rationality postulates of neo-classicism (McKean, 1975). To the chagrin of the older generation institutionalists, James Buchanan has been dubbed a "leader of the 'new institutional economists'" and Ronald Coase the "founder of the 'new institutional economics'" by Edwin G. Dolan in his book *Basic Economics*.[9] Dolan's designation was inspired by the fact that conservative economists like Buchanan, Coase, and others who work in the public choice, property rights and law/economics interfaces take the emergence of property, customs and laws as objects of analysis and not as 'givens'. This is one of the distinguishing marks of institutional economics, as stated by Walton Hamilton in the charter of 1918.

Institutional economics has made the passage in ideological development from left to right. In that sense, as in others, its career closely parallels that of the ideologies that have been studied by Lewis Feuer. In almost all respects it has followed Feuer's theory of the life-period of typical ideological movements: its origin in the generational revolt against the established orthodoxy of economics; its repetition of the Mosaic myth (in the life of its father-figure and founder, Thorstein Veblen); its charter or manifesto that defined itself in its separation from the older generation (issued by the first inheritor of Veblen's mantle, Walton Hamilton); its attachment to basic philosophical and scientific premises (of John Dewey through the influence of Clarence Ayres); and finally, its traversing of the ideological spectrum from left-wing to right-wing over the course of its history. In those few academic enclaves where it still claims left-wing adherents, it has ceased to be a radical doctrine and has become dogma. But its force as an ideological movement has been spent.

Trinity University

NOTES

[1] See the summary of work on Veblen that was done after Dorfman completed his magisterial biography. Joseph Dorfman, 'New Light on Veblen' in the introduction to Thorstein Veblen, *Essays, Reviews and Reports* (Clifton, New Jersey: Augustus Kelley, 1973).

[2] See C. E. Ayres, 'Moral Confusion in Economics', *International Journal of Ethics* **45** (1934–1935), pp. 170–199; Frank H. Knight, 'Intellectual Confusion in Morals and Economics', *International Journal of Ethics* (Jan. 1935), pp. 200–220; C. E. Ayres,

'Confusion Thrice Confounded', *International Journal of Ethics* **45** (1934–1935), 356–358.
[3] For a recent appraisal, see James M. Buchanan, 'Methods and Morals in Economics: The Ayres-Knight Discussion', *Science and Ceremony: The Institutional Economics of C. E. Ayres*, edited by William Breit and William Patton Culbertson, Jr (Austin: The University of Texas Press, 1976), pp. 163–174. See also the exchange of letters between Ayres and Knight in Warren J. Samuels, 'The Knight-Ayres Correspondence: The Grounds of Knowledge and Social Action', *Journal of Economic Issues* **XI** (Sept. 1977), 485–525.
[4] Warren J. Samuels, "Technology Vis-à-Vis Institutions in the *Journal of Economic Issues*: A Suggested Interpretation" Journal of Economic Issues **11** (December 1977) pp. 871–895.
[5] An exception must be made in favor of Ben B. Seligman, *Main Currents in Modern Economics: Economic Thought Since 1870*. This work is partial to institutionalism contrasting it favorably with almost every other school of thought or approach. A lengthy chapter called 'Institutionalism and the Dissenting Spirit' treats Veblen, Commons, Mitchell, Clark, Hobson, Ayres, Hoxie, Perlman, Means and Galbraith as institutionalists. His conclusion is that contemporary economics outside of the institutionalists deals with cold analytical models and has a concentration on formalistic technique without serving as guidelines to policy. Seligman's extensive, favorable treatment of institutionalism is unique in the post-war period (Glencoe: The Free Press, 1962).
[6] C. E. Ayres, 'The Co-ordinates of Institutionalism', *American Economic Review* **XLI** (May 1951), 47–55. See also the papers by M. A. Copeland and by Abram L. Harris and the discussion by Joseph Dorfman and Kenneth H. Parsons in the same issue.
[7] See the remarks by Gruchy in *American Economic Review* **47** (May), pp. 13–15. See also the comments by Forest G. Hill, Frank H. Knight and Kenneth H. Parsons in the same issue. Also the closing remarks of Ayres, pp. 26–27.
[8] See, for instance, J. M. Clark, 'Soundings in Non-Euclidean Economics', *American Economic Review* **11** (March 1921).
[9] Edwin G. Dolan, *Basic Economics* (Hindsdale, Illinois: Dryden Press, 1977). For a recent attempt to reconstruct left-wing institutionalism, relying heavily on the work of Clarence Ayres, see Marc R. Tool, *The Discretionary Economy: A Normative Theory of Political Economy* (Santa Monica: Goodyear, 1979).

REFERENCES

Anderson, Karl. 1933. 'The Unity of Veblen's Theoretical System', *Quarterly Journal of Economics* **47** (August), 598–625.

Ayres, C. E. 1938. *The Problem of Economic Order*. New York: Farrar and Rinehart.

Ayres, C. E. 1944. *The Theory of Economic Progress*. Chapel Hill: University of North Carolina Press.

Ayres, C. E. 1952. *The Industrial Economy: Its Technological Basis and Institutional Destiny*. New York: Houghton-Mifflin.

Ayres, C. E. 1963. *'The Legacy of Thorstein Veblen,* in Joseph Dorfman, *et al. Institutional Economics.* Berkely Los Angeles: University of California Press,pp. 45 – 62.

Boulding, Kenneth E. 1957. 'A New Look at Institutionalism', *American Economic Review* **47** (May), 1–12.

Buchanan, James M. 1975. *The Limits of Liberty*. Chicago: University of Chicago Press.

Burns, Eveline M. 1931. 'Does Institutionalism Complement or Compete with Orthodox Economics?', *American Economic Review* **21** (March), 80–87.

Commons, John R. 1934. *Institutional Economics: Its Place in Political Economy*. New York: Macmillan.

Copeland, Morris A. 1931. 'Economic Theory and the Natural Science Point of View', *American Economic Review* **21** (March), 67–79.

Demsetz, Harold. 1967. 'Toward a Theory of Property Rights', *American Economic Review* (May), p. 347.

Dorfman, Joseph. 1934. *Thorstein Veblen and his America*. New York: The Viking Press.

Dorfman, Joseph. 1959. *The Economic Mind in American Civilization, 1918–1933*, Vols IV and V. New York: The Viking Press.

Dorfman, Joseph. *et al.* 1963. *Institutional Economics: Veblen, Commons, and Mitchell Reconsidered*. Berkeley and Los Angeles: University of California Press.

Feuer, Lewis S. 1975. *Ideology and the Ideologists*. New York: Harper Torch Books.

Gambs, John S. 1946. *Beyond Supply and Demand: A Reappraisal of Institutional Economics*. New York: Columbia University Press.

Gruchy, Allan G. 1947. *Modern Economic Thought: The American Contribution*. New York: Prentice Hall.

Gruchy Allan G. 1972. *Contemporary Economic Thought: The Contribution of Neo-Institutional Economics*. Clifton, New Jersey: Augustus Kelley.

Hamilton, Walton H. 1919. 'The Institutional Approach to Economic Theory', *American Economic Review* **9** (March; Supplement), pp. 309–318.

Harris, A. L. 1932. 'Types of Institutionalism', *Journal of Political Economy* **40** (Dec.), 221–249.

Hobson, John A. 1936. *Veblen*. London: Chapman and Hall.

Homan, Paul T. 1928. *Contemporary Economic Thought*. New York: Harper and Brothers.

Homan, Paul T. 1932. 'An Appraisal of Institutional Economics', *American Economic Review* (March; Supplement), pp. 10–17.

Jaffé, William. 1931. *American Economic Review* (March; Supplement), pp. 139–141.

McKean, Roland N. 1975. 'The Economics of Trust, Altruism, and Corporate Responsibility', *Altruism, Morality, and Economic Theory*, E. S. Phelps (ed.). New York: Russell Sage Foundation, pp. 29–44.

Spengler, J. J. 1931. *American Economic Review* (March; Supplement), pp. 135–136.

Taylor, O. H. 1931. *American Economic Review* (March; Supplement), pp. 136–137.

WILLIAM H. DRAY

GENERALIZATION, VALUE-JUDGMENT AND CAUSAL EXPLANATION IN HISTORY

I

Since the Second World War, English and American philosophers of history have had a good deal to say about historical explanation. To the extent that such explanation has been seen as causal – and the ultimate viability of any other kind has often been questioned – two assumptions have generally guided the analyses offered. The first is that what is to be explained must be shown to instantiate laws or universal generalizations for which reasonable empirical support can be found, and which show the predictability of what occurred. The second is that, in the determination of what constitutes the cause or causes of a given historical event or state of affairs, the value-judgments of the investigator have no role to play. In the accounts of historical explanation which these two assumptions have commonly generated, historians have often found it difficult to recognize anything much resembling their own modes of reasoning. Many have urged, in consequence, that before philosophers set about prescribing models of explanation for historical inquiry, they pay more attention to the way causal investigation in this field in fact commonly proceeds. Philosophers, of course, can scarcely be expected to renounce prescription altogether. The admonition is nevertheless one to which I think philosophers of history ought to pay some heed. I therefore propose, in this paper, to examine a sample of causal argumentation by an historian which, it seems to me, positively invites philosophical attention. No doubt the discipline of history has many mansions, and it cannot simply be taken for granted that what is found true of a selected causal explanation will prove true of all causal inquiries undertaken by historians in all styles and branches of the subject. The sample case I propose to examine nevertheless seems to me far from untypical. It seems typical, at any rate, of much of what, despite the various methodological "revolutions" that have been announced by *avant garde* members of the historical profession in recent years, still passes for reputable political, diplomatic and military history.

The sample case is an article by the American historian, Norman A. Graebner, entitled 'The Mexican War: A Study in Causation'.[1] This piece of historical writing has the advantage for present purposes not

S. Hook, W.L. O'Neill and R. O'Toole (eds.), Philosophy, History and Social Action, 137–155.

only of dealing with a set of events which are well known to American readers, but also of trying to be more explicit than is usual in traditional historiography about the principles and assumptions upon which its reasoning proceeds. Graebner's approach to his problem is that of a historian seeking to revise a received causal thesis. He does this less by calling upon hitherto unknown or unconsidered evidence than by reinterpreting what he takes virtually all historians of the period and most of their readers know. The received causal thesis – to quote a classic formulation of it by J. H. Smith which Graebner takes as his point of departure[2] – is that the war was "deliberately precipitated by the will and act of Mexico", it thus being imposed upon, and certainly not caused by the United States. In examining Graebner's case for rejecting this judgment, I shall pay particular attention to the way generalizations and value-judgments actually function in his argument, his text being unusually replete with both. The value-judgments, I shall argue, are far from the ornamental, functionless things that philosophers of history have so often implied they must be or ought to be when historians offer causal analyses. As for the generalizations, although many can be seen to serve clear polemical purposes, they seldom play quite the logical role, and indeed are seldom quite the logical entities, that the usual philosophical view of the nature of causal connection in history would lead one to expect.

The first thing to be clear about is precisely what, according to Graebner himself, was the cause of the war between Mexico and the United States which broke out in May 1846. While he never puts it in so many words, it would seem to be certain aspirations, policies and actions of the American president, James Polk. It was the latter's 'persistence' in pressing American demands upon Mexico, Graebner avers, diplomatic pressure and denunciation being reinforced by threatening military movements, that "led him and the country into war". Given Mexico's refusal to honor long-standing American claims for compensation for losses suffered by American citizens doing business in Mexico, or to cede territories now administered only with difficulty from Mexico City and long coveted by American expansionists – indeed, even to allow differences over national boundaries to become matters for discussion – Polk, according to Graebner, had three choices open to him, short of simply invading his southern neighbor. He could soften American demands with a view to eliminating confrontation; he could let things drift in the hope that the United States might achieve its expansionist ends later in more favorable circumstances; or he could escalate the crisis in the expectation that, in consequence, Mexican nerve would falter. In Graebner's view, it was Polk's opting for the third choice and then sticking to it regardless of how the Mexicans reacted, that caused the war. "What perpetuated Polk's escalation to the point of war", he declares, "was his determination to pursue goals to the end whose achievement lay outside the possibilities of successful negotiations." The Mexicans responded to the presence of American troops on the Texan frontier (or, as they saw it themselves, across that frontier, its precise location being in dispute) by firing on them; and the war began.

Graebner supports his own causal diagnosis with both positive and negative arguments: he gives reasons both for regarding Polk's contribution to the sum of what happened as cause of the war rather than anything done or attempted by the Mexicans, and against attributing causal status to certain other things that he thinks some historians, while perhaps loath to accept the traditional view as it stands, may want to point to instead. I shall look first at the way he claims, fairly early in his argumentation, to show the unacceptability of two such candidate causes of the war which he concedes to have a certain plausibility, drawing attention especially to the way generalizations function in the eliminative arguments he offers. I shall then look more closely at the positive argument he elaborates for his own position, which at the same time disposes of the traditional account as such. In this the role played by certain value-judgments will be found to be crucial.

II

If Graebner's negative argument is to apply the notion of causal connection as law-instantiation, it should throw doubt on the existence of any universal connection between either of the two candidate causes he wants to put aside and the outbreak of a war. To a considerable extent, it can be interpreted as endeavoring to do this. The way it goes about it, however, raises problems with regard both to the relevance and to the argumentative efficacy of various generalizations that Graebner himself conjures with as he proceeds; and certain asides and citations of supplementary considerations make one wonder whether he would really allow his case to stand or fall on the stated nomological considerations alone.

The first candidate cause to be eliminated Graebner sums up as "mutual animosities" which had developed over a considerable period of time. Historians, he notes, have often tried to explain the Mexican War by pointing to "the causes of friction in American–Mexican relations", these lying in "the vast discrepancies between the two countries in energy, efficiency, power and national wealth". Graebner doesn't deny the relevance of such considerations for any attempt fully to understand what happened. He nevertheless denies that they can be regarded as the causes of the Mexican War for the following reasons:

> Governments as divided and chaotic as the Mexican regimes of the 1840s usually have difficulty in maintaining positive and profitable relations with their neighbours; their behaviour often produces annoyance but seldom armed conflict. Belligerence toward other countries had flowed through United States history like a torrent without, in itself, setting off a war. Nations do not fight over cultural differences or verbal recriminations; they fight over perceived threats to their interests created by the ambitions or demands of others.

All of the reasons which Graebner here sets out assume the form of generalizations of one kind or another. The first amounts to the claim that the situation was of a kind that often causes something short of war – for example, annoyance – but seldom war itself. As a way of showing that the noted animosities did not cause war in *this* case, however, such a generalization is surely of limited value. The generalization itself implicitly concedes that *sometimes* such conditions cause war; and on the question whether this is one of the cases where they do so, the generalization itself can obviously tell us nothing. To put it another way, if we accept the generalization it follows that, since mutual animosities of the indicated kind seldom cause war, it was improbable that they would do so in this case. But as every historian well knows, the statistically improbable does sometimes happen, this being something one simply learns to live with in working with statistical generalizations. Citing the generalization (a vague statistical) thus leaves it quite open whether the cause of the Mexican War was the mutual animosities. If Graebner wishes to show that it was not, he will need to provide an entirely different kind of argument, one that deals more directly with the particular case.

The second generalization amounts to the claim that belligerence on the part of the United States had seldom caused war in the past. Clearly this is a generality of a quite different logical type from the first, and a much weaker one. It simply tells us what had been characteristic of

American history so far and implicitly invites us to extrapolate this trend into the future. Doubtless what has been characteristic of a certain historical individual up to a certain point in time, whether that individual is a person, an institution, or a nation, justifies a modest presumption about its future behavior. It is surely strange, however, to regard such a presumption as, in itself, effectively ruling out the possibility that American belligerence caused war in this case. In other words, one would expect a historian to move with considerable caution from the assertion of what he takes to be a true but purely *summative* generalization to the assertion of a corresponding generalization of extrapolative force, even when the latter, as in this case, still makes claims only about the way an individual historical entity behaves – claims expressible by what some philosophers of history have called 'singular hypotheticals' or 'limited laws'.[3] The way the purely summative generalization is formulated by Graebner in any case leaves it uncertain whether we are being told that American belligerence had *never* caused war or only that there had been at least some cases where it had not done so. Read in the second way it would be even less useful for his purposes.

The third generalization, unlike the first, is universal, not statistical or probabilistic in form: it purports to hold for *all* cases of cultural differences or verbal recriminations; and unlike the second it makes a claim about *all* nations, not just about a designated one, uniquely located in historical space and time. One may wonder whether the assertion that nations do not fight *over* differences of the sort instantiated in the Mexican–American case is equivalent to saying that war between them is never *caused* by their having such differences. But if this equivalence can be assumed, Graebner's third generalization, if true, effectively disposes of the candidate cause he has in view. An immediate problem, of course, is why anyone should think the generalization is true: it will surely appear dubious to many readers in the absence of very strong evidence in its support. But a further and theoretically more interesting problem arises out of its nature. If a historian were ever tempted to think himself justified in asserting such a generalization, it would presumably be because, in a considerable number of cases where he had discerned animositites of the indicated kind, he had found the war to have been caused, not by them but by something else. But his case for believing that animosities were not the cause in any of these particular cases would necessarily be far stronger than for any generalization he might eventually try to derive from them, his claim to know the generalization presupposing his having conducted successful causal analyses of the particular cases on which it rests. Graebner's appeal to it here is therefore rather puzzling: he seems to employ a mode of argument which presupposes his ability, as a historian, to reach independently the sort of conclusion which that argument is represented as justifying. Why not, then, apply that ability directly to the case under examination?

Some of the same considerations apply in appraising what Graebner has to say in setting aside the second sort of factor he thinks some historians may single out as cause of the war: "specific issues over which the two countries perennially quarrelled", like monetary claims, boundaries and the future of western territories. For the sake of simplicity, I shall look only at what he says about American monetary claims. Again Graebner's argument appeals immediately to generalizations. He writes:

> Nations have made claims a pretext for intervention but never a pretext for war. Every nineteenth-century effort to collect debts through force assumed the absence of effective resistance, for no debt was worth the price of war. To collect its debt from Mexico in 1838, for example, France blockaded Mexico's gulf ports and bombarded Vera Cruz.

The first of these generalizations, if we take it as stated, implies only that, if a war occurred, claims would not be the American pretext for fighting it. This would not rule out the possibility that they were still what caused it – the real reason why war broke out. In any case, although the generalization, if suitably reformulated in terms of causes rather than pretexts, would indeed support conclusively Graebner's negative causal thesis provided it can be taken as true, it is hard once again to see how he could claim to know its truth (a single instance is cited in which claims are said to have led to something other than war). If, in response to this kind of objection, Graebner were to reduce the generalization to a probabilistic or statistical one, the same difficulties that were noted previously about allowing mere probabilities to negate causal connection in a particular case would ensue. The second generalization, like the second one offered in arguing that animosities were not the cause, is a mere summary of a number of supposedly known cases in which claims did not cause war. The reader seems to be invited somewhat questionably to extrapolate to the present case without further ceremony.

Something significantly new by way of generalization, although added almost in parentheses, does nevertheless appear in this second quoted passage: the generalized value-judgment that no collection of debts was "worth the price of war". What is the function of such a generalization in Graebner's overall argument? Some indication of it may be found, I think, in his continuing as follows:

> United States claims against Mexico created special problems which discounted their seriousness as a rationale for war. True, the Mexican government failed to protect the possessions and the safety of Americans in Mexico from robbery, theft and other illegal actions. But Americans were under no obligation to do business in Mexico and should have understood the risk of transporting goods and money in that country. United States citizens suffered similarly at the hands of other people without provoking wars of retribution.

To this Graebner adds the further consideration that going to war with a foreign country simply because it could not pay its debts would show 'bad grace' given the fact that many American states were 'in the same situation', and the United States itself was 'deeply in debt to Britain'. Graebner sums up: "If Polk would take up the challenge and sacrifice American human and material resources in a war against Mexico, he would do so for reasons other than the enforcement of claims." What concerns him at this point, in other words, is finding adequate explanations of the actions of Polk and other actors in terms of the reasons which made their actions seem eligible to them. He sees the desire for the settlement of claims, given the circumstances, as offering an obviously insufficient reason for going to war with Mexico (among other things, Mexico simply couldn't pay, and the Americans well knew it). He may appear to move rather too easily from this to the conclusion that when Polk did go to war, it must have been for some other reason. But, in fact, he adduces independent evidence that Polk had no intention of fighting simply to settle claims, and his positive argument supposedly spells out further the latter's 'other reasons'.

Whether there is a fundamental difference between explanations in terms of laws and explanations in terms of agents' reasons for acting, and whether causal language in history is typically ambiguous in this connection, are controversial questions which cannot be considered in any detail here. It is of some interest to note, however, that, in raising the problem of the causal relevance of unpaid American claims at this point, what Graebner seems to be asking is not whether, given the truth of certain empirical generalizations or laws, it was predictable that the Americans would go to war when offered such a provocation, but whether Mexican non-payment made it sensible from their standpoint to do so. He therefore doesn't ask whether political leaders in Polk's position generally declare war when faced by unsatisfied claims. He does cite a certain type of generalization, but its function appears rather to throw doubt on the rationality of declaring war for such a purpose, and hence on any attempt to explain an actual outbreak by attributing such a reason to relevant agents. Appended to this quasi-normative (and certainly less than conclusive) reflection is the further generalization, merely summative in character, that the United States had in other cases not gone to war to secure payment of claims. One might see this perhaps as a first step toward asserting an extrapolatable generalization, although a singular one, lending some plausibility to a prediction that the United States would not go to war in the present case merely for claims. It seems just as plausible, however, to interpret it as reinforcing the value-judgment already noted by pointing out how widely accepted was the opinion that such a result was not worth fighting for – a reminder of conventional norms.

More will be said below about more direct ways in which value-judgments function in Graebner's overall argument. The consideration which most needs stressing with regard to his negative thesis, however, is that, if its goal is simply to question the adequacy of mutual animosities or unpaid claims as causes of the war by destroying any appearance of strong nomological connection between war and antecedent conditions of these kinds, he had no need to involve himself in dubious assertions of *contrary* nomological connections. In order to undermine a lawlike connection between two variables, all that is needed is to show that there are significant counter-instances. What is questionable about Graebner's negative argument is often traceable to his trying quite gratuitously to go beyond satisfying this requirement – perhaps too much influenced by the common complaint that traditional historiography is nomologically underdeveloped. He thus over-argues his case. Many of the conditions he adduces are nevertheless apposite, given the conception of causation as law-instantiation he appears to be working with. A further problem arises out of the way his negative argument often seems to rest on the false assumption that what historians normally call causes must be conceived as in themselves sufficient conditions of their effects, rather than (at most) sufficient in conjunction with other conditions. For without this assumption – without this expectation that historical causes will be more than especially important *necessary* conditions – merely showing that in some cases alleged causes occurred in the absence of their envisaged effects would not dispose of their claim to be considered operative in a case under examination. It is not without interest that this assumption quickly drops away as Graebner goes on to elaborate his positive thesis.

III

Graebner's argument for the positive thesis, like that for the negative one, makes reference from time to time to generalizations which he appears to regard as supportive. Here, however, the generalizations seem less intrusive; it is the author's value-judgments which catch the attention. The positive thesis, as indicated earlier, is that what caused the war was Polk's utter determination, largely motivated by personal political ambition, to settle outstanding differences with Mexico, and in so doing to realize American goals with regard to disputed boundaries and coveted western territories. He thought this could be done by pressure short of war. But when the Mexicans made it clear that it could not, he persisted in his course, keeping up pressure by sending American troops to the Rio Grande, where fighting soon broke out.

It is clear that, in pointing to Polk's persistence as cause, Graebner doesn't mean to represent it as a complete explanation of the war's outbreak in the sense of constituting, in itself, a sufficient condition of it. He knows perfectly well that it takes two sides to make a war, and that if the Mexicans had not resisted American pressure (they actually fired the first shots) there would have been no war for Polk's persistence to be the cause of. In fact, he stresses the crucial nature of the Mexican contribution to what eventuated and how easily things might have been otherwise. According to him, Polk was unable for political reasons to initiate hostilities himself, and was thus, so far as his own causal efficacy is concerned, quite dependent on the Mexicans obliging. When they obliged, however, it is his actions, not theirs, that he identifies as the distinctively causal factor in the situation. Such emphasis on the indispensability of the Mexican response makes it inappropriate for Graebner to try to ground his own position directly in some generalization like "Whenever politicians placed in certain circumstances (fill in a characterization of Polk's) are determined to pursue goals to the end which they cannot achieve without war (again characterize Polk's) then war follows" – and in fact he doesn't do this. He does, however, appear to seek support in a generalization of a rather different sort – one which, in effect, generalizes part of the particular argument he offers for his positive thesis, one aspect of which I summarized toward the end of Section I. This might be paraphrased as follows: "Whenever a stronger power on the make confronts a weaker one, the stronger gives only two choices to the weaker, capitulation or war, whereas the weaker offers the stronger three, moderation of its demands, drift, or escalation of the crisis." How close does calling upon such a generalization bring Graebner to the nomological theory of causation as generally understood, and how relevant is this generalization really to his own causal diagnosis?

Expressed in such woodenly quantitative terms, at least, such a generalization may well seem a rather odd one for Graebner to cite in support of this thesis. I think that the governing idea behind it can nevertheless be shown to be important for the concept of historical causation he is working with. What he wants to emphasize is that, in its quarrel with Mexico, the United States, being by far the stronger power, had what might naturally be called the greater 'freedom of maneuver'. Since the intense pressure exerted by the American government made it scarcely practicable for the Mexicans simply to stall, and the only other alternative to capitulating was to resist American demands by whatever means were available, it would not be unnatural to say that the Mexicans *had* to fight. Certainly they seem themselves to have seen their situation in these terms, and Graebner appears to agree with them. The Americans, on the other hand, operating from a position of strength, could not with similar plausibility say they *had* to fight. As Graebner rather quaintly puts it, the United States had no 'national obligation' to make war to extend its sovereignty. And since, on his view at least, New Mexico and California could both have been acquired without war (although not in time to serve Polk's personal ambition), the United States was even less obviously under any compulsion to acquire them in the mid 1840s. By contrast, according to Graebner, the Mexicans did have a compelling duty to defend their territory. Since 'national honor' and not just economic interest was at stake, they had this duty even if they were pretty sure – as they were – that they would lose much of it whether they fought or not.

But if, with Graebner, we are prepared to say that the Mexicans *had* to fight, it is important to be clear about the sense of 'had to' that is here being employed. It belongs not to the language of science but to that of morals. The question is not whether one or other party to the dispute was determined by antecedent conditions to act as it did, in a sense that made its so acting predictable in accordance with law. It is whether it was obligated so to act, whether it was coerced into so acting, whether so acting was the only tolerable alternative. Strictly speaking, what Graebner is talking about is thus not the mere *number* of choices available to the two sides, and certainly not their mere number as determined by psychological or physical criteria. What is at issue is their moral nature. For in a morally neutral sense, the Mexicans, too, could have done nothing. In abstraction from what it was reasonable to expect of them, they, too, had the choice of drifting, so that they, too, could be said to have had three choices. In fact, the way Graebner typically talks about the choices involved leaves it very much in doubt whether he thinks the Mexicans had even *two* choices in the relevant sense of the term. Accepting American demands, he argues at some points, was not a real option for them. Paying the American price for peace, given the height to which Polk had raised it, was, in a familiar moralistic sense of the word, no choice at all.

Since the moral nature of the choices available to the two sides is so important for the way Graebner assigns relative causal status to the contributions of the Americans and the Mexicans, it is hardly surprising that quite overt value-judgments enter so frequently into his characterizations of the situation and of what the various agents were saying and doing in it. Thus Polk's diplomacy is represented as 'aggressive', the behavior of the Mexican government not being seen in similar terms despite the stream of vituperation acknowledged to have flowed out of Mexico City and the stationing of Mexican as well as American troops to the frontier. Mexico is said to have been under no 'moral obligation' to alienate territory in order to ease diplomatic and military pressures, the task of easing them belonging to the United States. The two million dollars eventually offered by Polk for the Rio Grande is described as hardly a 'proper payment' for it, although it might well have been considered generous by anyone who thought of Mexican territory as just there for the taking. Again, the 'security interests of the United States' are said not to have been endangered by Mexico, a judgment that surely accepts the Mexican, not the American view of where the territorial destinies of the two countries lay. The same could be said of Graebner's opining that it 'mattered little' whether disputed boundaries were settled in 1846.

I have suggested that the function of such value-judgments is generally to represent the Americans as enjoying freedom of maneuver and the Mexicans as lacking it – to show that, morally speaking at least, the Americans did not have to do what they did whereas the Mexicans did. The actions of those whose decisions are seen as having been most free are then regarded as the more appropriate ones to single out as the cause of what ensued. It is worth noting that this idea that, when the explanatory conditions of an event include human actions, those performed freely are better candidates for causal status than those which are obligated or coerced does not come out of the blue. Proceeding on such a principle is not a Graebnerian idiosyncrasy: it expresses in a particular form a very common notion (it should perhaps be called a metaphysical notion) that a cause should be something 'active' – an 'agency'. Free action is customarily regarded as a paradigm of such agency, while action that is unfree is seen as enmeshed in a process. Of course, Graebner's argument gives the notions of being free and being forced a distinctively moral meaning, which, although common enough in traditional historiography (and in ordinary discourse) is not the only one they can legitimately possess. Graebner himself somewhat inconsistently moves toward others when, in effect, he tries to reinforce his selection of American actions as causes of the war by describing the Mexicans as 'distraught', as experiencing 'inward dread', as 'in despair', or as 'lashing out' – as if incapable of reasoned responses to problems which were too big for them, and to which they were therefore just reacting instinctively or emotionally. Polk and his colleagues, by contrast, are represented as deliberate and calculating (if sometimes also as *mis*calculating). Graebner's more usual way of characterizing Mexican unfreedom, however, applies notions like obligation and coercion, not lack of control.

It should scarcely be necessary to add that, in making moral judgment so relevant to causal judgment, Graebner is not *reducing* historical causation to mere moral appraisal. What is at issue here is which, of two sets of actions and attitudes, both judged to be indispensable to the coming of the war, is to be singled out as what caused it. As has already been pointed out, Graebner makes it clear that he sees the actions of both sides as necessary conditions of what occurred – as close to necessary conditions, at any rate, as one is likely to find in a subject like history. The associated judgments of necessity could appropriately have been supported by corresponding generalizations of the form 'Only if so-and-so does war follow' (by contrast with the 'whenever' sort generally envisaged in statements of the nomological theory of causation). If Graebner cites no such necessary condition generalizations, it is perhaps because he thinks that the particular judgments of indispensability he makes in the present case are obvious. Certainly it is his value-judgments which will appear to most readers as the more problematic.

But what can be said about the nature of and the need for the generalization which Graebner does cite, the generalization about the relative number of choices presented when powers of disparate strength confront each other? Some may be inclined to see this as virtually tautological – something merely excogitated from the very idea of a serious quarrel between a strong power on the make and a weaker neighbor. If so, it would have to be set alongside the other kinds which Graebner has already been seen to be employing – the probabilistic, the singular, the summative – as still a further variety of general claim that might enter into a historian's argument for a causal conclusion. It is surely not very plausible, however, to argue *a priori* that a stronger power *necessarily* deprives a weaker one of an option like drift (perhaps it depends on how much is read into characterizing it as 'on the make'). It may thus be more promising to treat the generalization as making an empirical claim, as holding that this distribution of 'real' choices is what in fact always or nearly always obtains when powers like those envisaged confront each other. Graebner himself seems to be thinking of it as empirical when, apparently with a view to offering *evidence* for it, he points to situations in American history which he believes offer parallels to the American–Mexican case. The United States, he declares, gave only the same two options to Spain over Cuban independence in 1898, to Germany over unrestricted submarine warfare in 1917, and again to Japan over the integrity of China in 1941. Thus empirically interpreted, the generalization could hardly be denied all validity, although one might doubt that its force would be more than probabilistic.

But how useful would it be, even thus interpreted, as support for the particular causal judgment that Graebner wants to make? Suggestive as it may have been for a consideration of the role of value-judgment in his argument, I cannot see that it really has much relevance *qua* generalization to the acceptability of his causal interpretation of the Mexican War. Surely all that he needs to claim in order to establish the relative degree of freedom attributable to the two powers concerned, and to select the causal condition accordingly, is that *in this particular case*, the weaker power, the Mexicans, gave the stronger, the Americans, more choices or greater freedom of maneuver than the latter gave them. Whether this is what *always* happens when a great power confronts a lesser one is a question that simply need not arise. Who gave whom how many 'real' choices in a particular case is entirely decidable in terms of what the historian knows about the case and the value-judgments he is prepared to make about the actions and attitudes concerned. In other words, as in presenting his negative thesis, Graebner over-argues his position, exposing it to needless objections. It might be observed, too, that, even if the truth of the generalization, empirically interpreted, were judged to be required for his causal conclusion, this support would be seriously incomplete from the standpoint of nomological theory. It would not render the outbreak of war in this case predictable, for example. The most it would justify our saying is that, in the present case, as in all similar ones, the Mexicans, unlike the Americans, would get only two 'real' choices, this rendering their response, whatever it proved to be, relatively unfree. It would not help at all to determine which of the two choices the Mexicans would in fact take. A nomologically grounded explanation of the war which covered this essential detail, would require still a further empirical generalization; and this Graebner makes no attempt to provide.

IV

The way Graebner argues for his causal thesis raises a number of other issues of philosophical interest, three of which I shall comment on briefly. The first concerns a certain exaggeration into which he sometimes falls in characterizing the attitude of President Polk which, in his view, was so important for the causation of the war. Graebner's official position is that Polk was reasonably confident until very late in the day that, if he stood firm, he would achieve his goals with regard to Mexico without war. He even offers reasons for accepting the opinion of most historians and of most of Polk's contemporaries that the American president's policy was not so much to make war as to obtain the fruits of victory without having to fight for them. He points out that Polk's political enemies, the Whigs, who were virtually pro-Mexican, had warned that war would surely ensue if an agressive approach was taken to matters like claims and boundaries, and especially to the possibility of annexing Texas. Polk, observes Graebner, would hardly want to make his opponents' unwelcome prediction come true. He also cites documentary evidence which suggests that Polk saw himself as simply taking a calculated risk, and not a very great one given the presumed military weakness and evident political disarray of the Mexicans. Yet, at times, he seems willing to castigate Polk's diplomacy as so reckless and so ill-conceived, so obviously unlikely to achieve its goals without war, that, in a sense, he must be said really to have wanted war all along.

Graebner never spells out precisely how we are to understand this supposed sense of 'really wanting'; he simply points to alleged parallels in American history – cases in which he thinks we are similarly inclined to speak of what was 'really' wanted or intended. Polk wanted war, Graebner contends, in the same sense in which his political successors wanted it with Spain in the 1890s when they over-reacted to the *Maine* incident; and again in 1917 when, in effect, they pushed the Germans into naval policies which they could use as an excuse to enter the war on the side of the Allies; and again in 1941 when, according to Graebner, a hostile American foreign policy drove Japan into attacking Pearl Harbor. In all these cases, he avers, Americans – or, at any rate, their leaders – 'really wanted' to take part in a war which it was not possible in practice for them to enter unless certain initiatives were taken by the other side. American governments did their best, even if this was not their admitted policy, to ensure that these initiatives were taken. Graebner never makes it quite clear whether his intention here is to attribute anything as esoteric as unconscious motives to the agents concerned. But some notion of such motives does seem logically required to make full sense of what he says.

I want to make only two remarks about this. First, it seems to me that, in all the cases Graebner cites, it is an abuse of language to speak of Americans as 'wanting' war. We already have a perfectly adequate way of describing what he appears to think actually happened in these cases, namely that American leaders knowingly pursued policies that *risked* war – how reprehensibly depending on how we should evaluate their problems and opportunities, morally and otherwise. To say of someone who deliberately does something that risks a certain result and perhaps even realizes that it does so, that he 'really wanted' that result is often a useful maneuver in moral polemic or attempts at moral persuasion. But if clear thinking about the subject-matter is the aim, it is surely questionable – a trick of rhetoric masquerading as analysis. The second observation is that, if speculation about unconscious motives is to be encouraged, Graebner's inclination to attribute to Polk a covert desire for war might itself, with some plausibility, be interpreted as unconsciously motivated. For it looks very much like an attempt to strengthen his own causal thesis by making it exemplify the same simple and powerful paradigm of human causation as does the traditional thesis he is combatting. The most elegant reply to the contention that the Mexicans *willed* the war, deliberately inciting the American response, would be that, on the contrary, it was the Americans who willed it. In his better moments, Graebner doesn't fall into this simplification. When he does, the result is strained and unconvincing – and, as should be clear in view of what was said in Section III above, quite unnecessary for establishing the causal conclusion he wishes to draw.

A second issue on which it may be useful to make a comment concerns the emphasis placed by Graebner upon the policies, attitudes and actions of a particular person, President Polk, in causally accounting for the Mexican War. His thesis may seem uncomfortably old-fashioned in this regard – just the sort of thing that has given traditional diplomatic and political history a bad name with modern historical methodologists who maintain that any adequate attempt to understand a substantial past event or state of affairs must trace the influence of large-scale social forces or conditions. Graebner's analysis may even appear to some as betraying acceptance of an outmoded 'great man' theory of history. Certainly, although he claims to consider the case for regarding as causes of the war some larger-scale candidates than the personal contributions of the American president, he quickly puts these aside. For example, he denies causal status to a generalized envy of things American on the part of the Mexicans, or a contempt for Mexican national self-consciousness on the part of the Americans, which might have entered into an explanation of the war as due to a conflict of cultures. And as we saw, he explicitly refuses causal status to chronic issues like boundary disputes and monetary claims, which might be thought of as transcending the particular individuals through whose actions and attitudes their causal influence worked itself out.

It is thus of some interest to note one passage in which Graebner seems inclined to desert what might be considered his extreme historical individualism, a passage in which, almost as an aside, he speaks of 'the pressure of American expansionism', or 'Manifest Destiny', as the real cause of the war. He appears thereby to look behind the actions and attitudes of Polk himself to that mental set of a whole nation that ensured effective support for his policies – perhaps even to the frame of mind that put him in the White House in the first place. Graebner seems to regard this as a mere elaboration of his more usual thesis, as if the two causal hypotheses, Polk and Manifest Destiny, amounted to the same thing. It must be insisted, however, that although Polk's particular policies and initiatives and the general climate of opinion referred to as Manifest Destiny are obviously related (one might represent Polk as the cutting edge of Manifest Destiny at this point), they are not the same thing considered as causes of the war. The whole burden of Graebner's argument is to claim for Polk a uniquely important and perhaps non-replaceable causal role in what took place. Without his intransigence and miscalculation, although Manifest Destiny might still have realized itself eventually in the extension of American sovereignty into nominally Mexican territories, Graebner sees no reason to think that this would have occurred at that particular time, and still less to think that there would have been a war with Mexico. This is hardly an expression of a 'great man' theory of history, if by this is meant a conception of historical events as under the *control* of unusual individuals. Graebner's analysis, however, like much traditional historiography, does assume that particular individuals, appropriately situated, can be crucial factors in the coming of events. He judges Polk to have been such an individual, so situated.

None of this, of course, means that in designating as cause of the war Polk's individual contributions rather than the thrust of Manifest Destiny or some other general social conditions, Graebner is denying the relevance of larger-scale factors for understanding why the war came, any more than he denies the relevance of what the traditional interpretation regarded as Mexican provocations. As has perhaps been somewhat labored already, his conception of a cause, as is usual in traditional historiography, is that of an especially significant necessary condition, not a sufficient condition of its effect. Graebner himself constantly reminds us of many background conditions – social, economic, cultural – without which his designated cause would have been powerless to produce its alleged effect. But the question which mainly concerns him is which, of all the conditions which are in one way or another relevant to a full explanation of the war, can be said to be what caused it. Although, for reasons already considered, he selects Polk's persistence as the crucial factor, this would not at all preclude his regarding social conditions as causes in circumstances where no relevant individuals were seen as acting sufficiently freely to be judged causes. The question arises whether, in cases where actions or attitudes of groups are selected, the same moralistic conception of cause as free action could be applied. It seems to me that it could, to the extent that it makes sense to talk of groups – for example, the American people – as being capable of action at all, and as being more or less subject to coercion and obligation. Had Polk been seen as no more than a mouthpiece of an American society obsessed with Manifest Destiny but without any real need to expand its frontiers, Graebner could consistently have selected the latter as cause of the war. For him, however, Polk was clearly more than a mouthpiece.

A third issue which perhaps deserves brief mention relates to the surprise Graebner expresses over the fact that historians find the causal explanation of the Mexican War such a controversial matter more than a century after the event, and the confidence he apparently feels that agreement should now be easy to obtain. "Although nationalistic bias will color the judgments of those who study war," he observes, "it seems strange that historical disagreement on a subject as remote and as well-documented as the Mexican War should be that elusive." What is required to change this state of affairs, he avers, is that historians pay less attention to "official rationalizations" and more to "the more universal causes of war which transcend individual conflicts and which can establish the bases for greater consensus". He evidently sees his own analysis as showing how this is to be done. Certainly his own argument repudiates some of the rationalizations of Polk and his colleagues, and there is much in it about general connections asserted or denied between stated variables. But Graebner offers no *explicit* generalization of the form "Given such-and-such, war always follows," nor of the form "War is always caused by such-and-such". And the explicit universal generalization he does apparently offer in support of his positive thesis, the three-versus-two-choice law, is, as we saw, insufficient to provide a nomological explanation of the actual outbreak of war.

Given the nature of his particular diagnosis, could Graebner perhaps be interpreted as holding at least *implicitly* that a universal cause of war is politicians commiting themselves to foreign policies whose aims cannot be achieved without fighting? For reasons which have been adumbrated in connection with other generalizations, I cannot see that his argument actually *requires* such a generalization. But in any case, as stated, it asserts too much. For in the kind of circumstances which the argument for his positive thesis envisages, a politician, no matter what his policies, is said to cause war only by acting freely or gratuitously rather than under obligation or coercion – moralistic qualifications which would, in consequence, have to enter also into the characterization of the supposed universal causes. This means that, although appeal to such causes might, as Graebner contends, usefully divert the historian's attention from 'rationalizations', it would not allow him to draw causal conclusions without making value-judgments. Clearly the disagreements that the relevance of value-judgments will almost certainly entail are not going to be resolved by mere completeness of 'documentation'. Nor is it easy to see why Graebner thinks that the 'remoteness' of the Mexican War should make consensus easier. The assumption seems to be that, as the past recedes, and vicarious practical interest in it becomes less tempting, the probability of our making *controversial* value-judgments about it, at least, declines. But such a tendency, if it exists, is easy to exaggerate, not least because the relevant sense of 'remote' is not just temporal – as is brought vividly to mind by some of the parallels Graebner claims to find between the slide toward war with Mexico and later American experiences like the military escalation in Vietnam. Unlike Graebner, therefore, I should not expect controversy over the causes of the Mexican War easily to disappear. Not, at any rate, unless historians stop using the traditional concept of causation which his analysis for the most part employs.

University of Ottawa

NOTES

[1] See *Pacific Historical Review*, 1980, pp. 405–426.

[2] *The War with Mexico*, New York, 1919, Vol. 1, p. 155. For a restatement of the traditional interpretation, Graebner refers the reader to S. V. Connor and D. B. Faulk, *North America Divided*, New York, 1971

[3] See, for example, N. Rescher, 'The Problem of Uniqueness in History', *History and Theory*.

ABRAHAM EDEL

THEORY AND PRACTICE: AN UNSTEADY DICHOTOMY?

The traditional problem of the relation of theory and practice runs strongly through Lewis Feuer's work. It surfaces particularly in specific studies where he gives us striking explanatory revelations. Throughout the decades of the now receding 'Age of Analysis' he pursued genetic inquiry into ideas, coming, by repeated researched examples, to grips with the dominant isolation of philosophical ideas as if they could be dealt with purely in their own theoretical terms without reference to sources, functions and consequences in the circumambient life. Much of this work is intellectually liberating. For example, in his early dealing with Marxian theory, in the midst of efforts to locate determinisms of ideas, he offered instead the analysis that the economic is the problem-generating factor, leaving ample room for other influences to bend the ideas proposed to resolve the problems at issue. In the case of Freud, when the philosophical atmosphere was most oppressive, Feuer not only looked for an emotional decision-base of philosophic ideas but even tried to work out a sense in which a philosophic idea might be genetically explained away. In writing of Spinoza, the philosopher so often worshipped for intellectual purity, Feuer skillfully used historical evidence to pinpoint the events of Dutch political turmoil that were almost isomorphic with the structure of Spinoza's free man and his intellectual love of God. Feuer's later detailed work on the emotional background of scientists produced the important historical generalization that a hedonist-libertarian ethic rather than a puritan-ascetic ethic lay back of the awakening and flourishing of science.

These works, irrespective of possible controversies over specific points and special hypotheses, open up new paths of inquiry by paying attention to the context and background and functions of practice. They help, thereby, in the general search for differentiating criteria of the authentic as against the ideological. This is properly a retail business; it comes in fine shades of more or less. Only in occasional works does Feuer lapse into the use of genetic inquiry as wholesale condemnation: for example, his *Ideology and the Ideologists* has an almost prophetic tone of setting apart the saved from the damned.[1] Perhaps the important

S. Hook, W.L. O'Neill and R. O'Toole (eds.), Philosophy, History and Social Action, 157–172.

lesson from the study of Feuer's work is that genetic inquiry into ideas is at its best when it looks not only backward but forward, when it does not stop with the source but goes on to set off the different potentials of the ideas that were produced, and evaluates the products as instruments in the advance of knowledge – in what precise respect the source threw up roadblocks, where it provided only a personally frustrating emotion to otherwise valuable work, where it may even have broken open an area insufficiently noticed in the usual emotional response.

This paper approaches the problem of theory and practice in another way and accordingly asks different questions. It puts aside the older concern with the 'purity' of theory for which the vital question was whether there was or was not a relation between theory and practice. It asks instead what kinds of relations may be possible, to what extent and under what conditions they are present, and in what contexts each kind may be useful. It is led to ask whether these notions have not themselves been fashioned for specific purposes under specific conditions. In pursuing this agenda the paper considers: familiar debates on the relation of theory and practice; theory and practice as themselves constructions; sample historical alternatives; types of relation (the heart of the inquiry); how theory and practice were related in twentieth-century educational philosophy (a case study from practice, illustrating the practical importance of the way in which theory–practice relations are theoretically construed); concluding reflections.

FAMILIAR DEBATES

Some advocate the separateness of theory and practice and take it to follow from their distinct natures. As activity theory is the thinking, and practice is the doing; the one lies in the realm of mind and the other in that of body. If we look to the object at which the activity is directed, then theory as intellectual formulation is distinct from practice as the product or application. (Ohm's law is a formula, not an electric circuit.) Others hold that theory and practice are correlatives, that operationally the meaning of a theory or concept is to be found in its practical or observational consequences; that the meaning of a social institution is to be found in the way it works out in practice. Still others argue normatively that whether or not they can be separated, they ought not to be; that new lines of inquiry, conceptual refinement and correction, in a word the growth of knowledge, depend on the interplay between theory and practical problems.

Matters are made much more complex by the ways in which theory and practice are relativized. Thus we could arrange a continuum in which what is practice or application with respect to what precedes may be theory with respect to what follows it. For example, consider: mathematics, physics, engineering, architecture, actual construction (literally digging); or in another field, physiology, medicine, theory of therapy, actual curing or operating. Of course how we treat such cases depends on the context, that is on the kinds of activities people are engaged in at a given time under given conditions.

Such general arguments, with occasional favorable illustrations thrown in, are by now boring. It is much more interesting to go to the blossoming studies of the history of science and technology. Nevertheless, the history of the general debate does reveal some things.

THE CONSTRUCTIONAL CHARACTER OF THEORY AND PRACTICE

To think of an idea as a construct (or construction) is to regard it as fashioned to serve a purpose in inquiry. It does not involve a commitment to conventionalism or to subjectivism. The history of the general debates indicated shows that *theory* and *practice* are themselves theory-laden, and when we unpack underlying theory (including any surplus) we can see what prompts the sharp separation or the greater closeness. Thus the dichotomy is sharp when coupled with a metaphysics that instates a mind–body dualism or with an economic theory that differentiates intellectual or skilled production and unskilled labor, brain and brawn. On the other hand, the distinctness of theory and practice tends to fade under the influence of a naturalistic psychology which stresses a controlling cognitive function (explicit and implicit) in virtually all behavior beyond sheer physical movement. Purposive doing in general implicates thinking where the overt behavior is construed or felt as a means to something beyond itself. Finally, theory and practice are firmly wedded under the influence of an epistemology in which ideas are seen as plans of action and knowledge is for the sake of action. Even on a less pragmatic epistemology, thinking in experiment science involves a reference to doing.

Perhaps we could also bring to bear on the consideration of theory and practice what can be learned from the history of dichotomies in Western thought. They usually enter on the scene with confidence, proclaiming their analytical necessity or metaphysical truth. Sooner or later phenomena appear in the middle ground which raise doubt about their sharpness. Eventually they are revealed as programs, and not always very successful programs at that. The hardened dichotomies are then turned into contextual distinctions, useful in some cases, misleading in others. It is experience that refuses their absolute partitioning and that determines their relative utility. Such has been the story of body and soul, human and divine, body and mind, sensation and thought, sensation and feeling, is and ought, fact and value, and innumerable others. Why should it not be the story of theory and practice?

HISTORICAL ALTERNATIVES

The twofold division into theory and practice is by no means the only possibility. Indeed it was not the first on the scene. In Aristotle's philosophy the concept of theory was narrower (*theōriæ* limited to a grasping of eternal truth), and instead of practice there were two notions – *praxis* or action and *poiēsis* or making. These are distinguished by whether the end or goal lies in the act itself (as taking a ride) or in a product separate from the activity (as in making a bridle). Now Aristotle's trichotomy obviously stands out as a conceptual construction. We can see what went into it. One strand was the belief in a dichotomy between the necessary and the contingent, itself identified with one between the eternal and the changing. Hence theory is concerned with what cannot be otherwise. And into the division between *praxis* and *poiēsis* there is incorporated at least the social distinction between the craftsman or producer who makes things and the leisured class who enjoy their use.

The constructional character of Aristotle's distinctions is further made manifest by the modifications insisted on in Puritan theology. Aristotle's view of *theōriæ* is condemned precisely for its assumption of an eternal world that has always existed, a pagan belief. Since God created the world, what we are learning in theory is the scheme of God's production; *theōriæ* thus becomes incorporated into a universal *poiēsis*![2]

There is no reason to think that our current distinctions are less constructed shaped by philosophical assumptions and traditional cultural attitudes. Certainly the dualism of mind and body is transparent in the dichotomy of theory and practice, and the social dichotomy of leisure and work is as much associated with it as with the Aristotelian distinction of *praxis* and *poiēsis*. It is interesting to note that *praxis* is now coming to take its turn at getting to the top. If Aristotle had at the pinnacle his conception of theory, and the Puritans their conception of making as creation, contemporary humanistic Marxism is attempting to fashion the concept of praxis to embrace human free activity in which theory is fruitful and production itself satisfying.[3]

TYPES OF RELATION

There are many possible kinds of relations between theory and practice. If we pursue them in detail, it is not mapping for mapping's sake, but in the hope that their varieties may suggest fresh ways in which theory may furnish guides to practice and the influence of practice on theory be more fully understood. When we think of theory as the activity of people theorizing and practice as their doing, the relations that come to mind are the ways in which, as a matter of psychology and history, the one has influenced the other in either direction. These are largely the genetic relations that Feuer has studied in broad scope.[4] When we think of theory as the body of knowledge gathered and systematized, and practice as the formulation of its application, then the relations that come to mind are likely to be logical, methodological, and instrumental.

This latter set of relations – the logical, methodological, and instrumental – require more detailed examination to see the factors that determine their utility. These factors will, of course, depend on how good the theory is and how rich the practice is. On the side of the theory, the important factor is the state of knowledge. It may run from the highly axiomatized and highly systematized down to the roughly empirical. A possible ambiguity in the term 'theory' needs to be noted here. The notion of theory has considerable stretch. Strict use would distinguish a theory from a scientific law; it would be some higher-level set of relations which unified laws. But in a broader use theory comes to include all established universal knowledge, and the theory of a subject would include the empirical generalizations as well, if they were not simply rule-of-thumb. But the borders of usage are shady. A discipline begins to use the term when it has worked out a schema for future theory, or else it has concepts refined in terms of which theories will be stated, or it has theories on the loose but none established, or it has bodies of data rapidly accumulating and the hope of theory is in the air. Soon every field has its theory, which is scarcely more than reflective inquiry or where practice will go for what help it might get to light its way. We are not called on here to approve or disapprove, but simply to note the breadth of usage. If we wished to exercise greater caution, we would speak in some domains of theory-surrogates, in others of theory-hopes, in still others of mere hypotheses.

On the side of practice, two factors important for determining the kind of relations readily emerge on inspection. One is *the purposive pattern* of the participants, or the established purposes of the institution or enterprise in which the practice is involved. Thus the practice of medicine has purposes of curing, adding to medical knowledge and making a living, while the practice of farming has the making of a living in addition to growing food, so that on occasion grain or coffee is burned in order to support prices. (In that instance the impact of economic theory on farming practice is enough to reverse the usual practice.) The purposive pattern of a practice may be one of established common aims or even of intense value conflicts. Now a second factor on the side of practice that is important for determining the kind of relations between theory and practice is more complicated than the purposive patterns. For any practice it is not enough to bring in some theory to be sure that it can be used. Particular conditions have to prevail or else the theory is irrelevant. It is no use bringing the advanced farming knowledge of the green revolution to a country that has no way of getting fertilizer, or advanced medical knowledge for complicated tests and surgery to a situation that has no relevant facilities or equipment. In the practice of teaching, the best theoretical knowledge of methods is of no avail if the class does not know the language of instruction or is hungry or is hostile. For brevity's sake we may call this factor on the side of practice the *situational conditions*.

Now if we put together the three factors – *the state of knowledge* in the case of theory and the *purposive pattern* and *situational conditions* in the case of practice – we have to ask what variations within them make possible what kinds of relations between theory and practice. In the state of theoretical knowledge, focus is on the degree of development of the relevant science and its specific methods of inquiry. Behind this, of course, lies the character of the actual phenomena involved: whether the particular subject-matter happens to be simple and fairly uniform, stable like the motion of the planets, or a complex and volatile one like the weather, or even worse, like the stock market or much of human behavior. Putting it abstractly, the extent to which we have systematic knowledge depends on both the advanced state of the science and on the stability and manageable complexity of the phenomena. The domain of the practice is a fresh area of phenomena and how stable it is depends in part on the extent of harmony within the purposive pattern, in part of the stability of situational conditions.

Inner conflict is not unusual in many of our professional practices. For example, the traditional medical aims of curing and advancing knowledge sometimes involve the doctor in a trade-off, and the modern journalist sometimes has to choose between his obligation to report the news and the loss of advertising revenue. In education, the devastating impact of the conflict between academic objectives and job preparation is a familiar source of difficulties. As for situational conditions, all fields were thoroughly shaken in the last few decades by the political and technological changes after the Second World War; even had theoretical knowledge remained unchanged, its impact and relation to practice would have been radically altered. Finally, since the three factors vary in a relatively independent way, the character of each of them may, in a given case, affect the kind of relations possible between theory and practice.

The kinds of possible relations are patterned on the variety found in science (where it is well worked out in the practice of prediction), in means–end action (which is a practice of goal achievement), and in reflective experience generally. It thus moves from logical relations to methodological and to instrumental relations and then to a kind of craft or skilled insight. Whether we have here a continuum corresponding to diminishing degree of orderliness and increasing complexity in the phenomena or utterly different capacities in the human makeup, is a controversial issue.

From the sciences we borrow the idea of relations that are either deductive or inductive. In the first place there are conclusions for practice deduced from statements of theory in that ordinary sense of deduction in which the occurrence of the eclipse at a given time is deduced from the laws of planetary motion and the positions and velocities of the relevant bodies. Engineering makes such deductions constantly in the course of its applications. The usual assumption has been that human affairs have too much variability (whether from 'free will' or from complexity) for such relations; they do not provide laws of sufficient strictness and regularity to be scientifically serviceable. Strangely enough, in matters of social practice, regularity may be secured through different means – by social fiat or determination. Large-scale institutions may be assigned a strictness of procedure so that conclusions and calculations may be made with a precision at least analogous to engineering deductions. Thus in an area in which all other features but a specified number could be ignored – for example, specific insurance policies – the 'laws' set up by the contractual relations that serve as the 'theory' can be programmed and when situational conditions are entered after inspection the payments can be revealed ('deduced') by computer. This type of relation is then worth keeping open as a possibility because larger areas of human affairs may come to be thus handled mechanically as a matter of social policy – for example, if a 'negative income tax' or a guaranteed annual wage were enacted. In human affairs, human fiat might thus create an order stricter than natural tendencies would manifest.

Inductive-experimental relations are familiar enough where prediction is involved. If theory–practice relations were so conceived, the practice would not be merely an application but testing whether the theory works or whether there is some hidden flaw or missing variable or sometimes whether the analysis of conditions of application has been wrong or incomplete; in any case, the theory–practice relation is here assimilated to the hypothesis–experiment relation. A mass innoculation program, primarily a practical application of a theory and not intended as just an experiment, occasionally shows this character.

Where the sense of theory is weaker, or the complexity of phenomena very great, the best that can be done is construe the relation so as to have theory furnishing a set of methods or instruments for assisting practice, but not pointing to a specific result. Different analogies are helpful. Some theories are like charts or road-maps; they guide explorations but do not warn of obstacles or disruptive factors or improbable occurrences. Some theories are simply theory-schemas; they are useful as instruments in that they tell the kinds of dimensions or components to look for in analyzing a situation, but not necessarily where to find them or how to put them together. In the social sciences, what is often called a theory of society turns out to be an analysis of just such factors. To divide the society in one way rather than another will provide a more systematic picture.

Finally, there is something which does not at first sight seem to be associated with theory at all, but appears to be a substitute for it. This is a kind of intuitive judgment or insight, not in the sense of beholding a self-evident truth, but rather grasping that something fits. It is the scientist's insight that a formula may fit the data or a curve is the one that will do the job, or a theory will unify materials. Perhaps there is only a difference of degree between this and the way in which an experienced sailor senses a coming storm or an experienced politician the first signs of a political rumble or an experienced teacher grasps where a lesson will take hold. The question here is whether these phenomena should be seen as embedded theory being translated into practice, or whether they express a distinct human capacity. They are often invoked in the claim that practice is an art, rather than akin to the family of theoretic science, but it is better to recognize a continuity with the other relations of theory and practice.

The kinds of relations given here are perhaps more than enough, expanded to suggest that when a distinction between theory and practice of the relative sort is emphasized it gives the freest rein to the different ways that theory can respond to the varying conditions and needs of practice. The variety of relations of theory to practice provides modes of guidance that should not be wasted. It is equally important to clear channels for getting what feedback there may be from the experience practice may have in the guidance of theory. Finally, there may be other cultural experiments that dispense with the whole theory–practice construction, and they are worth working out, but they would involve a major epistemological recasting, for they would have to carry out systematizations across our present theory–practice lines.

RELATING THEORY AND PRACTICE IN TWENTIETH CENTURY EDUCATIONAL PHILOSOPHY

Twentieth-century philosophy of education has treated rather cavalierly its own problem of relating theory (i.e. philosophy) and practice (i.e. educational practice). Unfortunately, the matter was debated about a categorical presence or absence of a relation. For example, it used to be said, when metaphysics and epistemology were seen as a set of conflicting 'isms', that a Platonic metaphysics meant an authoritarian classroom; a Lockean epistemology entailed an insistence on memorization, rote learning and the use of rewards; or that a Kantian ethics promoted rigidity and a pragmatism of the Deweyan kind, individual initiative. The objection was of course offered that the Platonist could be more Socratic than authoritarian, the Lockean's burden of homework could cause the pupil to catch fire in the concentration of his effort, the Kantian could stimulate moral insight and strengthen moral fibre, and Deweyan flexibility might turn into opportunism. Not only could the same philosophical theory be associated with different educational practices, but different philosophical theories with the same practice. Educational philosophers jumped to the conclusion that there was no determinate relation between theory and practice, precisely because they were assuming that it had to be of a single kind. Actually they should have gone on to examine the cases more minutely, to find out at what point the theory operates and how it produces what effect.

The challenge took the form of denying that philosophical theories have had any relation to educational practice, although philosophical discipline might be effective insofar as it increases sensitivity or sharpens intuition. For example, some time ago a sheriff in Texas advertised for a deputy, announcing that he would give preference to someone trained in ancient Greek philosophy; he explained that he had found such students to be better at analyzing situations sensitively. It would scarcely do to say that the deputy would be *applying* or *practicing* Heraclitean philosophy if he kept an eye on changing conditions in the town, or Platonic philosophy if he acted on the assumption that all criminals have a fixed nature. His contact with these philosophies and his struggles with their arguments would have quickened his spirit; his actions proceed from a quickened insight, not a theoretical learning, any more than the politician's strategies come from the study of political theory rather than accumulated experience.

Approaches such as these create an atmosphere of cleavage between theory and practice. One form it takes in the philosophy of education is to divide knowledge sharply into the theoretical and the practical, with the educational consequence that learning is sent in two different directions. In this manner, Gilbert Ryle distinguishes *knowing-that* and *knowing-how*.[5] The latter, as practical learning, comes through experience or in apprenticeship in which the experience of others is taken over. The former, as theoretical learning, comes through verbal instruction. Ryle's dichotomy set a direction which was richly mined in contemporary educational philosophy, but steered toward multiplying different modes of learning. Criticisms of it simply bid higher by making more distinctions, for example between *knowing-how* and *knowing-how-to* and differentiating other kinds of knowing which invoked still further human capacities, for example aesthetic, and so other modes of learning. Now on our strategy, instead of multiplying types of learning, we should look to the different conditions under which learning takes place, those factors we explored in studying the various kinds of relations. Thus the atmosphere of sharp cleavage is dispelled and continuities in learning may be restored. The absolute types should not have been created at the very beginning: Ryle's *knowing-how* has not that sharp a distinction from *knowing-that* as his selected illustrations suggest. It may look that way when we think of learning to ride a bicycle, but not when we think of learning to pilot a plane; of knowing how to win an election under simpler conditions, but not under contemporary conditions of large-scale organization; of sea-faring in olden days, but not for computerized space-travel. In short, we have not here absolute differences in kind, but relative differences in degree.[6] Apprenticeship is not a cognitive mystery, even though it is a very complex matter. Many tasks have been sufficiently unravelled not only to become matters of instruction but even to be mechanized so that a robot can do them. Others have been sufficiently understood so that they can be improved by instruction and the handicapped enabled to do them in normal fashion. Some are not yet fully fathomed, so familiar as reading a page of a book. Differences in conditions support differences in the relations of available knowledge and possible educational practice.

If this general outline of the relations of theory and practice is fruitful, the philosophy of education will not antecedently limit its direction of development by some fixed formula of their relations. Analysis of specific context of practice will be required to determine what is possible in what areas and in what ways at what times and where emphasis should fall.

CONCLUDING REFLECTIONS

There remains the (less than gnawing) worry whether we have succumbed to our present dichotomy, instead of rejecting it, by exploring its inner relations. Perhaps we would do better by reverting to a trichotomy, in Aristotelian fashion, or else by overcoming (dissolving, transcending, overriding) the dichotomy in some way that makes a unity out of its components. These possibilities – tinkering with our present dichotomy in one way or another, and unifying its two sides in one way or another – merit at least brief exploration.

The theory–practice dichotomy was set up earlier as including in theory the systematic knowledge available for the particular field of consideration, and in practice the purposive pattern and situational conditions invoked in the application of the theory. Now the simplest alternative if we are to preserve a dichotomy might be to include the purposive pattern in theory, leaving the situational conditions in practice. This is a ready model for human action and planning: all action involves on the one side some cognitive and value equipment (presuppositions, established habits and directions of effort), and on the other side the field conditions under which they are being exercised. Should this seem to threaten the purity of theory, we might set up a trichotomy instead, with theory in its abstract form as the first party, theory interpreted for action as the second, and application as the third. (Compare a pure geometry rendered abstract in Hilbertian fashion, such a geometry interpreted for a Euclidean space, and thirdly carpentry or surveying.)

A more complex alternative would be to assemble all the knowledge invoked on both sides of our original dichotomy and regard it as theory. Clearly there is considerable knowledge embedded in a purposive pattern and in the description of the situational conditions. The result might be rather hard on practice: all the sciences, laws, classified types, universals on the theory side, leaving nothing but bare existence or movement of the practice side. This brings us back to the crude dichotomy of all that is cognitive, all the thinking, on one side and the mere fact of spatio-temporal doing on the other. It may remind us of the Platonic treatment of universals as the real, monopolizing all account of sorts and kinds, with existence as simply the convergent reflection of a group of universals in some shadowy matrix. Or in a more even-handed way we could repeat with Kant that percepts without concepts are blind and concepts without percepts are empty.

Attempts to overcome the dichotomy in some unity of theory and practice take different shape in different philosophies. Some go in for transcendence in an ultimate unity of mystical experience; others trace the interaction of cognition and movement in human activity, or again the mutual involvement of the practical and the theoretical.

Of the efforts at unity by transcendence, the ancient neo-Platonic is the supreme example. It develops the Platonic duality noted above. Practice (in the sense of what goes on in the material world) is reduced to a bare resistance to the rational-theoretical. The full analysis of any productive process is found in the rationality it embodies. (Plotinus may yet come to be installed as a patron saint of an automatized and computerized production in which thinking and planning are all that really counts and the rest follows as a matter of course!) But – to complete the story – a dualism remains within theory itself, for it takes for granted the distinction between the knower and the known. This is what is ultimately transcended in the mystical-religious experience, which is itself indescribable.

More earthy attempts at finding the unity of theory and practice focus on human action. A commonplace instance is the perfection of a skill so that the *temporal* distance between cognition and movement is minimized. Take, for example, driving an automobile. The better the driver, the more rapidly the cognition involved in perception of the surrounding conditions, together with all the knowledge embodied in the theoretical interpretation of signs, signals, warnings, judgments of speed, knowledge of the habits of pedestrians and other drivers are translated into movements of the driver's wheel. In short, habit operating intelligently, is the unifier of theory and practice. Where, however, in a more elaborate operation theory takes the form of complex intellectual activity expressing a pattern of ideas while the practice is some construction of art presumably embodying those ideas, then the unity of theory and practice has to be found in the way the theory functions with respect to the practice. This brings us back to the path we dismissed in beginning the discussion of different types of relations – the psychological and social modes in which the theory operates. A familiar distinction here is whether it operates in realistic guidance or in ideological justification. (The theory of such relations is itself well developed in the twentieth-century analyses of ideology.) It is, of course, the closeness of real guidance that constitutes the stronger unity. Nevertheless, the quality of the guidance and the quality of the product may itself be a factor in the judgment of the unity achieved. For example, a writer discussing Wagner's own theories in relation to his music may say: "Wagner varied the philosophical, aesthetic and political theories he proclaimed in his writings entirely for the same of his musical dramas, which in the last analysis were the only thing that truly possessed him. The works are the key to the writings, not vice versa."[7] Or an architectural critic may complain that in the past "The critic and historian, quite properly, came after the creative act, with a particular knowledge and insight, to probe meanings and processes, discern norms and standards, evaluate the work of art, and relate it to the larger setting of time and place", but now the critic gets in ahead and expounds aesthetic standards to evaluate before the work is done.[8] Doubtless it is the fact that the writer does not approve of the outcomes of such influence that affects the judgment about the shift in the critic's mode of functioning. Such examples, which one comes on in all sorts of different fields, show how important in our day has become the relation of theory and practice and the assessment of the theoretician's role. In questions of public policy and political action it becomes particularly crucial. Whereas in the 'old days' the theoretician proposed an ideal that might be approached over the ages, so that some testing went on in the steps toward its achievement, nowadays the distance between theory and practice has diminished so that instead of ideals we have instant political programs. To some extent this is doubtless due to the rapid development of instruments of communication and agitation, as well as of course to the multiplication of serious social problems.

Such considerations suggest that perhaps the critique of the dichotomy of theory and practice is simpler than it appears in the many directions in which we may struggle with it. The very attempt to explicate each side of the dichotomy carries us into the other. No establishment of theory is possible without experimental practice and no guidance of practice is possible without some theory or theory-surrogates. How particular divisions are patterned is itself an experimental matter geared to the particular contexts and problems which that division is to serve. Every division has its limitations and its costs and is on constant trial to earn its keep.

In general, the distinction of theory and practice comes from the division of labor that gave rise to specialized bodies of knowledge, pursued with increasing independence, from the ancient Egyptian priestly studies of earth and heavens to predict the flooding of the Nile to contemporary separation of basic research in all fields from even research directed to technological application. It is not a distinction without social and intellectual costs: its history is replete with the privileges and snobbery of class and status, and with the illusions that have masqueraded under the slogan of the neutrality of science (invoking another dichotomy, that of fact and value).[9] The democratic attempt to diminish the costs by broader education, greater appreciation of science, and even increased participation in some ways in scientific ventures, points a direction for effort but scarcely amounts to overcoming the distinction. In sum, the present basis of the distinction in its most serious use is the social context of a relatively independent accumulation of systematic knowledge. Overcoming the distinction in this respect is logically possible: doubtless there was a time of largely self-sufficient practice in the form of rudimentary craft production in which the ordinary experience of Everyman was adequate for the daily round as practised. And in the contemporary context it would seem that Mao's cultural revolution in China (now abandoned) was in great measure prompted by the fear of the relation of specialized knowledge and a class monopoly of power. It looks as if, however, in the context of our contemporary world as it is likely to be for an indefinite time, the problem both for knowledge and action will remain the quality of relations between theory and practice rather than the dissolution of any distinction.

Under these conditions we may rest content with realizing the contextual rather than the absolute character of the division of theory and practice, with the vista of the many types of possible relations, and with establishing a practice that, when it has to face a division is some inquiry, will always look, in a normative stance, for the most suitable mode of relation in the materials and purposes provided by the context. Appreciating the instrumental character of the distinction, we can also be sensitive to the search for fresh modes of relation, if those at hand are not adequate or satisfactory. While such an outcome may not satisfy the metaphysical inclinations so often at the base of traditional dichotomies, it provides a fruitful replacement for a barren dilemma.

University of Pennsylvania

NOTES

[1] The book pins the label of ideologist on innumerable heroes of thought and civilization. Surprisingly Freud is wholly spared not by defense against the many critiques directed at him, but by total omission (even in the index).

[2] Thus what is regarded by us as science that contemplates would be really from God's point of view art that produces. Such continuity is found, for example, in the teaching of William Ames, a seventeenth-century Puritan philosopher working in the spirit of Ramus.

His views were influential in early Massachusetts controversies against the Aristotelian philosophy.

[3] See, for example, Mihailo Marković, *From Affluence to Praxis*, Philosophy and Social Criticism. Ann Arbor: University of Michigan Press, 1974.

[4] Cf. Abraham Edel, *Method in Ethical Theory*. Indianapolis: Bobbs-Merrill, 1963, Chapter XI (Genetic Inquiry and Truth Determination). Pages 273–279 examine critically Feuer's attempt to establish a basis in genetic inquiry for determining truth or adequacy of philosophic ideas.

[5] See Gilbert Ryle, *The Concept of Mind*. London: Hutchinson's University Library, 1949.

[6] Ryle was attempting to get rid of the mind–body dualism in the theory of knowledge but the outcome of his way of doing it was to substitute a theory–practice dualism. For a fuller critique of this whole episode in educational theory, see Abraham Edel, 'Analytic Philosophy of Education of the Crossroads' in *Educational Judgments*, James F. Doyle (ed.). London and Boston: Routledge & Kegan Paul, 1973, pp. 232–257. Also in *Educational Theory*, Spring 1972, pp. 131–152.

[7] Joseph Kerman, 'Wagner and Wagnerism' in *The New York Review of Books* **XXX**, 20 (Dec. 22, 1983), p. 27. Kerman is here quoting Carl Dalhaus, from *The New Grove Wagner* by John Deathridge and Carl Dalhaus (Norton, 1984).

[8] Ada Louise Huxtable, 'After Modern Architecture' in *The New York Review of Books* **XXX**, 19 (Dec. 8, 1983), p. 29. She declares that "The influences passing between print and practice are subtle and insidious, and the relationship grows increasingly incestuous."

[9] For analysis of the fact-value dichotomy in its various aspects, see Abraham Edel, *Exploring Fact and Value* (Science, Ideology, and Value, Vol. 2). New Brunswick, U.S.A. and London: Transaction Books, 1980. The appendix (pp. 338–363) outlines the sociological history of the dichotomy.

ROBERT G. GILPIN

DEVELOPMENT AND UNDERDEVELOPMENT: CONFLICTING PERSPECTIVES ON THE THIRD WORLD

In terms of the broad sweep of human history, the most pressing issue of international political economy in our era is the future of the so-called Third World. However one chooses to label it – modernization, development, industrialization, etc. – the intense desire of the greater fraction of the human race to escape its debilitating poverty and join the developed world is a determining feature of contemporary international politics. How this issue is resolved, if at all, will profoundly affect the future of the planet. It has given rise to bitter controversy regarding the causes of this poverty, who is responsible for it, and demands for a New International Economic Order.

The poor of course have always been with us. Poverty in fact has been the lot of all but a few members of the human race from its very beginning. A major change has taken place, however – revolution would not be too strong a term – with respect to the political and moral significance of human poverty. It is this change that has made the immense gap between the rich northern half of the globe and the largely impoverished southern half a new and explosive issue of international political economy. Among the reasons for this historic change, three are of primary importance and account for the international political significance of mass poverty.

In the first place, poverty is less tolerable in a world of instant communications. Even in the most remote parts of the globe, the transistor radio and an occasional television set has made people aware of the wealth of others and the benefits of material progress. Whole societies now want what only the rich could aspire to have in the past. The advanced nations have taught the rest of the world that escape from poverty is possible. This demonstration effect has made the desire for economic growth, modernization, and rapid industrialization the universal ideology of political elites in all countries.

In the second place, poverty is no longer regarded as somehow natural, the punishment of God, or one's Karma. Poverty and its consequences are held to be man-made and for this reason they have become less acceptable. Fewer people today resign themselves to pov-

S. Hook, W.L. O'Neill and R. O'Toole (eds.), Philosophy, History and Social Action, 173–208.

erty and accept it as their fate. Moreover, the progress that has been made in certain less developed countries and the immense distance yet to be traveled by most only reinforces this perception. It is almost a law of human behavior that the rise of human expectations outpaces the capacity of society to meet them. This revolution of rising expectations has become a universal feature of our age (Hirschman, 1981).

In the third place, the poverty issue has been internationalized (Carr, 1945). Until the modern era, the differences of wealth *within* societies was far greater than the differences of wealth *among* societies. In pre-modern societies everywhere a small wealthy elite sat on top of an impoverished mass. This situation is of course still too true in many societies. Today, however, on a global scale the differences of wealth within various countries is less striking than the difference of wealth among countries themselves; the individual at the poverty line of Europe and America is far wealthier than the overwhelming bulk of the human race living in the Third World. In the modern world industrialism and the welfare state, whether one is rich or poor has become increasingly a function of the particular society to which one belongs. As a consequence, the class struggle within societies – as Marxists would describe it – has become partially displaced, if not superseded, by the conflict among national societies over the international distribution of material wealth.

The effect of these changes in fact and in perception, has been to make economic development – and underdevelopment – a central issue today in international political economy. This universal concern over the distribution of global wealth is truly a novel concern of world politics. If one consults diplomatic histories, one will find scant interest in the subject. Though individual nations have always desired to improve their economies, the issues of economic development and the skewed international distribution of wealth were not on the traditional agenda of international diplomacy. The contemporary controversy over these matters is the subject of this article.

THE DEBATE OVER INTERNATIONAL POVERTY

The grossly uneven distribution of wealth in the world has become one of the principal and most divisive features of contemporary international relations. The immense gap between rich and poor countries in fact is an aspect of the modern world which distinguishes it from prior history. There has always been wealth and poverty but in the past the dividing line ran between elite and masses or social classes. Today, the line separates nations, races, and hemispheres. It sets the poor South against the affluent North and the Third World against the First World of the market economies and, to a lesser extent, the Second World of the command economies. This fact that the global poverty line runs along political boundaries has given the distribution of wealth an international dimension and has made it a major issue of world politics.

The rancorous debate over the so-called North–South issue centers on three sets of questions. The first is that of the cause of this situation and assessing the responsibility for the deplorable conditions in which much of humanity lives. Is it due to the operation of the world market economy and the evil practices of capitalism? Or, does the problem lie with more objective economic factors and the misguided policies of the poor countries themselves? The second question and the answers given to it flow directly from the first: What actions should be or are being taken to solve the problem of world poverty? Is integration in, or disassociation from, the world economy the best route to economic development? The third question relates to an assessment of the strategies actually followed by the less developed countries: Is anything working? In the following pages, these three issues will be explored.

Although the answers provided to these questions have proliferated in recent decades, they can be divided into two broad perspectives on the relative importance of internal and external factors in explaining development and underdevelopment. The first perspective is the development or diffusionist perspective, a position taken by liberal economists and classical Marxists. This position accepts the concept of the 'dual economy' divided between traditional and modernizing sectors; it stresses the importance of internal factors and the positive influence of the world economy for economy development. The second perspective is the underdevelopment or dependence perspective that is accepted by a large number of neo-Marxists and economic nationalists, especially among Third World intellectuals and politicians. This position accepts the concept of the modern world system in which traditional economies are linked to an exploited by more developed ones; it stresses the importance of external factors and the negative influence of the world economy for economic development. These developmental and dependency perspectives will be discussed below in broad outline. It must be appreciated throughout, however, that what follows is a general statement characterizing two huge bodies of thought and that no one adherent of either position would subscribe to each and every point.

THE DEVELOPMENT OR DIFFUSIONIST POSITION

Both economic liberals and classical Marxists subscribe to the dual-economy theory of the world economy. For both positions the evolution of the world economy must be seen in terms of the spread of the process of economic growth from advanced to more traditional economies. Through trade, technology, and investment flows, the less developed economies are incorporated into an expanding world economy and in the process are transformed from traditional to modern economies. These two positions differ, however, on the nature of this process. Whereas for liberals this process is generally benign and harmonious, for Marxists it is accompanied by conflict and exploitation. Despite this difference, however, the end result of this integrative process is the same: the economic development of the less developed economies.

The Liberal Perspective on Economic Development

According to the development or diffusionist position, the world economy is a beneficial factor in economic development. The interdependence and economic linkages of advanced economies and less developed economies ends to favor the latter societies. From the advanced economies, the less developed economies acquire capital, technology, and export markets. For this position trade is – or can be – an engine of growth. For these economists, however, the essential determinants of economic development are internal to the developing economy. While the world economy can help or hinder development through the diffusion process, the most important factors affecting economic development is the efficient organization of the domestic economy.

Although there exists a generally accepted liberal theory of international trade, money, and investment, there is not a comparable theory of economic development. The principal reason for this difference is that the body of theory regarding trade, money, etc., rests on the existence of a market or, at least, assumes that a free market does exist; economic theory is concerned with rational individuals seeking to maximize welfare under market conditions. For liberal economists economic development involves the establishment of an efficient market system (or, as in the case of socialist societies, its functional equivalent) in a traditional and inefficient economy. How this transformation is to be accomplished takes one beyond the economic sphere into the larger society and the realm of political and economic strategies.

The liberal perspective on economic development is above all concerned with the best route to the creation of an efficient market economy. In contrast to other areas of economics, liberal views on this issue cannot be restricted to conventional definitions of economic behavior and institutions. The study of economic development must of necessity be holistic and address itself to the analysis of social, economic, and psychological variables. Further, whereas conventional economics tends to assume a static framework of rules and institutions within which economic activity takes place, the purpose of the theory of economic development is to explain behavioral and institutional change (Davis and North, 1971). Given the magnitude of this task, it is hardly a wonder that the study of economic development has not produced in the fraternity of liberal economists a generally accepted body of developmental theory. Nonetheless, liberal economists are agreed on several points.

Although there are great differences of opinion and conflicting findings, liberalism holds that an independent world economy based on free trade, specialization, and an international division of labor facilities domestic development. Flows of goods, capital, and technology serve to increase optimum efficiency in resource allocation and transmit growth from the centers of growth in the developed nations to the less developed countries. The rate and direction of this spread effect are dependent upon a number of factors: the international migration of economic factors (capital, labor, knowledge); the volume, terms, and composition of foreign trade; and the international monetary system. Although economic progress is not uniform throughout the economy (domestic or international), the operation of market forces over the long term tends to develop all regions and nations. In time, according to classical doctrine, the economic system will achieve a more nearly just distribution of real income.

Liberal economists tend to focus their attention on the obstacles to economic development that exist in the less developed countries themselves. Among these constraints are the preponderance of subsistence agriculture, the lack of technical education, the low propensity to save, and the weakness of the financial system. The assumption of all liberal thinkers is that once these bottlenecks are removed and a market begins to function, the society will commence its escape from its condition of economic backwardness. As Lewis has put it, any economy can develop it if it has three simple ingredients: adequate rainfall, a system of secondary education, and sensible government. For the liberal, the question is not why the poor are poor. Poverty been the lot of the human race for millennia. The real question, as Adam Smith phrased in *The Wealth of Nations*, is rather why certain societies have overcome the obstacles to development and become rich (Smith, 1776). The answer given is the establishment of an efficient market.

From the perspective of this position the integration of a less developed economy into the world economy is generally beneficial. There is in fact a harmony of interest between developed and less developed economies. The flow of commodities and resources between the two types of economies is a stimulus to growth in both areas. This is the notion that trade serves as an engine of growth. The less developed economy gains capital, technology, and access to world markets. The developed economies obtain raw materials and outlets for their manufactured goods. As the less developed economies have smaller markets, the opening of trade with advanced economies actually serves to benefit them relatively more. Moreover, as the factors of production flow to those areas where they receive the highest rewards, the less developed economy (frequently with a surplus of labor and a deficit of savings) receives infusions of foreign capital that accelerates its growth. This liberal position adheres, thus, to a diffusion theory of economic growth in which the requirements for economic development are diffused from the advanced core of the world economy to the developing economies in the periphery.

For this position the causes of economic backwardness are found principally in the less developed economies themselves (Bauer, 1976). The failure to develop is due to domestic market imperfections, inefficiencies, and rigidities. These nations are held beck by political corruption, a parasitic social structure, and a failure to make appropriate investments in education, agriculture, and other prerequisities for economic development. The fault lies in improper public policies fostered by burdensome bureaucracies and economic nationalism. Although the advanced economies can hinder the progress of the less developed economies by restrictive policies of their own such as protectionist policies against Third World exports and can accelerate development through foreign aid, the principal focus of responsibility in liberal theory lies with the less developed countries themselves.

From this perspective, the foremost task of development is to increase the domestic rate of saving and thereby accelerate capital accumulation. Although the advanced economies can and perhaps should assist in the process of capital formation through loans, foreign investment, and international assistance, the ultimate responsibility is said to rest with the less developed countries themselves. As W. Arthur Lewis, a sympathetic student of the problems of the less developed countries has argued, "no nation is so poor that it could not save 12 per cent of its national income if it wanted to, and this amount would be sufficient to put it firmly on the path of economic development."

In defense of their position, the proponents of this perspective point out that those economies that are doing best among the developing countries are precisely the ones that have put their own houses in order and participate most aggressively in the world economy: Hong Kong; Singapore; South Korea; and Taiwan. These newly industrialized countries, or NICs, have received infusions of capital and technology from the advanced countries and have flourishing export markets. The least integrated economies such as Albania and Burma, on the other hand, are among the most backward. Even Communist China has come to realize its need for Western assistance. Eastern Europe as well as the Soviet Union itself seek Western capital and technology. A primary reason for this continued dependence on Western technology, liberals would argue, is the failure of the socialist economies to remove the domestic impediment to economic growth and technological innovation.

Beyond this general agreement on the primacy of internal factors, liberal development theories differ profoundly among themselves on the appropriate strategy that a less developed economy should follow. In the first place, they disagree whether there exists a set of rather definable stages through which a developing economy must progress, or whether there are as many routes to development as there are national experiences. Second, they stress either balanced or unbalanced growth as the proper means for breaking out of historic poverty. Third, they differ on the degree of emphasis to be given to agriculture or industry. Fourth, they have differing views on the nature and the extent of the contribution developed economies should or can make to the efforts of the developing countries. Fifth, they place different emphases on efficiency and equity in the process of economic development. And, sixth, they differ on the role of the state in the process of economic development. These and similar issues constitute the subject of economic development as treated by liberal economists.

In summary, due to the absence of a commonly accepted body of theoretical ideas, the debate over economic development among liberal economists centers upon a set of strategic choices and alternative routes to economic development. That is, the intellectual differences among these economists revolve around the question of the appropriate set of economic policies to achieve an efficient economy. In contrast to underdevelopment theorists, however, they share a conviction that the two foremost causes of international poverty are (1) national politics and (2) inadequate incorporation of the less developed countries into the world economy. The proposed solution is the creation of an efficient domestic economy and participation in the international division of labor.

The Marxist Perspective on Economic Development

Karl Marx and Frederick Engels were first and foremost theorists of Western economic development; the bulk of the corpus of their work was devoted to the transition of European society from feudalism to capitalism to socialism and the elaboration of the inherent laws of capitalist development. In several places, however, they formulated what one can consider a theory of economic development applicable to the less developed economies. Subsequently, Lenin and later nineteenth-century Marxists extended these nascent ideas and formulated the Marxist theory of capitalistic imperialism.

Marx saw capitalism as a world-wide economic process. Its distinctive feature in contrast to prior economic system was its dynamic and expansive nature. It had originated in eighteenth-century Great Britain and by the middle of the nineteenth century had expanded to Western Europe. In time through imperialistic expansion it would incorporate into itself the entire world and would bring all societies under its mode of commodity production. Its historic mission was to develop the forces of production throughout the world. When this task of transformation was completed, capitalism would have fulfilled its assigned role and its rationale would be extinguished thereby giving way to its successor, the socialist system.

The evolution of Western civilization, according to Marx, passed through a set of relatively well-defined stages. Initially, there were the ancient economies of primitive commodity production such as characteristic of ancient Greece. This stage had been followed by the feudalism of the Middle Ages. Next, came the capitalist mode of economic production to be followed by socialism and eventually communism. The driving force at each stage had been the class conflict between the owners of the means of production and the dispossessed. It was the dialectics of this class conflict that moved history from one stage to its successor.

When Marx turned his attention outside the European continent to Asia, the Middle East, and elsewhere – as he was forced to do in response to increasing colonial clashes – he discovered that his theory of European development did not apply. In these immense agglomerations of mankind the pre-capitalist stages did not exist; there appeared to be no stages identifiable with the ancient and feudal mode of production. These civilizations, moreover, were void of any internal mechanism of social change. There was no class conflict that could drive them from one stage of social development to the next. They were in fact stuck historically.[1]

In order to account for this anomaly, Marx innovated the concept of the 'Asiatic mode of production'. What characterized this mode of production was (1) the unity and relative autarchy of agricultural and manufacturing production at the village level and (2) the existence at the top of society of an autonomous and parasitic state separated from the rest of society. This conservative social structure was held to be responsible for the millennia of social and economic stagnation suffered by these non-Western societies. There existed no internal forces to move these societies forward historically. An external force, Marx believed, was required. This force was Western imperialism.

The historic mission of capitalistic imperialism, Marx argued, was to smash the conservative Asiatic mode of production that held back the modernization of what we would today call the Third World. In an essay, *The Future Results of British Rule in India (1853)*, which contemporary dependency theorists might read with profit, Marx argued that British imperialism was necessary for the modernization of Indian civilization.[2]

Through the introduction of modern technology, imperialism creates in the colonialized areas a set of opposed classes and thereby implants the mechanism that will move the society toward economic development. Once the Asiatic mode of production was eliminated the forces of capitalist accumulation and industrialization would be released to do their work in transforming the society and put it on the track of historical evolution. Thus, the creation of a railway system by the British in India, Marx believed, was the "forerunner of modern industry".[3] Without Western imperialism the economies of Asia and Africa would remain in their state of stupor forever. Imperialism was immoral, Marx believed, but it was also a progressive force.

In his attack on the evils of capitalist imperialism, Lenin carried further this classical Marxist view that Western colonization was progressive. Like Marx, and unlike contemporary neo-Marxist theorists of imperialism and dependency, Lenin regarded colonialism and neocolonialism as progressive and necessary for the eventual modernization of less developed countries. The export of capital, technology, and know-how to colonies and dependencies, he argued, would greatly further their economic development. In equal measure, as the advanced capitalist countries exported capital and technology to their colonies their home economies would become rentier economies and their industrial technological bases would stagnate (Lenin, 1917):

> The export of capital greatly affects and accelerates the development of capitalism in those countries to which it is exported. While, therefore, the export of capital may tend to a certain extent to arrest development in the countries exporting capital, it can only do so by expanding and deepening the further development of capitalism throughout the world.

Lenin argued that the inherent contradiction of capitalism was that it develops rather than that it underdevelops the world. The dominant capitalist economy plants the seeds of its own destruction in that it diffuses technology and industry, thereby, undermining its own position. It raises up against itself foreign competitors that have lower wages and can out-compete the more advanced capitalist economies in world markets. The intensification of the economic competition between the declining and rising capitalist powers leads to economic conflicts and imperial rivalries. Such he believed was the fate of the British-centered liberal world economy of the nineteenth century. Similarly, many Marxists today argue that as the American economy is increasingly pressed by rising foreign competitors, a similar fate threatens the U.S.-centered twentieth-century liberal world economy.

In summary, orthodox Marxism from Marx to Lenin believed that capitalism develops the world. But, in contrast to liberalism, it does not do so evenly, continuously, and without limit. It is a process laden with conflict, inherent instability, and that one day would reach its limit necessitating a transition to socialism and communism. Lenin firmly believed, however, that capitalist imperialism would give the 'coloured races' of the world the tools for their emancipation; the incorporation of non-Western societies into the world economy through trade and investment would lead to their economic development. The fact that this has not yet happened and the colonies, decades following their political emancipation, remain economically dependent and underdeveloped has given rise to another perspective: dependency or underdevelopment theory.

THE DEPENDENCY OR UNDERDEVELOPMENT POSITION

The proliferation of dependency or underdevelopment theories has been a response to the fact that, although the former European colonies have achieved political independence, they are seen as not developing and as remaining economically subordinate to the more advanced capitalist economies. Contrary to the belief that political independence and decolonization would automatically bring economic development, most countries in black Africa, Asia, the Middle East, and Latin America continue to be economically and technological dependent. They continue in their traditional role of exporting commodities and raw materials in exhange for manufactured goods. Many have become penetrated by large multinational corporations. Rather than progressing into higher stages of economic development, certain of these countries are in fact slipping backward and increasing their reliance on advanced economies for food, capital, and modern technology. Whatever the condition of particular less developed economies, underdevelopment theory places the responsibility for this situation on the external world economy and not on the less developed countries themselves.

The essence of all underdevelopment theories is that the international capitalistic economy operates *systematically* to underdevelop and distort the economies of the less developed countries. The argument is not simply a matter of the activities of one developed economy or multinational corporations inflicting damage on a less developed economy through its trade policies or other actions; this occurs frequently in international economic relations. It is rather that this damage to and exploitation of the less developed economies is an inherent feature of the normal, everyday operations of the world market economy. The very nature of the world economic system itself, according to underdevelopment theory, is detrimental to the interests of the poorer countries. From this perspective it follows that the rich who control the world economy are responsible for the poverty of the Third World and, therefore, have a moral obligation to alter their policies and make amends.

The attempts to explain the profoundly disappointing and frequently deteriorating economic performance of the less developed economies have undergone an important transformation over the past several decades. The initial efforts to account for the lack of Third World progress were associated with the researches of scholars such as Ragnar Nurkse, Gunnar Myrdal, and Hans Singer and especially with the work of the United Nations Economic Commission for Latin America (ECLA) under the leadership of Raúl Prebisch.[4] Their *structuralist* theory of underdevelopment focused on the features of the contemporary world economy that allegedly restrict the development prospects of less developed economies, in particular the deteriorating terms of trade for the commodity exports of the less developed countries. The proposed solution of these structuralists was a reform of the international economy and the adoption of development strategies based on import-substitution – that is, the proper path of development for the less developed economies was to industrialize rapidly and to produce for themselves products formerly imported from the more advanced economies. The subsequent failure of this import-substitution strategy and the onset of the Vietnam War led in the late 1960s and early 1970s to the rise of present-day dependency theory, a far more radical interpretation and proposed solution to the problems of the less developed countries.

The most concise and frequently quoted definition of dependence remains that of the Brazilian scholar, Theotonio dos Santos (1970):

> By dependence we mean a situation in which the economy of certain countries is conditioned by the development and expansion of another economy to which the former is subjected. The relation of interdependence between two or more economies, and between these and world trade, assumes the form of dependence when some countries (the dominant ones) can expand and be self-sustaining, while other countries (the dependent ones) can do this only as a reflection of that expansion, which can have either a positive or a negative effect on their immediate development.

Although there are many varieties of dependency theory, they all combine elements of traditional Marxism and economic nationalism. From Marxism, dependency theorists take their analysis of capitalism, particularly the Marxist theory of capitalistic imperialism. From the theorists of economic nationalism they take their political program of nation-building. Although different dependency theorists lean in one direction or another – toward Marxism or nationalism – they all tend to have a common set of assumptions and explanations regarding the causes of and the solution to the problems of less developed countries. This position is captured by the statement of Andre Gunder Frank "that it is capitalism, both world and national, which produced underdevelopment in the past and which still generates underdevelopment in the present".[5] As Thomas E. Weisskoff has put it, "the most fundamental causal proposition [associated] with the dependency literature is that dependence causes underdevelopment".[6]

Whereas liberals define underdevelopment as a *condition* in which a nation may find itself because it has not kept up with the front runner, dependency theorists see it as a *process* in which the less developed economies are caught because of the inherent relationship between developed and underdeveloped nations. For the latter, development and underdevelopment constitute the two parts of a system that generates economic wealth for the few and poverty for the many, or what Frank (1970) has called "the development of underdevelopment". In contrast to liberals who stress the dual nature of domestic and international economies – i.e. the contrast between the modern sectors integrated into the national and international economies and the backward, isolated, and inefficient sectors – dependency theorists argue that there exists one functional and integrated whole. The underdeveloped periphery is backward and underdeveloped because it is systematically exploited and kept from developing by international capitalism and its domestic reactionary allies in the Third World economies themselves.

This functional or organic relationship between the developed and underdeveloped countries, it is argued, was first created by colonialism that restricts the colonials to commodity exports and limited industrialization. Since the achievement of formal political freedom, a neocolonial and impoverished status has been maintained by economic and technological forces. These forces operate to concentrate wealth in the metropolitan countries rather than to diffuse it to the less developed nations in the global periphery. Contrary to the liberal view that a time lag underlies the phenomenon of underdevelopment and that the gap between rich and poor will disappear as Western economic methods and technology diffuse all over the world, the essence of this position is that "underdevelopment followed upon and did not precede development".

Dependency theory arose in the mid-1960s, in part as a response to the seeming failure of the structuralist analysis and prescriptions of Prebisch and others. In particular, dependency theorists argue that the import-substitution industrialization strategy of the structuralists failed to produce sustained economic growth in the LDCs because the traditional social and economic conditions of these countries have remained intact; the neo-colonialist alliance of indigenous feudal elites and international capitalism was in fact reinforced by the strategy. The results, according to dependency theory, are an increased maldistribution of income in the dependent economy, domestic demand too weak to sustain continued industrialization, and ever greater dependence on the multinational corporations that took advantage of the import-substitution policies. As a consequence, the less developed countries have lost control over their domestic economies (denationalization) and are ever more dependent on international capitalism. The solution to this situation, dependency theorists argue, is a socialist and nationalist revolution that would promote an equitable society at the domestic level and an autonomous nation at the international level.

Despite its many variations, there are essentially three major components to dependency theory. The first is an analysis of the nature and dynamics of the capitalist world system. The second concerns the relationship of advanced capitalist and less developed countries. The third focuses on the internal characteristics of the less developed countries themselves. Each of these aspects of dependency theory will be considered in turn.

The first ingredient of dependency theory is the original Marxist critique of capitalism set forth by Lenin and others. The laws of the motion of capitalism and the contradictions existing in a capitalist economy force it to expand into the less developed periphery of the world economy. The falling rate of profit, the overproduction of goods, and the requirements of accumulation necessitate that advanced capitalist economies acquire vents for surplus capital and markets in the Third World. The survival of capitalism forces it to hold the less developed countries in economic bondage. International capitalism creates a hierarchical structure of domination between the industrial core of the world economy and the dependent, exploited periphery. Thus, imperialism and exploitation today, as in the past, remain inherent features of the capitalist mode of production.

Dependency theory differs, however, from traditional Marxist analyses of capitalist imperialism in three important respects. The first is the substitution of economic for political means of subordination; whereas Lenin believed that political control was the principal feature of capitalist imperialism, in dependency theory formal colonialism has been replaced by economic neo-colonialism and informal control. The second difference is that dependency theorists reject the classical Marxist view that imperialism develops the 'colonized' economy to the point where it can cast off its bonds; even if development does take place, as long as an economy is dependent, it cannot escape its shackles as predicted by Lenin's law of uneven development.[7] And, third, in the contemporary world economy the principal instrument of capitalist domination and exploitation has become the multilateral corporation. The great corporations of the world are said to have replaced *haunt finance* and the colonial governments that dominated the less developed countries in the past.

The second principal component of dependency theory involves the relationship of the advanced capitalist economies and the less developed economies. Although world capitalism and external factors are held to be responsible for underdevelopment, the advocates of dependency theory differ on the precise mechanism that has brought about this condition. In seeking an answer to the causes of underdevelopment, three general positions can be found in the dependency camp: the 'exploitation' theory, the doctrine of 'imperial neglect', and the concept of 'dependent or associated development'.

The 'exploitation' theory holds that the Third World is poor because it has been systematically exploited (Amin, 1976). Through the mechanisms of trade and investment, a transfer of wealth has taken place from the less developed to the developed capitalist economies. The underdevelopment of the former is functionally related to the development of the latter. The operation of the modern world system has functioned to permit the advanced core to drain the periphery of its economic surplus. As a consequence, dependence does not merely hold back the full development of the Third World. The effect of dependency is actually to immiserize the less developed economies and make them worse off than they would have been if they had been left alone and had been free to develop on their own.

The 'imperial neglect' position takes a decidedly different view regarding the effect of the world economy on the less developed economy and perhaps should not even be classified as dependency theory (Brown, 1979). The problem of the less developed economies and most certainly of the least developed ones, it argues, is that they were bypassed by the forces of capitalist imperialism. The expansion of world capitalism through trade, investment, and European migration created an international division of labor that favored some lands and neglected others to their detriment. In particular, nineteenth-century capitalist imperialism laid the foundations for industrial development through the stimulus of international trade and infrastructure investments (port facilities, railroads, and urban centers) in a privileged set of less developed countries, most notably, the so-called 'lands of recent settlement'. Elsewhere, capitalism's penetration and impact were insufficient to destroy archaic modes of production and thereby open the way to economic progress.* The lament of those bypassed is "Why didn't they colonize us?" Even today the investments of multinational corporations (MNCs) bring industry to a relatively few less developed countries and completely neglect the great majority. Thus, the world capitalist economy is held to be ultimately responsible for underdevelopment because of the patterns of trade and investment it has fostered have had a differential impact on the global periphery.

The 'dependent or associated development' school is the most recently evolved interpretation of dependency theory (Evans, 1979). In the light of the rather spectacular economic success of the newly industrializing countries such as Brazil, South Korea, and Taiwan, this position holds that dependency relations under certain conditions can lead to rapid economic growth. It argues, however, that this dependent-type of economic growth, characteristic of the so-called NICs, is not *true* development because it does not lead to national independence and has very detrimental effects on the less developed country.

Economic dependency as a conditioning situation for development is said by one author or another to have the following evil consequences:

1. Overdependence upon raw materials exports with fluctuating prices causes domestic economic instability.
2. Dependent development leads to the maldistribution of national income. It creates in the elite inappropriate tastes for foreign,

luxury goods, and neglects the true needs of the masses. Social inequalities and cultural imperialism fostered by dependent development thus reinforce the domination of external capitalism.

3. Manufacturing investments by MNCs and dependent industrialization have the effect of creating an inefficient branch plant economy with high production costs, destroying local entrepreneurship and technological innovation, and bleeding the country of repatriated profits. These foreign firms gain control of key industrial sections and crowd out local firms in capital markets. They bring into the country inappropriate technology, i.e. capital-rather than labor-intensive. An international division of labor is created between the high-technology industries of the core and the low-technology industries of the periphery. As a consequence, dependent development prevents autonomous or self-sustaining development based on domestic technology. Moreover, as the MNCs pay higher wages than domestic employers, they distort the local labor market causing waste and unemployment. And, finally, reliance on foreign capital tends to encourage authoritarian-type governments that cooperate with and give these corporations the political stability they demand. Thus, whatever its immediate economic benefits the costs of dependent development to domestic welfare and national autonomy are much too great.

Dependency theorists thus argue that this type of dependent or associated development is not *true* development. It results in income inequalities, cultural imperialism, and other evil effects. It does not lead to the economic and political independence of the country.

The third major component of dependency theory is a quasi-Marxist analysis of the dependent economy; it is this aspect of dependency theory that distinguishes it most from what its adherents regard as the reformist, bourgeois position of the structuralists. The latter, dependency theory holds, do not go far enough in their analysis of or prescriptions for the dependent economy. Specifically, dependence theory holds that the dependent country is fastened to the world economy by a transnational class linkage. An alliance of convenience and common interest exists between the centers of international capitalism and the clientele class that wields power in the dependent economy. This parasitic or feudal-capitalist class is composed of agrarian interests, the military, and the indigenous managers of the MNCs who have a vested interest in maintaining the links with international capitalism and in preventing the development of an independent and powerful industrial economy through social and political reform; even the strategy of import-substitution promoted by the structionalists, it is argued, served only to increase further the foreign hold over the economy. Tariff barriers and other policies encouraged direct investment and the establishment of domestic subsidiaries by multinational corporations. This co-opted local elite, dependency theorists argue, resists the loss of their privileges and is kept in power by the forces of world capitalism.

The crux of the attack by dependency writers on established elites in the Third World is that the cooperation of the latter with international capitalism and the integration of the society into the world economy thwarts sound reform, the economic development, and political independence of the society. The structuralists, they argue, failed to appreciate that a social revolution is required to eliminate the 'soft state' and create a 'hard state' capable of resisting foreign domination. A socialist regime is necessary to produce the necessary social discipline for economic development and, at the same time, through the redistribution of wealth, to create an equitable society capable of self-sustaining economic growth and national independence.

The conceptions of development and underdevelopment held by dependency theorists are in reality political rather than economic ones. What they seek is not merely the economic growth of the economy but the transformation and development of the society in a particular economic and political direction. Their objective is to create an independent, equitable, industrialized nation-state. From their perspective, the only true development is autonomous national development.

The solution to the problem of dependence underdevelopment advocated by the proponents of dependency theory is to the destruction of the linkage between international capitalism and the domestic economy. This requires the emergence of a revolutionary national leadership that will overthrow the clientele local elite and replace it with one committed to the autonomous development of the economy. This new elite will dedicate itself to the industrialization of the economy, the prompt eradication of feudal privileges, and the achievement of social and economic equity. Though the replacement of capitalism by socialism and the course of self-reliant development, the new elite would create a just society and strong state. In summary, the theory of dependence is first and last an ideology of nation-building. It falls within the statist traditions of eighteenth-century mercantilism and nineteenth-century economic nationalism.

AN EVALUATION OF DEVELOPMENT STRATEGIES

Every theory of poverty, however elaborate and sophisticated it might appear, can be reduced to one or a combination of three formulations: The first is that the poor are poor because they are inefficient; this is essentially the position of economic liberalism. The second is that the poor are poor because they are powerless or exploited; this is the argument of most contemporary Marxists and dependency theorists. The third is that the poor are poor because they are poor; that is, they are caught in a vicious cycle of poverty from which they cannot escape; this is the view of traditional Marxists and present-day structuralists (Nurske, 1953). For traditional Marxists, the function of capitalist imperialism was to break this cycle and place the less developed economy on the road to economic development. For structuralists, the reform of international institutions is to perform the same function. In brief, the development strategy advocated for a less developed country rests largely on which of these several interpretations one believes to be correct.

In this article, these conflicting interpretations have been analyzed primarily from the perspective of the significance of the world economy for the distribution of wealth and industry among countries. Liberals argue that the integration of an economy into the world economy is beneficial to its long-term economic growth and development because the flows of trade, capital, and technology increase its efficiency. Contemporary Marxists and dependency theorists, on the contrary, believe that integration is detrimental and causes the underdevelopment of the less developed economies. Structuralists believe that the rules and institutions of the world economy must be reformed if it is to operate to the advantage of the poor countries of the globe.

An evaluation of these theories is very difficult if not impossible for several reasons. They are all much to imprecise and more in the nature of prescriptive rather than scientific statements; each can argue that its prescriptions have never really been tried and therefore not adequately tested. Moreover, we are dealing with too short a span of time to declare the success or failure of various development strategies. The most serious obstacle, however, is that these theories have very different objectives and definitions of economic development. If taken on its own terms, each theory must be judged by a different set of criteria. Thus, whereas liberalism tends to define economic development as an increase in wealth per capita regardless of how that wealth is generated and distributed or its implications for national autonomy, dependency theorists and structuralists tend to define economic development in terms of self-sustaining industrialization and increased national power.

The fundamental question of concern here is the relevance of these theories and the development strategy each entails for the power and independence of the newly emerging nation-states. This article accepts, if you will, the position of dependency theorists and structuralists that the 'name of the game' is what it was to Alexander Hamilton, Friedrich List, and other economic nationalists, namely, that of state-building. Thus, the appropriate question is the following: Under the conditions of the late twentieth century and based on the limited evidence available, what has been the best strategy for a less developed economy to pursue, either singularly or in alliance with other countries, in order to become a unified and powerful nation?

The purpose of the following discussion, therefore, is to analyze and evaluate the development, or what underdevelopment theorists might call the emancipation, strategies that less developed economies have in fact pursued over the past several decades. If one excludes those few countries such as Burma or Liberia that appear to have opted out of the game of development altogether, these strategies range from that of self-reliant or autonomous development advocated by dependency theorists to aggressive participation in the world economy. Obviously a brief discussion of each must be very incomplete and judgments rendered are tentative at best. After all, the historical drama of state-creation among the less developed countries is only just beginning.

Autonomous or Self-Reliant Development

Both structuralists and dependency theorists have advocated a development strategy based on self-reliance. For structuralists this has meant an emphasis on import-substitution, industrialization, high tariff walls, and a reform of international institutions. Dependency theorists go further and argue that national autonomy, and self-reliant development require a social transition from a feudal-capitalist society to a socialist one. Has this strategy worked out in actual practice?

Import-substitution industrialization was first begun in Latin America and certain other less developed countries during the Great Depression of the 1930s, and was accelerated by the Second World War. With depressed prices for their commodity exports and cut off from manufactured imports from the industrial countries, many less developed countries began to develop their own manufacturing industries. Although this strategy has led to varying degrees of industrialization, its results in certain important respects have been disappointing. For a number of reasons, it has resulted in the creation in most countries of an inefficient and high-cost industrial structure; governments encouraged the establishment of too many industries in which their countries had no comparative advantage.

The reasons for the failure of this strategy were numerous. The relatively small size of national markets led to the creation of uneconomic plants. Excessive protectionism further weakened incentives to improve the quality of production. Of equal importance, the need to import industrial technology and capital goods caused massive balance of payments and debt problems. As a consequence, by the 1980s, most LDCs have come to appreciate that a strategy of industrialization based on national self-sufficiency is inadequate and emphasis must be placed on efficient industries capable of competing in world markets (Strange, 1985).

The alternate route of autonomous development via domestic transformation advocated by dependency theorists have been chosen at one time or another by Cuba, Tanzania, and China. Self-styled socialist or communist countries, they sought to minimize their involvement in what they regarded as the hostile, imperialist world capitalist economy. Whatever one may think of the internal politics of all three, Cuba and China at least have achieved a degree of social and economic welfare that is noteworthy among less developed countries, although not more equitable than Taiwan or South Korea. The economic success of all three, however, can only be described as disappointing. Cuba's economy has changed little since it broke with the West; its exports continue to be mainly sugar, tobacco, and other commodities. Its economy is highly subsidized by the Soviet Union for political reasons; in effect it exchanged one set of dependency relations for another. Tanzania's economic performance is dismal to say the least; it lags behind its neighbor, Kenya, which has chosen a more openly capitalist route to development. Tanzania has in fact become highly dependent on South Africa. It is in China, however, that one must look for an evaluation of the strategy of autonomous development because of its economic scale and strong state.

Although China received in the 1950s and 1960s substantial infusions of Soviet aid, the Chinese under Mao Tse-tung committed themselves to a course of self-reliant development. They would modernize their economy outside the framework of the capitalist world economy, mobilizing the capital from their own labors, and creating their own technology. Chinese industrialization would be based on home-ground, labor-intensive technology for a mass market. This self-reliant strategy was launched by Mao Tse-tung with the Great Leap Forward (1958–1961). Sympathetic Western observers had nothing but great praise for the backyard iron works that came to symbolize this massive effort to modernize China. Enthusiasts proclaimed the wisdom and the success of what they labeled the Chinese model of economic development and recommended it to others who wished to escape the yoke of international capitalism.

The Great Leap Forward proved to be a stumble for the Chinese economy. This self-imposed disaster was further accentuated by the onset of the Sino-Soviet split and the Russian effort to sabotage the Chinese economy by removing their technicians and eliminating all aid to China. There then followed the Cultural Revolution that devoured itself, causing further damage to the economy and the scientific-technical foundation of the country. For two years China tore itself apart and slipped backward. Alienated from both East and West, the leadership that emerged after the death of Mao came to realize that China could not do it alone.

It is much too soon to know what the effects will be of China's re-entry into the world economy. The opening of China to Western investment, the transfer to it of modern technology, and the enlargement of trading activities are really just beginning. But the moral of the tale is that the strategy of autonomous development advocated by the more extreme of the dependency theorists thus far has been demonstrated to hold little promise for the less developed economies. If a China with its advantage of a strong state, abundant resources, and a relatively large internal market cannot be self-reliant, what hope is there for a Tanzania? Even the Soviet Union, it should be remembered, had a well-developed industrial base prior to the Revolution and infusions of Western technology continued under the New Economic Policy (NEP) of the 1920s. As the Yugoslav writer, Milovan Djilas, once put it to the present writer, no communist society has or can fully develop without the assistance of capitalist economies. More generally, in varying degrees all development is dependent development and no society can develop without acquiring at least the productive technology of the most advanced economies.

Economic Regionalism

A second strategy which has been employed by developing economies, and others as well, is economic regionalism. Several approaches have been tired or proposed. First, through economic cooperation and alliance a group of countries in a geographically restricted area seek to improve their overall position with respect to more advanced economies. This cooperation may take several forms of which the following are the most important:

(1) the formation of a free trade area or customs union that increases the scale of the internal market and, at the same time, protects domestic producers against outside competitors;
(2) the enactment of investment codes and agreements that strengthen the bargaining position of the members *vis-à-vis* developed economies and especially their MNCs, such as the Andean Pact;
(3) the development of regional industrial policies that rationalize and concentrate local, fragmented companies into regional champions (public or private) in such fields as textiles, steel, and automobiles.

In response to the disappointment with the strategy of import-substitution, the United Nations Conference on Trade and Development, under the leadership of Prebisch, began to push for a regional approach to the problem of the less developed countries. In order to overcome the problem of small national markets and improve their bargaining position *vis-à-vis* the multinational corporations, it was argued that the less developed countries should form regional monopolies in important industrial sectors, create a regional division of labor based on specialization, and formulate a set of rules toward outside multinational corporations.

Without exception the various attempts to establish regional economic entities among the less developed countries since the end of the Second World War have failed. These attempts have taken place in

West Africa (Kenya, Tanzania, and Uganda), the Caribbean, Southeast Asia, Central America, and most significant of all, the Andean region. These efforts to create a common market have invariably been torn apart by regional conflicts and economic rivalries. Intra-regional competition for foreign investment and trade have destroyed the united front against multinational corporations. Attempts to rationalize and concentrate industries in order to create a regional division of labor have been undermined by the desire of each country to have the regional champion one of its own. In brief, the very forces of economic nationalism that prompted the initial commitment to regional cooperation have led to its destruction as each nation has sought to advance its own national interests.

There have been in fact only two relatively successful examples of economic regionalism: the European Economic Community (EEC), or Common Market, and the COMECON in Eastern Europe. They have resulted in a high degree of economic integration. Yet the unusual circumstances surrounding both endeavors and the limited nature of their success restricts their usefulness as models for the less developed countries. In each case one or other of the superpowers has played a significant role in its formation, and security motives have been of paramount importance. Furthermore, it should be appreciated that even the EEC has been unable to advance much beyond its common external tariff and common agricultural policy. Although the Soviet Union has forced its Eastern bloc members to specialize in a "socialist international division of labor", this effort has been resisted and, where possible, these economies have sought economic openings to the West. Thus, in Europe as in the less developed economies, economic nationalism plays a strong constraint on regional integration.

A second form of regionalism is the creation of special trading relations between developed countries and particular groupings of less developed countries (LDCs). The Lomé Conventions, various proposals for a Pacific basin economic zone, and President Reagan's Caribbean basin initiative are examples of this type of regionalism in which one or more advanced economies extend preferential trading and other benefits to selected countries. For example, the Lomé Conventions between the European Community (EEC) give approximately fifty African, Caribbean, and Pacific states privileged access to the Community for their commodity exports and certain types of manufactures. Without exception, however, these arrangements are interlaced with restrictions on agricultural and manufactured exports from the LDCs. In particular, they restrict exports that compete against local European products thereby limiting this type of regionalism as a vehicle of industrialization and means of escaping the dependency relationship.

In recent years, a third type of economic regionalism has been promoted by scholars and governments. This is the concept of delinking trade between developed and less developed economies and forging trade links and a division of labor among the less developed countries independent of the more advanced economies (Lewis, 1980). This concept of intra-Third World trade is an attractive one and may yet succeed. For the moment, however, the developed countries still constitute the engine of the world economy and are the major importers of LDCs exports of all types. Moreover, this delinking strategy suffers from the general weakness of economic regionalism. Individual less developed countries seek to advantage themselves at the expense of others and when possible continue beneficial trading and investment relations with more advanced economies. Frequently, they form alliances with foreign multinationals in order to acquire capital, technology, and access to foreign markets. By giving the multinationals a monopoly position in its own closed markets, a particular country seeks to draw upon the corporation's resources and enhance its own economic position. Few less developed countries are willing to sacrifice their perceived national interests to the good of Third World solidity.

Commodity Cartels

A third strategy seemingly available to certain states in the Third World is to emulate the Organization of Petroleum Exporting Countries (OPEC) and form a commodity cartel that can force a dramatic improvement in the terms of trade for Third World raw material and food exports. Such cartels have been proposed in copper, zinc, bauxite, and other commodities. In the wake of the initial OPEC success, there was much talk along these lines. In response, two schools of thought arose in the developed countries. On one side were those who spoke of the threat from the Third World, foreseeing the proliferation of southern commodity cartels that could reign havoc on the North. On the other side were those who argued that 'oil is the exception' and that no such threat existed (Krasner, 1974). The available evidence suggests that the latter position has been very much vindicated.

The success of OPEC in quadrupling the price of petroleum was due to a peculiar set of favorable circumstances. Both demand and supply factors were ripe when in 1973 the third Arab–Israeli war caused Arab oil exporters to impose an embargo on the West; this move permitted the Shah of Iran to take advantage of the situation by raising massively the price of petroleum exports. During the months just prior to the outbreak of the war, demand for petroleum and other commodities had increased greatly at the same time that accelerating inflation had reduced the real price of oil. On the supply side, no excess capacity existed that the West could tap to make up for the Arab induced short-fall as it had in earlier wars.

The success of a commodity cartel requires the existence of a producer who has excess capacity and can control world prices through increases or decreases in the aggregate supply. In 1973, this pivotal position shifted from the Texas Railway Commission (the agency that controls American domestic production) to the King of Saudi Arabia. Subsequently, the Saudis have largely dominated world energy markets. Through increasing or decreasing their production they have maintained the existence of the cartel and influenced the setting of the world price. They have operated the cartel to their national advantage and, of course, to the advantage of other producers. Despite its relative success, however, OPEC has had its own internal difficulties due to market conditions and the frequently conflicting interests of its members.

There does not appear to be any other commodity in which a situation similar to that of petroleum exists. Unlike petroleum, substitutes are readily available, albeit at a cost, with respect to almost all other commodities. With the exception of a few metals, the United States or one of its allies can produce the commodity. But more importantly there is not a single producer such as Saudi Arabia that can control the supply and hence the price. Ironically, the only two commodities that might be cartelized in the future are uranium and wheat, and in both cases the United States is one of the primary producers. Finally, it should be appreciated that although cartels may benefit certain less developed countries, as in the case of petroleum, they do so at the expense of everyone else. Thus, it does not appear that cartels in scarce commodities are a promising route to improve the lot of the less developed countries as a whole.

The Demand for a New International Economic Order (NIEO)

The failure of alternative strategies (import substitution, self-reliance, and economic regionalism) and the demonstrated success of OPEC led to the launching of a new strategy in 1974, at the Sixth Special Session of the United Nations General Assembly. At that session, a group of less developed countries (the Group of 77) led by three OPEC members (Venezuela, Iran, and Algeria) adopted a Declaration and Program of Action for the Establishment of a New International Economic Order (NIEO) that included the right of the LDCs to form producer associations, the linking of commodity export prices to the prices of manufactured exports from developed countries, the right of LDCs to nationalize foreign enterprises and gain sovereignty over their natural resources, and the formulation of rules to regulate the multinational corporation. On December 12, 1974, the General Assembly adopted these demands in the form of the Charter of Economic Rights and Duties of States.

Although this demand for a NIEO was profoundly influenced by radical and dependency critiques of world capitalism, it was generally in the spirit of structuralism. The goal of industrialism and economic development could be achieved within the framework of the world economy; it was not necessary to overthrow the capitalist system. What was required was a set of policy and institutional reforms that would make the international economic system operate to the advantage of the LDCs and enlarge their role in the running of the system. Among the most important demands for changing the terms on which the LDCs participated in the world economy were the following:

(1) Measures that would increase Third World control over their own economies, especially in natural resources;
(2) agreements to maintain *and* increase the purchasing power and terms of trade for their raw material exports;
(3) enactment of a code of conduct increasing their control over the MNCs;
(4) reductions in the cost and increased availability of Western technology;
(5) increases in the flow and liberalization of foreign aid;
(6) alleviation of their growing debt problems;
(7) preferential treatment and greater access for their manufactured goods in developed markets;

(8) greater decision-making power in the IMF, World Bank, the UN, and other international organizations, thus making these institutions their more responsive to the development needs of the less developed countries.

The essence of the initial proposal for a New International Economic Order and of subsequent reformulations is that the operations of the world economy should be made subordinate to the perceived development needs to the less developed economies. To this end various commissions and reports have advocated changes in the rules governing international trade, the monetary system, and various aid programs in ways that would benefit the LDCs. International regulations have been drafted to control the multinational corporations. In particular, the LDCs have advocated changes in international organizations – the United Nations, the World Bank, and the IMF – that would give them greater influence over the management of the world economy (Krasner, 1985).

Following an initial disarray and conflicting responses among them, the Western powers have in fact capitulated to few of these demands. The debate and the discussion, however, continue over this most concerted and significant attempt of the less developed countries to change the international balance of economic and political power. The reasons for this continuing lack of success to implement the NIEO program include the following:

1. Despite rhetorical and marginal differences in their positions, none of the developed economies is willing to make any significant concessions. This resistance has been led principally by the United States which regards the proposals either as unworkable or as contrary to its commitment to a free market economy. Although other Western countries have been more accommodating in spirit, they have supported in substance the American stance.
2. Contrary to their statements and the expectations engendered by them, the OPEC members have been unwilling to put their power and wealth at the services of other Third World states. For example, they have not used their economic resources to finance a general commodity fund or the development efforts of other than a very few countries. Instead, they have used their newly gained economic wealth to support their own nationalistic interests and have invested their financial surplus in that most capitalist of institutions, the Eurocurrency market.
3. The rise in world petroleum prices has had a devastating impact on non-oil producing countries, particularly those in the Third World.

> In addition to burdening them with high import bills, it triggered global recessions in 1973 and 1979 that undercut the rising world demand for their commodity exports. In effect, the success of OPEC undercut the bargaining power of the LDCs and in doing so blunted their demands for a New International Economic Order.

In short, the history of the NIEO demonstrates the fundamental dilemma of the less developed countries. In the name of nationalism they seek to counter the operation of the world market economy and improve their own relative position. Yet, this same nationalistic drive undermines their own efforts to cooperate and form an economic alliance against the developed countries. When one gets beyond the realm of rhetoric to the specific, as in the case of the negotiations over the law of the seas, the conflicting interests of the LDCs come to the fore. The process of nation-building, it must be remembered, is set in a competitive system of states with conflicting national interests. The less developed nations of the Third World are not exempt from the logic of this situation.

The NIEO has failed to produce the reform desired by its proponents; yet, this fact obviously does not invalidate either the basis of LDC grievances or the desirability of certain changes in the relationship between North and South. Certain of the NIEO proposals could become the basis for reforms that would improve the operation of the world economy and be of mutual benefit to both the developed and the less developed economies. For example, although the developed countries are loath to accept proposals that would raise the real price of commodities beyond their market value, it would be in their interest to stabilize the export earnings of the LDCs. One can envisage similar mutually beneficial arrangements in other areas such as those of debt relief and foreign aid. Most important, the developed economies must maintain open markets for the manufactured exports of the LDCs. It would be foolish to expect in present circumstances, however, the enactment of a set of sweeping reforms that would change the overall position of the LDCs in the world.

The Process of Uneven Development

In the light of this discouraging experience, at least from the perspective of the less developed countries, what is one to conclude regarding the validity of the argument of underdevelopment theories that the capitalist world economy operates systematically to thwart the development strategies of the Third World? One should begin by observing that a single, independent variable – the functioning of the international economy – is being used to explain three quite distinct types of phenomena found today in the Third world: underdevelopment, marginalization, and dependent development. From a simple methodological point of view, something is wrong with any theory in which a single independent variable seeks to explain three actually opposed sets of dependent variables. Let us consider, therefore, each of these alleged consequences in turn.

The first charge of dependency theory is that a number of countries have been integrated into the world economy as commodity exporters, are systematically exploited, and are held in this position by an alliance of international capital and a corrupt, local elite. This charge, it must be admitted, is not completely without foundation as there do exist many examples of this type of dependency relationship in the world today. But it should also be noted that corruption is rife throughout the Third World in every conceivable type of society and has little to do with external factors. It should also be noted that dependency relationships are characteristic of the socialist Soviet Union and its clients in the Third World such as Cuba, Yemen, and Vietnam. Further, multinational corporations, witness Gulf Oil in Angola, have no difficulty working with 'socialist' governments; political stability and not a sold-out, clientele elite is all that international capitalism requires. Most importantly, the terms of trade for commodities have shown no secular tendency to decline and many commodity producers have fared well economically. There is little evidence for the charge of the dependency theory that the international economy operates *systematically* to the disadvantage of the LDCs.

The second criticism leveled by underdevelopment theorists against the world market economy is that it has neglected and bypassed many countries in the Third World. This is correct. The process of global economic integration and development that began apace in the latter part of the nineteenth century through the expansion of trade and investment among the developed and the then less developed countries has been a highly uneven one. The simple fact is that nineteenth-century imperialism and the operations of twentieth-century multinational corporations have left much of the world's traditional economies untouched because they found too little there to be 'exploited'. This marginalization of destitute areas (the Fourth and Fifth Worlds) such as the Sahel and much of tropical Africa, however, constitutes a sin of omission rather than of commission as the dependency theorists allege. As such, the moral obligation of helping these peoples better their lot lies with the socialist bloc, the oil producers, and the well-off everywhere, and not just with the capitalist West.

The third argument of underdevelopment theorists that dependent or associated development such as characterizes the newly industrializing countries (NICs) of Brazil, South Korea, and other countries today is not 'true' development is, of course, a tautology. However, even if one accepts the position of these theorists that the objective of development is national independence and autonomous industrialization, the evidence to support their contention is weak. As certain Marxist writers have effectively demonstrated, dependent development has begun a process of sustained industrialization and economic growth in an increasing number of less developed countries. Thus, recalling the position of Marx, Lenin, and other classical Marxists that the function of capitalism is to develop the world, Bill Warren (1973) writes:

> If the extension of capitalism into non-capitalist areas of the world created an international system of inequality and exploitation called imperialism, it simultaneously created the conditions for the destruction of this system by the spread of capitalist social relations and productive forces throughout the non-capitalist world.

Thus, he returns to the thesis of classical Marxism. One can readily apply to the most successful of the NICs – South Korea, Taiwan, and other Asian countries – Marx's observation regarding the role of British imperialism in India. The relevant imperialism, however, is that of the Japanese. Just as Great Britain performed the dual role of smashing the traditional economies of these countries and implanting in them the seeds of modern industrial society, Japanese imperialism may be said to have performed a similar function in those societies. Japanese imperialism swept away their traditional economies and, on the foundations of their ancient Confucian cultures with its ideal of the subordination of the individual to the state, commenced the creation of the modern development state (Cummings, 1984). In the contemporary world, economic ties with Japan and the United States have further stimulated these economies.

In these development states of the Third World, the two prerequisites of economic success can be found (Johnson, 1982). The first is the existence of a strong state and an economic bureaucracy that set national priorities and implement a coherent industrial policy. The economic managers of these societies impose themselves between the developing economy and the powerful outside forces of the market. They deal with the multinational corporations and other foreign supplies of capital and technology. It is their task to make trade, investment, and other commercial arrangements serve the national interest. As these states have shown, the issue is not independence or interdependence but the terms under which the domestic economy is integrated into the larger world economy, and these terms are largely the result of negotiations between the state and foreign entities. But the experience of Libya, Cuba, and other Third World states reveals that a strong state in itself is not enough to assure economic development. One needs also a well-functioning market that spurs individual initiative and promotes economic efficiency. Both a strong state and an efficient market are the earmarks of the development state in the Third World and constitute the necessary ingredients of national development in the contemporary world (Hofheinz and Calder, 1982).

We are witnessing in these NICs the operation of what Leon Trotsky called the law of combined and uneven development. As Trotsky observed in his analysis of Russia's late industrialization, what occurred earlier this century in Japan, and is now taking place in the NICs, is that a strong state has encouraged the importation of foreign technology and has combined it with traditional social forms. These contemporary development states have further been benefited by a rapid growth of international trade and the openness of the world economy over the past several decades that have enabled them to pursue export-led growth strategies. Whether these favorable conditions will continue in the future will profoundly influence their ultimate success and the availability of an export-led growth strategy for other less developed countries. A strategy available to a few countries may not work if pursued by a relatively large number.

In summary, we return to the theme that the workings of the capitalist world economy develop the world, but it does so, as Marx and Lenin first noted, unevenly. In the nineteenth century, this growth process spread from Great Britain – the first industrialized country – to Western Europe, Japan, and the New World. Today, the newly industrialized countries (NICs) – Taiwan, South Korea, Hong Kong, Singapore, Brazil, Mexico, etc. – are joining the ranks of industrialized countries.

Although their development strategies have ranged from export-led growth to import substitution, contrary to dependence theory, the operation of the world economy has been a positive factor in each case. But what has been of equal importance is the capacity of the state to order its priorities and guide the forces of economic development.

CONCLUSION

If one means by dependence a conditioning factor that profoundly affects the development strategies of developing economies, then the fact of dependency can hardly be denied. Every less developed economy is certainly dependent upon fluctuating world market conditions; they must import capital, technology, and industrial know-how. Export markets are very difficult to penetrate given the advantages of powerful, established exporters and protected markets in the developed countries. These aspects of dependency surely exists. A continuum of dependency exists in which every country is more or less dependent upon others, albeit some are more dependent than others. If, however, one employs this condition of dependence as an explanation of underdevelopment, then the arguments of the dependency theorists lose most of their force. There is a tendency, unfortunately, to confuse these two meanings of dependence, even by the same writer and to make the *fact* of dependence the *explanation* of economic underdevelopment.

The less developed countries have a high degree of dependence because they are underdeveloped rather than the other way round. They are the weak in a world of the strong. They are dependent because they are underdeveloped. It is not the condition of dependence that holds them back but primarily the lack of an effective and appropriate development strategy to overcome their lack of development. Those less developed countries that have put their houses in order and have created efficient domestic economies have succeeded in achieving very rapid rates of economic growth.

Throughout the Third World, an increasing number of societies have established the political stability and social discipline that are the prerequisites for economic development. Modernizing elites in the public and private sectors have learned to exploit the opportunities provided by trade, foreign investment, and technology imports to attain a rapid rate of economic and industrial growth. As Marx predicted, capitalism tends to develop the world but, as Lenin amended this prediction, it does so in accordance with the law of uneven development. As in the nineteenth century, a process of uneven development is taking place among the less developed countries, profoundly altering their positions in the international system and their relations with one another.

What is happening today in the Third World is that the competitive nation-state system with all its capacity for good and for evil is spreading there and transforming it. The concept of the Third World evolved in response to the bipolar Cold War; its leaders rejected both the Soviet and American blocs and sought to develop themselves independently and preserve their unity as a third force. There subsequently arose among them such various pan-movements and regional organizations as pan-Arab and the Organization for African Unity. Inspired by structionalism and dependency theory, they formulated autonomous and cooperative routes to economic development and nation-building. The two ideals of political non-alignment and Third World internationalism would characterize their new world order.

Today, the Third World no longer exists, if it ever did. The notion of a homogeneous and united bloc of less developed societies is breaking down. In its place, one is witnessing a differentiation taking place in the standing and the policies of those countries. In every region, particular nation-states are forming and emerging as centers of power: Brazil, India, Mexico, Venezuela, Nigeria, Saudi Arabia, Indonesia, Vietnam, China, etc. These emergent powers are pursuing foreign policies designed to further their own particular national interests. Differences in interests and ambitions are producing conflicts and intense wars among these newly emergent powers.

In effect, the modern state system is forming in what was once regarded as the amorphous Third World. These newly organized nation-states are becoming part of the global system. Both the structuralists and dependency theorists are wrong in their belief that the less developed countries would forever remain underdeveloped and would have to cooperate to emancipate themselves. Development is taking place but it is doing so as it did in the developed West, some states becoming strong while others remain impotent. These emergent states are becoming part of the first truly global system of international relations in the history of the world.

Faculty Associate at the Center of International Studies

NOTES

[1] Shlomo Avineri (ed.), *Karl Marx on Colonialism and Modernization*. (Garden City, N.Y.: Anchor Books, 1969) is an excellent collection of Marx and Engel's writings on this subject.
[2] Reprinted in *Ibid.*
[3] *Ibid.*, p. 136.
[4] A good summary of this position is Ian Roxborough, *Theories of Underdevelopment* (London: Macmillan, 1979), Chap. 3.
[5] Quoted in Anthony Brewer, *Marxist Theories of Imperialism: A Critical Survey* (London: Routledge and Kegan Paul, 1980), p. 158.
[6] In Thomas E. Weisskopf, 'Dependence as an Explanation of Underdevelopment: A Critique'. Center for Research on Economic Development, University of Michigan. Unpublished, 1976, p. 3.
[7] Lenin was aware of what neo-Marxists today call 'dependency' relations. In *Imperialism*, p. 85, he noted the dependence of Argentina on Great Britain. He apparently did not believe, however, that this type of economic relationship was very important in contrast to formal political annexation. The classical Marxist view that capitalist imperialism develops the colony was amended in 1928 at the Sixth Congress of the Communist International in favor of the contemporary dependence theory formulation. See Jay R. Mandle, 'Marxism and the Delayed Onset of Economic Development: A Reinterpretation', *Journal of Economic Issues* **14** (1980), p. 736.

REFERENCES

Amin, Samir. 1976. *Unequal Development: An Essay on the Social Formations of Peripheral Capitalism*. New York: Monthly Review Press.

Bauer, Peter T. 1976. *Dissent on Development*, rev. edn. Cambridge: Harvard University Press.

Brown, Michael B. 1979. *After Imperialism*. New York: Humanities Press.

Carr, Edward H. 1945. *Nationalism and After*. London: Macmillan.

Cummings, Bruce. 1984. 'The Origins and Development of the Northeast Asian Political Economy: Industrial Sectors, Product Cycles, and Political Consequences', *International Organization* **38**, p. 8.

Davis, Lance E. and North, Douglass C. (with the assistance of Calla Smorodin). 1971. *Institutional Change and American Economic Growth*. Cambridge: Cambridge University Press.

Evans, Peter. 1979. *Dependent Development: The Alliance of Multinational, State, and Local Capital in Brazil*. Princeton: Princeton University Press.

Frank, Andre G. 1970. *Latin America: Underdevelopment or Revolution*. New York: Monthly Review Press.

Hirschman, Albert O. 1981. *Essays in Trespassing: Economics to Politics and Beyond*. New York: Cambridge University Press, p. 3.

Hofheinz, Roy Jr and Calder, Kent E. 1982. *The Eastasian Edge*. New York: Basic Books.

Johnson, Chalmers. 1982. *MITI and the Japanese Miracle: The Growth of Industrial Policy, 1925–1975*. Stanford: Stanford University Press.

Krasner, Stephen D. 1974. 'Oil is the Exception'. *Foreign Policy*, No. 14, pp. 68–84.

Krasner, Stephen D. 1985. *Structural Conflict: The Third World Against Global Liberalism*. Berkeley: University of California Press.

Lenin, V. I. 1917. *Imperialism: The Highest Stage of Capitalism*. New York: International Publishers, 1939, p. 65.

Lewis, W. Arthur. 1980. 'The Slowing Down of the Engine of Growth', *American Economic Review* **70**, pp. 555–564.

Lewis, W. Arthur. 1970. *Theory of Economic Growth*. New York: Harper and Row.

Nurkse, Ragnar. 1953. *Problems of Capital Formation in Underdeveloped Countries*. New York: Blackwell, p. 4.

Santos, Theotonio dos. 1970. 'The Structure of Dependence', *American Economic Review* **60**, p. 231.

Smith, Adam. 1776. *An Inquiry into the Nature and Causes of the Wealth of Nations*. New York: Modern Library, 1937.

Strange, Susan. 1985. 'Protectionism and World Politics'. *International Organization* **39**, p. 252.

Warren, Bill. 1973. 'Imperialism and Capitalist Industrialization', *New Left Review* **81**, p. 41.

ELI GINZBERG

OCCUPATIONAL MOBILITY: A PERSONAL PERSPECTIVE

1. INTRODUCTION

My contribution to this major testimonial volume celebrating the life and work of Lewis Feuer addresses a theme which he has discussed with broad and deep scholarship on more than one occasion. I hope that this personal retrospective will add a little additional understanding to a theme that has commanded the attention of the American polity for more than a half century since Lewis and I graduated in 1927 from DeWitt Clinton High School in New York City.

From the time when A. Philip Randolph first threatened to lead a march on Washington in early 1942 and President Roosevelt responded by issuing an Executive Order as his part of the bargain to get Randolph to cancel his plans, the issue of discrimination in the workplace has held a prominent place on the nation's agenda. Let there be no misunderstanding: I do not claim that anti-discrimination has commanded or received continuing attention and action from the three branches of the federal government, from other levels of government, and from the non-profit and private sectors during all of the years since Randolph made his threat. I am suggesting the more modest hypothesis that during the past four decades and more, the subject has repeatedly engaged the attention of employers, workers, and the public. During this time, government and the courts have paid attention, demonstrated initiative, and taken actions to force changes in the status quo as the legislators and administrators read the tea-leaves in the political arena.

My plan is to reconstruct, highly selectively, the environment that various minorities – religious, ethnic, racial, and sexual have encountered in higher education and in the world of work in the late 1920s and 1930s up to the Second World War; the economic and social forces that contributed to the marked reductions in the barriers they encountered in the ensuing decades; to end with a few reflections on what this retrospective can contribute to an understanding of the probable course of events that will ensue in the years that remain in this century and the period immediately beyond.

S. Hook, W.L. O'Neill and R. O'Toole (eds.), Philosophy, History and Social Action, 209–219.

2. THE NEW ERA AND THE GREAT DEPRESSION

If one follows the current national debate about the shortcomings of the American educational system, one might be inclined to believe that the schools in generations past performed satisfactorily, even at a high level of efficiency. But nothing is further from the truth. As my friend, John Fischer, the former president of Teachers College, Columbia University pointed out, the public schools never did a good job for poor youth – white majority or black minority. My rough estimate is that when Lewis Feuer and I graduated from DeWitt Clinton, the diploma recipients accounted for between one-third and two-fifths of the members of the entering class of four years earlier.

Those who stayed and earned their diplomas comprised the following to a disproportionate degree: good students who liked school and were headed for college and a minority of others who came from families that did not require that they become early earners. The majority who dropped out contained a large proportion of the whole spectrum of students – Jewish, Irish, Italian, blacks and WASPs. The labor market of that day did not require or reward an applicant because he had acquired a high school diploma. The major reason for a young person to stay the course was that he was planning to enter college.

Our classmates had four choices: Columbia, New York University, City College, or an out-of-town college. A few, but a very few, from upper income homes chose the last. Most of our class, including Lewis, entered City College. A few opted for New York University which had two campuses: one with a definite restriction on the number of Jews who would be accepted; and the other unrestricted. But each required a relatively high tuition in the range of $300–400 per annum.

Columbia College, following in the footsteps of Harvard College but without any fanfare or adverse publicity, had moved to an admissions policy which aimed at a student body broadly representative of the nation. Accordingly the percentage of Jews was set at around 10 percent and only part of this implicit quota was available to applicants from the New York City area. Columbia College did offer 10 Pulitzer scholarships, awarded primarily on the basis of scores on written examinations, and Jews represented a high proportion of the winners. I believe that Lewis never took these examinations, but the reasons are unclear to me. On the basis of his high school performance he surely would have won a Pulitzer fellowship.

I have speculated whether, had Lewis gone to Columbia College, his later development would have been significantly different. I doubt it. He simply would have had less competent instruction in philosophy, for at City College Morris R. Cohen taught undergraduates, but no one with his philosophical and pedagogic skills was lecturing to Columbia undergraduates.

Among the almost 500 men who entered Columbia College in the fall of 1927 there was one black student from DeWitt Clinton, George Gregory, who went on to lead the Columbia basketball team to many victories. If we set up as an example a simple dichotomy – WASPs versus all others – the total number of 'all others', including the 10 percent of Jewish students plus a few blacks and Asiatics and a limited number of sons of Irish, Italian and Eastern European parents, could not have exceeded 25 percent.

Columbia University had a small undergraduate college; its reputation rested on the strength of its graduate departments and professional schools. In general, admission to the graduate faculties was not controlled on the basis of race or similar criteria, but in the Medical School and, to a lesser degree, in some of the other professional schools, the number of minorities accepted was definitely limited.

So much for Columbia University admissions. What about academic staffing and employment opportunities for college graduates and others with advanced degrees?

To oversimplify: even in a public institution such as City College, tenured Jewish professors such as Cohen and Klapper (later the first president of Queens College) were exceptions. At Columbia and the other Ivy League institutions the phenomenon of tenured Jewish professors was even more rare. My rough estimate is that, at the end of the 1920s, the number of Jewish professors with tenure in all of the divisions of Columbia did not exceed 10. I am well nigh certain that the university had no black tenured professor and I do not recall a woman professor. I studied with Ruth Benedict in 1930–1931 but despite the support of Franz Boas, her mentor, it was years before she received tenure.

To oversimplify once again: the external employment market had the following characteristics. Jewish graduates of the College of Physicians and Surgeons had great difficulty in obtaining an internship in any but a Jewish or public hospital. The major law firms in the city did not interview and surely did not hire Jews no matter how high their ranking in law school. Engineering positions were largely closed to Jews, which explains why so few pursued an engineering curriculum. The large corporations other than those in retailing, communications, international banking and a few other service sectors also were closed to Jews.

We know that the same exclusionary policies were directed against women and blacks and to a somewhat lesser degree against Irish, Italian, and other minorities. The corporate world remained a preserve of the WASPs.

One important change that occurred during the Great Depression of the 30s was the considerable number of Jewish college graduates and graduates with advanced degrees who sought and obtained employment in either the rapidly burgeoning federal bureaucracy or in several sectors of state and local government in New York, particularly in the large educational system but also in other departments such as the expanding new agencies involved with welfare.

Another impact of the depression years that had a delayed effect were the considerable numbers of talented young Jews who prolonged their higher education on the assumption that they would be better positioned for employment if they acquired additional degrees. This strategy paid off when mobilization for the Second World War got under way. Many with scientific and engineering training were able to break into the corporate world because of the much enlarged R&D and related activities that were generated by our preparation for, and later entrance into, the war.

3. THE SECOND WORLD WAR AND AFTER

The war itself presented a frontier of opportunity for all of the groups which had been the victims of discrimination. Those who gained most were married women, who were avidly recruited to take the place of men who were drafted. For the first time blacks were able to apply and be hired for jobs from which they had previously been automatically excluded. Roosevelt's Executive Order and the exigencies of the marketplace with its shortage of workers speeded the process. In my view, 1942 was the first year in the economic emancipation of blacks. The 1940 Census still showed most of them (4 out of 5) living in the South, living and working as tenant farmers, crowded at the very bottom of the income distribution.

In the wake of the nation's struggle against Hitler many prior-existing attitudes and practices against Jews, Catholics, and other ethnic minorities began to erode. I had earlier been the beneficiary of such a development when the authorities at Columbia obtained the agreement of the William Bayard Cutting family to remove the criterion of 'preference to children of American parentage' for one of the university's prestigious traveling fellowships.

The performance and experiences of the Negro soldier is recounted in my *The Negro Potential*, Columbia, 1956. Discrimination in the Armed Forces was not substantially reduced until after the re-election of President Truman in 1948. Since I played a role in speeding the process in Europe, as Consultant to the Secretary of the Army, we can date the crucial year as 1951–1952 when the generals accomplished in about 10 months what they had earlier assured me would take between 25 and 50 years. Once the Commander of the U.S. Army in Europe headquartered in Heidelberg learned by communicating out of channels with the Pentagon that the message which I delivered was real, he decided to move expeditiously to desegregate fully all aspects of his command from living quarters to swimming pools; he redistributed among white units all previously segregated black soldiers and sergeants as well as the few black officers who had been commissioned.

Other developments of the war and early post-war years revolved around the new prosperity that provided important avenues for economic betterment, including new and good opportunities for small businessmen as well as for salaried employees. The fact that the birth rate had dropped precipitously during the depressed 1930s proved a boon in the 1950s for the generation ready to enter college and professional schools. The large decrease in the age cohort made it relatively easy for qualified Jews and members of other minority groups, blacks and women still excepted, to gain entrance to prestigious institutions from which they had earlier been excluded. The most striking change came in the much greater ease of ethnic and religious minorities in being accepted by leading medical schools.

The booming economy, the substantial migration of blacks out of the South, the economic invigoration of many southern cities, all opened important new employment opportunities for blacks in blue-collar, white-collar, and service jobs, reinforcing the substantial gains they had made during the war years. This is not to say that job discrimination against blacks had disappeared by the mid-1950s – far from it. However, as I pointed out in an address (1956) to the directors of the

National Urban League, the opportunities in the labor market were opening up at a rate faster than qualified blacks were coming out of the high school and college stream.

As far as Jews were concerned, the job environment in academe had taken a major turn for the better. In the scramble for quality, especially in the sciences where objective criteria had more sway, prestigious institutions were willing to hire and promote Jews into tenured positions at a rate many times greater than in the pre-Second World War period. I have long believed that the booming 1950s to 1960s presented so many good opportunities to WASPs that they were able to realize their occupational and income goals without having to worry about competitors from less favored groups. There was more than enough room for both.

But the boom was not sufficiently prolonged and by the early 1960s there had not been substantial broadened opportunities for blacks and women, surely not at the higher levels of the occupational hierarchy. It is worth recalling that when President Kennedy sent his proposed Civil Rights bill to the Congress in 1963 it did not contain a section on employment discrimination. A section was finally included because of the work of interested legislators, especially Senator Joseph Clark of Pennsylvania. I recall with pleasure and satisfaction my presentation before his sub-committee wherein I, among others, emphasized the urgent need for this section.

It is also worth recalling that the extension of the anti-discrimination legislation which covered blacks to women came about as the result of a legislative maneuver in the House. Judge Smith of the Rules Committee thought that he could kill the bill by adding women to the groups to be protected; he thus inadvertently assured the required votes for passage.

Legislation, court decisions, and administrative actions all contributed to reducing discrimination in the labor market with respect to both initial hiring and promotion. But in my opinion it was the strong and continuing demand for labor at all levels that was responsible for much of the change that occured. Moreover, I would give substantial weight to the behavior of senior executives of many of the nation's leading corporations who concluded that the time had come to reposition their enterprises to take full advantage of the enlarged pool of applicants and at the same time to do what was right in the eyes of the law and the new directions of national opinion.

The steadily growing participation of women in the labor force, from one-third of all adult women at the end of the Second World War to over one-half today, inevitably assured that the initial national effort to reduce discrimination against blacks would become intertwined with the intensified demands of women who accounted for three of every five new workers, for equal opportunity. With the federal and some state governments pressuring employers vigorously to lower their discriminatory barriers, it was not surprising that many corporations found it easier and more congenial to open professional job categories to women while lowering the barriers against blacks.

It was much easier for leading medical, law, and graduate business schools to admit a larger number and proportion of qualified women candidates than to find significant numbers of equally qualified blacks, many of whom were still handicapped because of their earlier inferior educational experiences. In addition, most of the women candidates came from middle-class or affluent homes and did not require heavy or any scholarship assistance.

Faced with the need and/or desire to act, educational and employer institutions opened their gates to both women and blacks but they found many more qualified women than blacks.

This formulation recognizes that, in addition to objective considerations, matters of performance and taste contributed to giving white women a big edge over black men and black women: after all, white women were the wives, sisters, daughters of the white males who controlled the centers of power in the United States and most males found it easier to associate, work, and socialize with them than with blacks.

4. A FEW TENTATIVE GENERALIZATIONS

The term 'personal' in the title is my excuse for bringing my highly selected and abbreviated account to an end with the aim of directing the remaining space to a consideration of some of the lessons that can be extracted and to venture a few prognostications.

First as to the lessons, the most important from my perspective are as follows:

1. The single most important societal development which contributed to the marked reduction in employment discrimination since the onset of the Second World War was the long and sustained period

of economic growth which created a great number of educational, job, and promotional opportunities for hitherto discriminated against groups – Jews, other ethnic minorities, and particularly women and blacks. Without the sustained burgeoning of the economy the rate of progress in removing discriminatory barriers would have been much slower.

2. The second contributing factor was the change in social attitudes and behavior, reinforced by governmental and voluntary action which sought to reduce policies and practices that were inherently inequitable or which indirectly reduced the chances of specific groups to participate fully in developing and utilizing their potential and their talents and thus to enjoy rewards similar to those which accrued to white males as they advanced up the occupational scale.
3. The increasing shift of the economy from the production of goods to the production of services – more than 7 out of every 10 jobs are now in the service sector – placed a higher premium on young people entering the workforce with more and better education. The explosion in higher educational facilities, mostly in the public sector, greatly facilitated the opportunities for persons from low-income families to secure the educational credentials and competences they need to start on the track to better jobs and incomes. However, the poor quality of many inner-city schools, where large numbers of blacks are concentrated, proved a handicap to their fully exploiting the educational route to better careers.
4. From many vantages the opening up of the previously controlled domains in higher education and in the corporate world to Jews and other minorities was striking and much faster than any optimist would have anticipated a half century ago.
5. The transformation of the role of work in the lives of women and the participation of women in the world of work has been the most revolutionary of all changes.
6. There are two ways to read the record of the blacks. On the one hand, their economic progress since 1942 has probably been as fast as that of any previously discriminated-against group in a comparable time period – roughly two generations. But since all blacks were concentrated at or close to the end of the queue in 1942, a disturbingly high percentage are still in serious plight, poorly educated, and with only a marginal adjustment to the labor

market. The cumulative burdens of racism which represent a weight much greater than the burdens of religious or ethnic discrimination are not moot. Almost a century and a third after the Civil War the black person is not yet a full member of our society and polity.

5. A LOOK AHEAD

Shortly before his death *The New York Times* quoted Martin Luther King, Jr as saying that he did not expect discrimination based on race in the labor market to disappear until after the beginning of the next century, a forecast that I found at the time to be very pessimistic. I regret to say that I now find it unduly optimistic for the reasons that I will identify as follows:

1. A disturbingly high proportion of all black youngsters are growing up in one-parent households with marginal income, living in segregated areas rife with crime, drugs, and other pathologies; attending schools where effective instruction is rare; getting signals from the larger society that they are second-class citizens. They drop out of school inadequately prepared for the world of work and unless there is a return to the exceptionally high levels of employment that prevailed during the Second World War, they are unlikely to make more than a marginal adjustment to work and to society. The shortcomings from which they suffer will be projected onto many in the next generation, and this explains why King's hope and expectation of the disappearance of job discrimination by 2005 is likely to be wrong. Of course, a mainline economist will point out that there is no job discrimination if the lowest qualified are at the end of the queue. Admitted – but what about the panoply of social forces that contribute to the poor preparation of so many blacks? One can insist that the appropriate term should be social or societal, not job, discrimination but that doesn't change the outcome.
2. The outlook for blacks growing up in better positioned families is more favorable. If they graduate from high school and particularly if they graduate from college, the odds are good that they will be able to make a satisfactory adjustment to the world of work. They will still encounter barriers in seeking to move toward the top,

among other reasons because of the racism that continues to govern relationships off the job, and further because of the lack of support from a critical mass of black peers and supervisors. But these disadvantages notwithstanding they should do relatively well.

3. The numbers of blacks with strong academic credentials who seek to pursue careers on the campus and to make their marks in their disciplines are strikingly few, reflecting, in my opinion, the cumulative weaknesses in their environmental and educational backgrounds. And I do not see an early or substantial increase in their numbers. The large number of Jews who made their way into academe two generations ago are not being replicated by blacks.

The outlook for women is definitely more favorable for the reasons alluded to earlier and also because, with the retirement of the older generation, men who obtain leadership positions are likely to be more relaxed about having women colleagues.

The continuing shift of the economy in the direction of services, which is likely to be accompanied by a still further weakening of the internal labor market characteristic of the manufacturing sector, will reduce one of the significant avenues that helped many black blue-collar workers to get reasonably good jobs and to earn reasonably satisfactory incomes. Henceforth one's educational preparation will exercise more influence on his/her occupational mobility.

Reference must also be made to the second largest minority group, Hispanics, which by the turn of the century may outnumber blacks. Primarily located in the faster growing regions of the country, the Southwest and the West, Hispanics have enhanced their occupational advancement to a point midway between whites and blacks; the most recent data, reflecting recovery from the long recession, show Hispanics as among the most important gainers – more than double the average for the nation – in terms of the percentage increase in total employment. In addition to the large numbers of Hispanics in the Southwest and West there are three other centers: Miami, where the early immigrants from Cuba have been doing very well; New York City, where the Puerto Ricans have been doing quite poorly; and selected communities in the mid-West, such as Chicago, which has become home for large numbers of relocated Chicanos.

The final group that warrants attention is the large number of immigrants and refugees who have been entering the United States mostly legally, but some illegally, amounting annually to three-quarters of a million to more than a million. The immigrants are persons who believe that they can better themselves by coming to the United States for a short, long, or permanent stay. The refugees are also in some measure a selected group. Willing to undergo considerable hardships after they arrive in order to get a foothold and then to improve themselves, the evidence points to the fact that many make considerable progress in a relatively few years often much faster than blacks who have had roots in this country for over three hundred years. Several factors help to give the immigrants an edge: self-selection, superior education, family networks and lower short-term expectations, but long-term optimism that they will be able to alter their circumstances for the better. While the issue is not clear cut or simple some blacks are probably further handicapped by this large immigrant stream.

6. SALUTE TO LEWIS FEUER

Lewis' own career, which involved appointments at City College, Vassar, University of California (Berkeley), Toronto, and the University of Virginia, speaks to a considerable reduction in the bias against Jews, even those whose approach and ideology were somewhat off the narrow path. During these years, Vassar was transformed into a co-educational institution and the University of Virginia admits considerable numbers of blacks and has added some blacks to its faculty.

Lewis has directed considerable of his scholarly energy and talent to exploring diverse issues related to the transformation of modern societies, including a special interest in the changing role of Jews within the halls of academe. This piece has been written with a sensitivity for Lewis, my New York City high school classmate, who became a student of both Marx and Freud, but whose interests were more deeply grounded in the dynamics of societal change. I hope he finds this modest offering to his liking.

Columbia University

CHARLES HANLY

FROM ANIMISM TO RATIONALISM

In this paper, I propose to explore a facet of one of the great cultural achievements of the ancient Greeks. The psychoanalytic vantage point of this paper is one that has informed many of Lewis Feuer's contributions to the field of applied psychoanalysis. The transition from animism to rationalism in ancient Greece can be illuminated from various perspectives: economic, geographic and environmental, technological, political, sociological. Each of these perspectives can provide knowledge essential to a comprehensive understanding of this important transition, but the psychology of the process is no less fundamental. For the study of the history of ancient Greek culture leaves one with the clear impression that there gradually took place a quite far-reaching development in the psychic organization and functioning of these remarkable people.

It is my thesis that the cultural transition from animism to rationalism came about in part as a result of a gradual improvement in individual psychic functioning. During a relatively short time (roughly of the duration of the Renaissance to this century) the ancient Greeks advanced from an animistic culture in which borderline personalities were normal to a rationalistic culture in which neurotic personalities were normal. This statement is not intended to imply a unidirectional causality from changes in individual psychology to changes in culture. On the contrary, the relationship is reciprocal. Individuals inherit the accumulated experience of their ancestors not through the genetic transmission of their acquired characteristics [Freud's archaic inheritance (Freud, 1915, 1923, 1939)] but by means of childhood identifications with the parents and their later surrogates (Freud, 1923). As a result of his discovery of the fundamental importance of identification in psychic organization and functioning, Freud no longer needed his Lamarkian hypothesis. Identification provides for the transmission of pre-reflective cultural values and modes of conduct culturally via the parents and their generation. Thus to emphasize the importance of improvements in individual psychic functioning is also to emphasize the importance of the pre-cognitive cumulative experience of previous generations as reflected

S. Hook, W.L. O'Neill and R. O'Toole (eds.), Philosophy, History and Social Action, 221–234.

in the actual lives of each parent generation and, thereby, made available to the next one.

Since the concept of a borderline personality is being used in a technical and non-pejorative sense, some definition may be useful. Borderline personalities have a relatively tenuous grasp on reality; they are subject to rapid and extreme mood shifts from omnipotence to terror, from ecstatic hope to despair, from elation to depression; they are impulse driven; their super-ego functions remain precarious and unreliable; they are vulnerable to paranoid forms of perception and thought; they rely on projection, denial and splitting for instinct mastery (see Fenichel, 1945; Kernberg, 1970; Frosch, 1970).

In order to gain an empathic grasp of the borderline personality and its typical forms of experience, it is only necessary to remember emotionally significant experiences of our childhood. For we were, each of us, once borderline personalities. Children may have hallucinatory experiences without being psychotic. They may have 'spirit' companions with whom they converse and play. Paranoia (e.g. fear of a malignant being in the dark) is normal in childhood when dangerous non-existent animals, witches and bogey men often appear under beds and at windows. It is normal for children to perceive inanimate objects as having souls: "Bad table to hit Johnny." The parallels between ordinary childhood behavior and adult borderline behavior are there to be observed by anyone. And just as animism is the natural world view of modern children, so too animism was the world view of the Homeric Greeks (Freud, 1901). Animism is the culture that is natural to the borderline personality.

However, it must also be recognized that the comparison is incomplete. An adult Homeric Greek was different physically and psychologically from a modern child. Also he was different psychologically, although not physically, from a modern, adult, borderline personality in the clinical sense. The differences between the adult Greek and the modern child are obvious. The difference between the Greek and the modern borderline personality results from the fact that the ancient Greeks lived in a culture that was congruent with and congenial to the borderline personality. A modern general would be stripped of his command and hospitalized if he advised a council of war to consult a prophet, priest, or dream interpreter as a means of finding out why a campaign was going badly. Achilles' advice to do these things was considered to be perfectly sensible by his Homeric contemporaries. The ancient Greeks were not surrounded by a social environment in which borderline personalities were at a disadvantage and which reminded them constantly that there was something amiss with them. Hence, they were able to act with the force, confidence, and dignity of a twentieth-century rational adult even though their actions and their reasons for taking them were dangerously out of touch with reality by any modern standard.

At the same time it must also be appreciated that remnants of animism not only continued to play an active part in the lives of the more rational fourth-century Greeks and in the lives of twentieth-century children, they still have their place in the lives of contemporary adults. For example, some of our contemporaries, faced with the uncertainty which remains after consulting scientific water tables, turn to water-witches to make a decision on where to dig their wells. Mackenzie King, Canada's most successful politician (if success is measured by the length of occupation of the highest public office) regularly consulted a medium and held seances with the dead, both before and during his lengthy tenure as Prime Minister. However, he was sufficiently uncertain about the public response to his eccentric method of reaching decisions that he confided the practices only to his diaries and to his trusted associates. Moreover, it is likely that his seances with the dead were limited in their function to dealing with residual uncertainties and doubts concerning matters of state. The substance of his decisions would have been the result of normal consultations and realistic considerations. These practices appear to have served the same function in the life of Mackenzie King that water-witching serves among country people. The practices are vestiges of animism preserved in our century to patch over the anxieties left by gaps in knowledge and technology. The greatest of these vestiges of animism is, of course, religion.

In two respects individual and collective life are similar. Just as an individual retains the residues of his infantile modes of psychic functioning (e.g. in dreaming) so too civilizations retain the residues of their earlier forms. Just as individuals may, under conditions of severe anxiety, regress to more infantile modes of functioning, so too a society, however advanced its culture, acquires no immunity against regression to more primitive forms.

However, these factors notwithstanding, there remains a clear sense in which civilization has advanced. I am of the opinion, and this study as a whole is an attempt to corroborate it, that the cultural anthropology of

Levi-Bruhl (1910) which allows for real and fundamental cultural change (without denying its reversibility) is truer to the human condition than the cultural anthropology of his successor, Levi-Strauss (1962) which proposes that there is no fundamental cultural change. My thesis is that a basic cultural advance was accomplished by the ancient Greeks in whose debt we moderns still remain in so far as we are its beneficiaries. I shall then propose an explanation of the psychological factors which were at work in this change. My focus will be on the psychological changes that were connected with the cultural change. For this purpose I shall consider a series of contrasts between the Homeric Greeks and their fifth-and fourth-century descendants.

The perceptual life of the Homeric Greeks was not able to penetrate beyond the global projections that invested it to the inanimate world of physical objects. As a result they lived in a world which was systematically overdetermined. No objects were simply material. No forces were simply physical. No events were simply mechanically or chemically produced. When the string of a Homeric hero's bow broke, he (Teucer) replaced it with another carefully selected for its physical strength, but he would also experience the occurrence as the consequence of a psychic intervention: a recalcitrance on the part of the bow-string's animate being, the intervention of a daemon or a god (*Iliad*, 28, see Dodds' (1963) comment, p. 12). A Homeric Greek might simultaneously use all his acumen as a bow man to select a sturdy replacement and care for it properly and also undertake the ritual precautions needed to make it obedient to his purposes. (Teucer, in fact, is described as having abandoned his bow for a spear.) Of these precautions the magical ones would have been deemed by the Homeric Greek to be the most important.

The central action of the *Iliad* is initiated by a plague. In order to solve the problem of the plague Achilles recommends a consultation with a dream interpreter; he speculates that Appolo has been offended by a broken vow or a ritual failure. And Achilles was found to be wrong only in details, not in principle, for it was an insult by Agamemnon to a priest of Appolo that caused the plague, in Homer's plot. In his account, no consideration is given to any realistic efforts to contain the plague; thought is given only to ritual and magical solutions. It is reasonable to suppose that when knowledge was scarce and anxiety was high, recourse to omnipotence of thought and magical actions increased.

The whole of nature was overdetermined for the ancients. They suffered, as it were, from hypergnosia (Waelder, 1926). Natural events had an ominous character. For example, during the Trojan counter-attack, the Trojan soldiers observe an eagle with a snake in its talons flying over their positions. Suddenly the snake manages to free itself and it falls among the soldiers. They draw back from it in horror, not primarily because of a physical fear of it, but because they see it as an evil omen. That it *is* portentous they have no doubt; all they need is a priest to tell them what it foretells. The priest tells them that the eagle represents Hector, and the snake Agamemnon; the failure of the eagle to keep its prey in its talons means that Hector will fail in his efforts to take the Greek camp (*Iliad*, pp. 226–227). The natural world and its events were experienced as if they were invested with the hidden purposes of fate-disposing gods.

The psychic experience involved in such an episode is not difficult to reconstruct. Made anxious by their own fear of failure too much to be able to experience it in themselves, they projected it onto a suitable natural event in which they could become aware of it as something ominous that endangered them. By means of a culturally sanctioned interpretation the meaning of their fear could be made known and, at the same time, legitimated. This elaborate detour by way of projection is evidence of an ego life lacking in stability and strength, and dominated by the terror of shame; shame, that is, at being seen to have fears or anxieties, deemed cowardly by the soldiers. This type of projection should not be considered as itself the source of the animism in these primitive men; projections of this kind exploited, and were facilitated by, a more profound global structure of experience, of a paranoid type.

This paranoid structure is evident in the mental events, character, and volition of the Homeric Greeks as well as in their experience of natural events. It dominates their perception of each other and their self-awareness as well. Psychic interventions experienced as alien to the ego were experienced as the source of special emotional states, of dreams, of exceptional achievements, follies, and failures. Hector contemptuously rejects the prophecy of his failure to destroy the Aecheans because he has received in a vision, or visionary dream, sent from Zeus a promise of victory (*Iliad*, p. 227). Agamemnon has already had a similar dream of opposite import. An *ate* (which means some state of mind varying from momentary inattention or carelessness to madness) was experienced as the handiwork of an alien psychic agency: Zeus, *moira* or the Erinys. It is an *ate* that causes Odysseus to fall asleep and thus to allow his retainers to kill a tabooed oxen (see Dodds, 1963, pp. 6–8). A *menos* is an exceptional emotional state which gives rise to extraordinary deeds. Diomedes, wounded by Pandarus, receives a *menos* from Athena so that he fights with unusual ferocity and bravery to revenge his injury and to wreak general havoc upon the Trojans (*Iliad*, p. 95). Hector is inspired with a *menos* of manic triumph: he foams at the mouth, his eyes burn with a strange fire when he puts on the armour of Achilles after having killed Patroclus (*Iliad*, pp. 321–322). Aias is similarly inspired by Poseidon on the Achaean side and is certain that the priest Calchas who had encouraged him was an apparition of the god; for it was "not hard to recognize a god" (*Iliad*, p. 236). Thus the highly volatile shifts of mood, attitude and behavior characteristic of Homeric man were attributed to interventions by gods and daemons over which they had no control.

> But the thoughts of Zeus outstrip the thoughts of men. In a moment the god can make a brave man run away and lose a battle; and the next day he will spur him on to fight (*Iliad*, p. 3ll).

Thoumos (an interior voice seated in the chest or midriff of Homeric man which counsels him) was commonly experienced as being independent of the ego even when it was not felt to be acting under the influence of a god. The inner as well as the exterior landscape of Homeric man was experienced as being regularly subject to alien psychic intervention.

The struggle with strong conflicting passions which in modern man takes place within the self could be experienced by Homeric man as a struggle between himself and a constraining god. When Agamemnon punishes Achilles for his arrogance by insisting on taking Briseis from him for himself, the enraged Achilles is about to kill Agamemnon, but Athena appears to Achilles and persuades him to desist. The description is of an hallucination. Achilles can see and hear the goddess, but the others cannot (*Iliad*, 28). Evidently, the super-ego of Homeric man was not yet a secure, internal acquisition. In conditions of crises, ego ideals made themselves felt as hallucinations: the projections of the archaic parental images. We do not have to look further than these hallucinatory experiences in order to find the origins in individual experience for the Olympians who were the collectively edited and elaborated versions of the arhaic parental images projected onto an outer scene.

Among classics scholars there has been a debate between those who see Homer's descriptions of the gods as an allegorical poetic machinery without substantive roots in the life experience of ancient men and those, like Dodds (1963) and Cornford (1957) for example, who treat Homer's descriptions as having been, in their essentials, literal for the Homeric Greeks. Psychoanalysis supports the later position. No other understanding can do justice to the antiquity of Homeric man nor appreciate their achievement. If the Homeric Greeks really were no different, psychologically, from the Greeks of the classical period and, hence, scarcely different from modern man except for the lack of science and an advanced technology, then one of mankind's greatest cultural achievements, the gradual conquest of animism in favor of rationalism would have to be attributed to some yet earlier people. However, the evidence that would establish the transition as belonging to an earlier people would have to be the same sort of evidence we find in the Homeric and Hesiodic legends. And the same arguments that these accounts are merely literary devices would equally apply, *ad infinitum*. Finally, this argument establishing the gods as 'literary' rather than 'real' for the Greeks would lead us to postulate the creation of the original *homo sapiens* not essentially different from ourselves. Is the 'literary' theory based on a narcissistic denial that *homo sapiens* as he now is was once a non-rational animal who, as a consequence of evolutionary pressure and genetic accident, became the precariously civilized, although inventive, creature of today? This denial would protect us against the anxiety provoking realization that since what has been accomplished by 'civilized man' is not sanctioned by instinct, genetic endowment, environment, or any divine protection, it can also be undone and, therefore, needs to be assiduously preserved.

From Homer's description in the *Iliad*, it is evident that the Greeks were subject to severe shifts of mood. Agamemnon goes from exultant confidence to tearful despair. Achilles goes from arrogant challenging of Agamemnon to tearful grief at his public humiliation (when he is obliged to accede to Agamemnon's demand for Briseis) and, then, to sullen, suspicious, vengeful rage. It is also clear that the Homeric heroes are lacking in impulse control. Agamemnon's realization that the safety of his enterprise, let alone its success, depends upon Achilles and his men is lost in his determination to humiliate Achilles. Realistic considerations of state or of military strategy are constantly at the mercy of impulsive responses carried immediately into action. These mood shifts are sometimes occasioned by the course of the battle between the Achaeans and the Trojans but more often the outcome of the fighting is determined by mood shifts from despair to manic triumph in the leading warriors now on one and now on the other side. These 'inspirations' are always perceived as the handiwork of a god. To the fear of defeat there is added the dreaded of the loss of divine favour; to the hope of victory there is added the exultation of divine sanction and support.

Homeric man's morality was governed by five factors: fear of superior physical force, fear of superior guile, lust after wealth, the pursuit of sexual pleasure, and fear of the loss of reputation. To the extent that Homeric Greeks were able to master impulse, affect, and behavior, the mastery was accomplished by shame reactions – by the fear that the contemplated action would result in the loss of esteem in the eyes of others. It is to these same reactions that Achaean and Trojan commanders appeal in their efforts to get their friends and the ranks to perform their duties on the field of battle. Consequently, in the absence of the fear of being recognized no actions were constrained. The fear of the loss of self-respect which is the more reliable source of a sense of duty did not play a significant part in the psychology of Homer's Greeks. The gap created by the lack of an internalized conscience was filled by physical intimidation and psychic projection. The gods, of course, acted no differently than men. When Zeus was absent from Olympus, it was the fear of his unquestionable physical superiority that constrained the gods from disobedience. When the mortal commanders were unable to compel their troops to action by invoking shame and the prospect of triumph, booty, and women, they threatened them with death.

By the fifth century B.C. some fundamental changes had occurred. The psychic structure of the characters who populate Plato's *Dialogues* make themselves known as being fundamentally akin to ourselves. The archaic mastery and grandeur of Homeric man's paranoid, narcissistic psychic organization had been brought under cultivation and subjected to maturation. For example, Plato attempted to demonstrate the correctness of his tripartite division of the soul by appealing to the universality of psychic conflict. The nature of this conflict he illustrated by the example of Leontius, a man who, having chanced upon the bodies of some recently executed criminals found that, despite his best efforts, he could not restrain a wish to look with an excited, morbid fascination at them. Leontius is inescapably aware that it is his own necrophilia impulse that drives his conduct, curse his eyes for the lust as he may (Plato, *Republic*, 4, 439–440). Even unwanted, inconvenient, guilt provoking and calamatous impulses are accepted as belonging to the self. To be sure they are alienated from the rational and moral self and assigned to the work of a degraded and degrading body. Nevertheless, the instinctual, affective, and moral activities of the psyche have been largely withdrawn from the world into the self which has been transformed in the process.

In contrast, Homeric Agamemnon, as part of his reconciliation with Achilles, found it necessary to solemnly declare that it was Zeus who had caused him to act vengefully toward Achilles. In place of remorse or self-criticism, there is the experience of having been a passive vehicle for the machinations of an external psychic agency, the god Zeus, against whose force no mortal could successfully struggle.

> It was Zeus and Fate and the Fury who walks in the dark that blinded my judgement that day at the meeting . . . What could I do? At such moments there is a Power that takes complete command, Ate, the eldest Daughter of Zeus, who blinds us all, accursed spirit that she is . . . flitting through men's heads, corrupting them, and bringing this one or the other down (*Iliad*, p. 356).

Numerous other illustrations of the changed experience of self and nature can be found in Plato's dialogues alone. Although Plato's version of the crime of the ancestor of Gyges the Lydian is replete with magical descriptions and symbolism, at its core is the idea of a character whose Oedipal criminality is the expression of his own hidden desires (*Republic*, II, 359–360; see also Hanly, 1977). When Socrates (*Republic*, IX, 572) wanted to establish that evil impulses exist even in the most decent among us, he appealed to the evidence of dreams of patricide, incest, and cannibalism which give expression to latent carnal wishes (Plato, *Republic*, 9, 572). No longer, in Plato, are dreams so easily experienced as divinely visionary and prophetic. They have been integrated into the self as an important register of its reality. Lest we suppose that the achievement of rationalism was a peculiar achievement of Plato, it is not difficult to find evidence of an advance along a broad front.

In the *Poetics* Aristotle asserted that the tragic poet is able to gain a more fundamental insight into the human condition than the historian because he need not concern himself as the historian must with the accidental and adventitious. Aristotle also asserted that *Oedipus Rex* was the most perfect example of tragedy. There is, in these assertions, an implicit translation of the machinery of the gods back into its sources in the instinctual vicissitudes of men and women. Thucydides' account of Pericles' funeral oration to the grieving Athenian parents of the heroes of the first year of the Peloponnesian War reveals a profound acceptance of a cruel reality unmitigated by denials or projections (Thucydides, 2, 44ff).

Later, Epicurus conceived of an atomic universe which, among its other considerable virtues, held out for Epicurus and his contemporaries the principal value of binding the gods and exiling them to a remote province of the universe where they could live with such pleasure that they would have no interest in creating mischief for men out of envy for their poor and temporary happiness. Epicurus' explanations of natural phenomenon (eclipses, storms, earthquakes, plagues, etc.) in terms of the mechanical action of inanimate bodies were aimed at the events in nature which, in the earlier animistic culture, had been the objects of the superstition which created projections (Epicurus, *Letter to Heroditus*, 766–777).

But it is essential not to oversimplify or dichotomize. Progress in civilization is intrinsically precarious. No satisfaction or consolation once enjoyed is altogether renounced. Even Socrates, the questioner, when faced with death bethought himself of a dream in which he had received an injunction from the gods to make music and, uncertain as to whether philosophy altogether qualified, he sought to console himself and to satisfy the gods by writing poetry. There is still an element of projection in Aristotle's antology. The relationship between form and matter from which the cosmos is born remains a sexual one on the analogy with a biologically misconceived procreative relationship between men and women. (Aristotle believed the male sperm to be the source of the rational nature or essence of the offspring while the woman provides only the matter out of which the foetus is formed.) In Plato's metaphysics there is a no less narcissistic overestimation of the power of ideas which are envisioned as archetypes according to which the cosmos is constructed. The purposive character of human mental processes were unquestioningly projected by both Plato and Aristotle into inanimated nature. And although Democritus, Epicurus and later Archimedes, Euclid, and no doubt, the ingenious artisans of the time (I am persuaded that Bacon (1620) was correct in attributing to them a better grasp of reality than was common among the philosophers) had constructed a fundamentally adequate concept of matter, stripped of earlier anthropomorphism, for Plato and Aristotle a correct understanding of the nature of matter remained inaccessible. Nevertheless, the ontologies of Plato and Aristotle removed the gods from nature, and their epistemologies where reality bound even if the nature of reality was not adequately conceived. The ethics of Plato and Aristotle demanded the test of values of reason and implied individual responsibility for conduct. The experience of self, society, and nature by the fourth century had changed in a fundamental way. Among these fourth-century Greeks the gods had died, were dying or, at least, had 'departed'.

In stating this, I am not asserting that fourth-century Greeks did not have religious beliefs or that they did not consult oracles or engage in cult observances. I am asserting, rather, that religious belief and practice had by then acquired approximately the same psychological character as it has among believers today. Religion had become a question of belief rather than one of experience as it had been before. Perceptual experience had undergone an adaptation to reality. We would like to be able to understand the psychogenesis of this adaptation.

Fundamental to the transformation we are considering is a modification of the psychic organizations and processes involved in self-awareness or self-experience. If we reflect once more on the description of Achilles' struggle with his wish to kill Agamemnon, we see, at once, that the conflict between the murderous drive demand and the inhibiting agency was experienced by him as a conflict between himself and an admonishing, hallucinatory figure who was able to exercise an influence upon him as if it were another person. Psychoanalysis has made us familiar with the fact that small children (before the resolution of the Oedipus complex) form a narcissistic projective identification with their parents which gives to the parents a god-like quality. Parents, as a consequence of this identification, are able to act as surrogate egos for their offspring with respect to both morality and the definition of reality. Even if a child is unable to experience the world as his parents describe it, he 'knows', because of this identification, that it is as it is said to be and not as it appears to him to be. This psychic dependency is necessary and adaptive because of the incompletely developed capacities in the child for instinct mastery and reality testing. We may assume, therefore, that the psychic structure of Homeric Greeks had reached a rudimentary stage of development but that maturation did not yet provide for any significant advance beyond this stage. My hypothesis is that the mature Homeric Greek continued to live, to a large extent, on the basis of projective identifications. Hence, the transition from narcissistic dependency on parents to psychic independence and autonomy was never quite accomplished because it was achieved, not by the development of new psychic structure within the self which could provide the basis for a genuine rational autonomy, but by repetition of the projective identifications and the substitutions of hallucinatory figures of the gods for the parents who could no longer serve this function. Thus premonitory and admonitory hallucinations occupied the place in psychic life that was later to be occupied by conscience. The psychic organization that Homeric man had not yet adequately developed was an effective super-ego.

The formation of the super-ego provides for more than instinct mastery. It facilitates and consolidates the functions of self-observation and self-criticism. If thus allows for self-objectivity, autonomy and responsibility (Waelder, 1934; Hanly, 1979) all of which are psychic qualities upon which rationality depends. The capacity for self-objectivity brings in train a much more refined and powerful ability to differentiate what is subjective from what is objective in sense experience. Hence, the ability to 'objectify' objects is facilitated by the ability to objectify the self. Thus the development of effective super-ego functioning was an important step in de-mystifying, de-divining and de-anthropomorphizing perceptual life.

Similarly, the super-ego provides for autonomy in so far as it brings self-evaluation and self-regulation within the grasp of the individual himself. The price to be paid for this acquisition is a consolidated realization of one's own motives and doings as well as a more just and accurate recognition of self-limitations. The exercise of the autonomy that is consequent upon the formation of a stable super-ego is incompatible, on the one hand, with the narcissistic grandeur, frenzy, mania, and elated triumph of the Homeric Greeks and, on the other, with their dread, terror, and despair.

Responsibility depends upon a sagacious and tenacious self-observation which is able to undo the need to alienate from the self acts the consequences of which may range from inconvenient to calamitous. Self-observation allows for a more correct attribution of the motivation and causation of attitudes, wishes and actions. It brings into existence a personality with a real capacity for self-honesty, whether or not that capacity is exercised, as well as the capacity to react with guilt to failures of self-honesty. Responsibility made possible by the formation of a stable super-ego causes the individual to be accountable to himself no less than to others. However, it must be understood that Homeric man is not being accused of dishonesty. He was honest to his experience of self, others, and nature. To accuse him of dishonesty or inauthenticity would be like accusing a child, who is actually experiencing the dark as menacing, of lying when he says that he is afraid.

My hypothesis is that at the heart of the ancient Greek creation of more rational forms of individual life and culture was the gradual bringing into existence of individuals who developed stable, effective, internalized super-egos as part of their maturation. This hypothesis, if true, is only one step toward a better understanding of the historical origins of modern man in ancient Greece. It raises a series of questions concerning how this change in the process of psychological maturation came about. The questions can only be posed here. But two general points can be made.

On reliable theoretical grounds, we are entitled to wonder whether an essential element in this historical process must have been some real reduction in the amount of anxiety to which individuals were subjected. Since there is no reason to assume that the subjective sources of anxiety deriving from constitutionally determined vicissitudes of instinctual life had declined then, or have declined even now, we must assume that the ancient Greeks, exploiting the precarious protection offered them by their archaic religious life, managed to bring about real improvements in their social and natural environment so that they could progressively abandon their projections and begin to take up the quest for self-understanding and self-mastery. The achievement of the psychological conditions for a more rational life is reversible. And this achievement was reversed historically for many centuries, until during the Renaissance propitious conditions once more arose. It was no accident that the men who moved civilization forward again did so by returning to the Greeks for inspiration.

One merit that the hypothesis put forward in this paper can claim is that it is consistent with certain insights of the greatest of the Greek philosophers. I shall conclude by citing four of these. Plato believed that, in order to know, the soul must be just, i.e. harmoniously ruled by reason. Epicurus valued his atomic materialism because it banished the gods from nature. By rescuing man from religious anxiety, materialism offered him a better prospect of living the only life he has with greater pleasure. Aristotle taught that tragedy, by bringing about a catharsis of otherwise debilitating emotions, protected a homeostasis of psychic life necessary to a rational and civilized existence. Socrates believed that in order to know reality it is necessary to know oneself. No statement could better summarize the argument of this paper.

University of Toronto

REFERENCES

Bacon, F. 1620. 'Novum Organum' in *The English Philosophers from Bacon to Mill*, E. A. Burtt (ed.). New York: The Modern Library (1939).

Cornford, F. M. 1957. *From Religion to Philosophy*. New York: Harper and Brothers.

Dodds, E. R. 1963. *The Greeks and the Irrational*. Berkeley: University of California Press.

Fenichel, O. 1945. *The Psychoanalytic Theory of the Neurosis*. New York: Norton.

Freud, S. 1901. 'The Psychopathology of Everyday Life', *Standard Edition*, **6**.

———. 1915. 'Instincts and their Vicissitudes', *Standard Edition* **14**, 117–140. London: Hogarth Press.

———. 1923. 'The Ego and the Id', *Standard Edition* **19**, 3–66. London: Hogarth Press.

———. 1939. 'Moses and Monotheism: Three Essays', *Standard Edition* **23**, 7–137. London: Hogarth Press.

Frosch, J. 1970. 'Psychoanalytic Considerations of the Psychotic Character', *Journal of the American Psychoanalytic Association* **18**, 24–50.

Hanly, C. 1977. 'An Unconscious Irony in Plato's *Republic*', *The Psychoanalytic Quarterly* **46**, 116–147.

———. 1979. *Existentialism and Psychoanalysis*. New York: International Universities Press.

Homer, *The Iliad*, transl. E. V. Rieu. Harmondsworth: Penguin Books (1950).

Kernberg, O. 1970. 'A Psychoanalytic Classification of Character Pathology', *Journal of the American Psychoanalytic Association* **18**, 800–822.

Levi-Bruhl, L. 1910. *Functions mentales dans les sociéties inférieures*. Paris: Alcan.

Levi-Strauss, C. 1962. *La Pensée Sauvage*. Paris: Plon.

Plato, *Republic*, transl. B. Jowett. New York: The Liberal Arts Press (1948).

Thucydides. *The Peloponnesian War*, transl. T. Hobbs (1629), D. Grene (ed.). Ann Arbor: University of Michigan Press (1959).

Waelder, R. 1926. 'Schizophrenic and Creative Thinking' in *Psychoanalysis: Observation, Theory, Application*. New York: International Universities Press (1976), pp. 42–56.

———. 1934. 'The Problem of Freedom in Psychoanalysis and the Problem of Reality Testing', in *Psychoanalysis: Observation, Theory, Application*. New York: International Universities Press (1976), pp. 101–120.

SIDNEY HOOK

TOWARD GREATER EQUALITY

The concept of equality has moved into the center of intellectual and cultural interest in recent years. In some quarters it has become a verbal fetish, used synonymously with terms like 'democracy' and 'social justice'. Reflection will show, however, that when it stands by itself, 'equality' or the demand for equality is an incomplete and ambiguous expression. If we want to move 'toward greater equality' we don't know what it is we want unless we can specify *what* we want in greater equality. In a country in which members of group A have the vote while members of group B have not, we can move toward greater equality either by granting members of both groups the vote or by *denying* it to both groups. If we stress only equality we could not distinguish between saying that "justice consists in treating all persons alike" and "justice consists in mistreating all persons alike". At best, then, equality is a necessary but not sufficient condition of a just society – since human beings can be equal in poverty or affluence, equally enslaved as well as equally free. That is why a just rule or law must go beyond mere impartial application. The rule or law itself must be morally valid.

My starting point, which I am assuming is common for all of us, is that we are interested in furthering greater equality of participation in democracy as a way of life. The commitment to democracy as a way of life does not entail a belief that all men and women are physically and mentally equal, but the postulate that, whatever their differences, they are *morally* equal. The recognition of the moral equality presupposed by democracy as a way of life is expressed as *an equality of concern* for all members of the community to develop themselves to their full capacities as human beings. This makes it obligatory for the democratic community to equalize opportunities for its citizens regardless of race, color, religion, sex and national origin. Toward greater equality, then, means toward greater equality of opportunity, for persons in all areas, especially education, housing, health and employment, required for individuals to develop their best potential as human beings.

It is obvious that as long as the family and different home environment exists, as well as extreme genetic variations in capacities, *absolute*

S. Hook, W.L. O'Neill and R. O'Toole (eds.), Philosophy, History and Social Action, 235–243.

equality of opportunity is unattainable. But this is no more a justification for abandoning the continued quest for equalizing the conditions of opportunity than our failure to achieve absolute health, wisdom and honesty undercuts the validity of these ideals. Properly understood, equality of opportunity commits us to programs of continuous reforms. It is the revolution *en permanence*.

But is equality of *opportunity* enough? Are competence and merit, which determine awards and rewards when careers are truly open to talents, sufficient? It is sometimes argued that equality of opportunity necessarily converts social life into a race or battlefield in which the consequence of the victor's triumphs is invidious defeat of his competitors. This is a non-sequitor. If we remember that our controlling moral postulate is equality of concern for all persons to reach their potential, not all forms of competition are desirable; neither are all forms of competition undesirable; not all forms of competition need result in disaster for those who are not winners, and especially for their children. After all, we do not tolerate or encourage opportunity no less equal opportunity, for many kinds of anti-social actions. Our moral postulate commits us at the very least to see that the basic human needs for food, clothing and shelter are gratified for all, regardless of the outcome of social programs geared to merit. It commits us to provide a floor or level of life, whose extent and nature is a function of our technological capacities, beneath which human beings should not be permitted to sink through no fault of their own.

It is further objected: if careers are to be open to talents, what about the fate of the untalented? To which I reply: it is a tolerable fate provided that there are opportunities of employment for all; provided that no persons, regardless of their vocations, are regarded as second-class citizens of the political community; and provided that all citizens have the assurance that the community in diverse ways will seek to equalize the opportunities open to their offspring.

So I cheerfully acknowledge that equality of opportunity is not enough because it does not absolve us from the responsibility of responding to basic human needs; and because, as Jefferson realized, not all the possible consequences of equal opportunity, especially extreme disproportions of wealth and power, are acceptable, particularly if their operation undermines through monopolies of press and other strategic goods and services, equal opportunities of development for subsequent generations.

Nonetheless because equality of opportunity with all its institutional reinforcements is not by itself adequate for an enlightened social philosophy, it does not in the least follow that we must supplement it or substitute for it, or use as a test of its presence, the principle of equality of *result*. It is strange logic that argues from the undesirability of *some* extreme inequalities of result to the desirability of equal results in every form of social distribution, or seeks to justify equality of results as essential to social justice or to genuine democracy. In a world in which men and women vary in all sorts of ways, how can equality of concern eventuate in equality of result except by chance or ruthless design? The expectation is as absurd as the presumption that equality of medical concern for those who are ailing necessitates the same medical prescription or regimen for them all – the unfailing sign of a medical quack. Equality as a moral postulate envisages the moral equality of the different and not merely of the same; the application of an equal desirable standard to those relevantly situated, regardless of their differences. It was none other than Karl Marx who reminded us that since a right by its nature consists in the application of an equal standard, applied to different individuals, "and they would not be different individuals if they were not unequal. . . . every right in its content is one of inequality." It is a gross misconception of the socialist ethic to interpret its principle of distributive justice as aiming at equality of result or situation regardless of merit, variations in need, responsibility, and social utility. After basic needs have been met, with respect to goods and services that are scarce, i.e. relative luxuries, it is not unfair to distribute these on the basis of desert. This makes it more likely that the availability of such goods and services will increase.

How shall we apply these general considerations to the concrete situations we face in the United States today? I start with the assertion that among the very highest items on the agenda of unfinished business of American life is the final elimination of all practices of civil, social and economic discrimination against any citizen on grounds of race, color, sex, religion or national origin. We can move toward this objective by vigorous enforcement of all federal, state and local statutory prohibitions against discrimination, not only in the field of civil and political rights but in public and private employment with respect to remuneration, promotion and tenure, and in access to education, housing, recreation and other areas that affect the development and fulfillment of human capacities.

It goes without saying that vigorous enforcement of existing laws against discrimination will not by themselves achieve the objective of equal opportunities. Where there are no opportunities, there is little solace in the assurance of equal opportunity to fill them. If decent housing is not available, equal opportunity of access to it is literally equal opportunity of access to nothing. It is therefore incumbent upon the Federal Government to commit itself to a program of full employment at a decent wage level. For those able and willing to work there must be posts for all who are qualified, and also for the unqualified, granting that the posts will naturally be different. In all areas, if private enterprise is unable to meet public need, public agencies must move to meet them. In education, particularly, special supplementary programs should be established to permit all *individuals* who have been disadvantaged in the past to make up for opportunities lost or denied to them. The basic strategy for realizing equality or opportunity is a combination of rational public and private policies, whose detailed nature we cannot explore here, that will expand opportunities for all. If there are not enough places in law school for all *qualified* students, where that has been determined on objective grounds, then instead of preferential racial or religious selection, let us establish more law schools. The same goes for medical schools. What we must avoid is, first, policies that discriminate *in favor* of any group at the cost of discriminating *against* some other group on grounds irrelevant to the objectives of the policy and, second, policies that fail to consider the rights and needs of the individual person, considering him or her only as a representative of a group. We must avoid them precisely because it has been such policies that in the past have resulted in gross injustices to racial and religious minorities and women. We are inconsistent, as well as insincere, if in attempts to rectify the arbitrary and invidious discrimination of the past we practise arbitrary and invidious discrimination in the present. Morally illicit proposals for such types of rectification have been made by partisans of preferential hiring, of quota systems, of schemes which require that "numerical goals and timetables" be mandatory for all institutions that have contractual relations with the government.

Let us consider this issue which currently divides those who are opposed to the patterns of prejudice and discrimination of the past.

It is sometimes denied that it is arbitrary to give preferential treatment to someone who has been the victim of preferential treatment. And with this we can agree. Any particular person who has suffered in consequence of injustices toward him or her is certainly entitled to compensatory treatment as an individual. But morally it is altogether different to say that since *other* individuals of his or her group have suffered in the past because their qualifications were disregarded that therefore individuals who belong to that group in the present, who were not themselves discriminated against, should be judged not on the basis of their qualifications, but on the basis of their group membership, even if this means barring individuals who are better qualified.

I say that it is altogether different because it violates the very principles of equity that justify compensatory treatment for the individual victimized by past discrimination. What I am asserting is that granting the evils of past and present discrimination, the remedy in *not* bureaucratically prescribed 'numerical goals' and 'time-tables', a quota system, imposed on educational institutions and other organizations. Such remedies are advanced on the basis of a highly questionable assumption, to wit, that a society of equality is one in which all the various groups and sub-groups within that society are represented in all disciplines and professions in precise numerical proportion to their distribution in the total pool of the population, or in the community pool, or in the pool of those potentially utilizable. The additional assumption behind such remedies is that any statistical evidence that reveals a marked variation from these proportions is proof of a policy of discrimination.

Not only are these assumptions highly questionable, we never make them in other situations where abuses have been checked or abolished. No one would reasonably argue that because many years ago blacks were deprived of their right to vote and women denied the right to vote that today's generation of blacks and women should be compensated for past discrimination against their forebears by being given the right to cast an extra vote or two at the expense of their fellow citizens or that the male descendants of some prejudiced white men of the past be deprived of their vote. Take a more relevant case. For years, blacks were shamefully barred from professional sports, until Jackie Robinson broke the color bar. Would it not be absurd to argue that, therefore, today in compensation for the past there should be discrimination against whites in professional athletics? Would any sensible and fair person try to determine what proportion of whites and blacks should be on basketball or football teams in relation to racial availability or utilizability? Do we not want the best players for the open positions regardless of the percentage distributions in proportion to numbers either in the general population or in the pool of candidates trying out? Why should it be any different when we are looking for the best qualified mathematician to teach topology or the best scholar to teach medieval philosophy? If we oppose all discrimination, why not drop all color, sex, religious and social bars in an honest quest for the best qualified – no matter what the distribution turns out to be? Of course the quest must be public and not only fair, but must be seen to be fair.

One might even consider the situation with respect to athletics as a paradigm. Since there should be equal opportunities for health and recreation, the community has the responsibility for providing facilities for all citizens to engage in sport and other exercise and, in fact, many communities are moving in this direction. There are still areas of discrimination here with respect to women. There is a neglect of their legitimate athletic interests, especially in competitive sports with men, and they have not yet cracked the sex bar in professional athletics. The law can properly prevent some types of discrimination in budgeting appropriations to enlarge the field of opportunity for women. The rest ultimately depends on them. In time, as more and more women participate in sports, and the social stereo-types about women in this field disappear – as they have in tennis and swimming – I am confident that opportunities will open to women in professional sports as they have to blacks. Were a woman to appear who bats almost as well as Hank Aaron or who steals bases as well as Lou Brock, it would not be long before she broke into the big leagues. Where there is a vested interest in victory, one standard for everyone will sooner or later prevail. In a free and open society one standard for everyone should prevail, regardless of any vested interest.

We may even draw an analogy with equality of educational opportunity. On the basis of the principles I have outlined, we can support programs of open enrollment, in the sense that the community should provide opportunities for all young men and women to enter some institutions of higher or tertiary education, and beyond that to *compete* for entry into *any* educational institution. However, the right to receive an education, and to the schooling required for that education, does not carry with it the right to enter any specific institution at any specific time. But it does carry with it the right to be to be judged and evaluated by the same standards as all others who have been admitted.

There are some institutions that seek to staff themselves with the most talented faculty available, with the aim of serving the ablest students, engaging themselves in an educational experience and in a process of learning and discovery of inestimable social value and human benefit to the entire community. Provided that one standard of excellence is applied to all applicants, when all the programs of remedial and special training and coaching to enhance the skills of those who need this support to compete have been concluded, the percentage distributions of students in various ethnic groups or among men and women are irrelevant.

But of course these are not the only kind of students that should be of educational concern to us. The ordinary students, even the dull students, have the same right to the best educational experience for their aptitudes as have the superior students. That is why there must be a provision for various types of educational institutions for students of various interests and capacities. That is why there must be provision for continuing education. After all, not all teachers can be the most talented nor all students the ablest. There is a uniqueness about every student that must be respected. That respect is perfectly compatible with the application of a single standard of achievement or award in any given institution. We may and should guarantee the basic needs of food, shelter, education and, whenever possible, vocation, too, but we cannot guarantee anyone against educational failure.

It should be obvious that my whole approach is based on the belief that it is the individual who is the carrier of human rights and not the ethnic or national or racial group. Once we disregard this universalist approach which is blind to color, deaf to religious dogma, indifferent to sex where only merit should count, we practically insure the presence of endemic conflicts in which invidious discriminations are rife and tensions mount until they burst with explosive force. The pluralist society then becomes a polarized society. A pluralistic society is one in which the *individual* is a member of many different associations, in which he does not stand stripped of all ties and parochial loyalties before the power of the all-encompassing State. A pluralistic society is *not* one in which individual rights are bestowed by different contending groups or possessed by virtue of membership in these groups. Such a society, whether it is called multi-ethnic or multi-racial, whether it be Malaysia, Northern Ireland, Cyprus, South Africa, Uganda and other areas where persons do not enjoy equal rights unless their mothering groups possess equal power, is in an incipient state of civil war.

There are many cultural and psychological side-effects that derive from employment practices based on membership in groups rather than from individual merit. First, there is a tendency to dilute and debase academic and professional standards in order to accommodate members of groups who, it is feared, would not make the grade. There is something offensively patronizing in assuming that members of some groups, even when given educational opportunities commensurate with others, cannot, without preferential treatment, compete on equal terms with their fellow citizens. Second, a double standard of status begins to emerge among the personnel of institutions recruited in this way. Invidious distinctions are drawn between those who have made it strictly on their own merits, selected by their professional peers in open competition, and those who have made it by virtue of membership in some group. Qualified members of disadvantaged groups who have succeeded without benefit of preferential selection have been quite eloquent on that score.

I want to conclude by presenting a program of non-discriminatory hiring for institutions of higher learning – the institutions I know best – and by recommending the underlying principle, *mutatis mutandis*, to all other institutions of our society.

Let us assume for the moment that institutions of higher learning have been asked and have agreed to comply with the following rules:[1]

(1) that they publicize in the most open and evenhanded way all their academic and other job openings;
(2) that they recruit applicants from *all available* sources;
(3) that they maintain fully non-discriminatory hiring procedures and keep full records of such items as interviews;
(4) that they comply fully with non-discriminatory promotion and tenure policies;
(5) that they abolish all rules and regulations which are discriminatory with regard to pay, leave or fringe benefits.

Let us also assume that our colleges and universities:

(1) open up fully their respective institutions to all qualified student applicants;
(2) recruit their student body evenhandedly from *all* secondary schools and other possible preparatory channels;
(3) maintain vigorous remedial programs for entrants who wish to remove deficiencies;
(4) maintain comprehensive counseling and other auxiliary programs to facilitate the entry of disadvantaged students into the mainstream of academic life.

Let us further assume that there be maintained a *simple*, *speedy* and *effective* complaint and grievance mechanism inside and outside the academic institutions (lower and appellate levels) to promptly handle complaints involving alleged discrimination on grounds of race, sex, or creed.

Let there be academic and mixed academic/non-academic study groups and standing commissions to investigate continuously the employment *possibilities* and *practices* and, when necessary, recommend censure and the withholding of government funding.

Here is a practical program of positive commitments, remedial measures, and monitoring bodies that constitutes a plan as comprehensive as any in existence to combat discrimination – without demanding 'numerical goals', 'time-tables' or quotas of any kind.

I ask: wherein is it lacking? In what respects is it unfair? Why will it not work if it has the active support of all persons of conscience and good will?

New York University

NOTE

[1] These are substantially the proposals made by University Centers for Rational Alternatives, through its Executive-Secretary, Prof. Miro Todorovitch, to HEW.

IRVING LOUIS HOROWITZ

LEFT-WING FASCISM AND RIGHT-WING COMMUNISM: THE FISSION–FUSION EFFECT IN AMERICAN EXTREMIST IDEOLOGIES

The work of Lewis Feuer provides a kaleidoscopic view of many tendencies and trends within contemporary social science. Above all, he illustrates the fission–fusion trend in intellectual currents as exemplified in the work of Spinoza, Marx, Freud, and Einstein. He not only writes brilliantly on these figures now appended with 'isms', but his work is permeated by socialist, physicalist, and psychoanalytical explanations of political events.[1] In this sense, Feuer is an old-fashioned thinker in the best sense, someone who combines in his person radical persuasions with democratic practice in a variety of fields. His work is not so much anti-communist, as it is anti-totalitarian. I dare say he has been singled out as a Soviet *bête noir* precisely because he has so artfully disentangled radical rhetoric from totalitarian reality[2] (Feuer, 1979).

I am struck by the obviousness of this in reviewing his superb work on Marx and Engels on Politics and Philosophy. What emerges is an individual who believes fully and firmly that the Marxian kernel can be rescued from its Stalinist and Soviet protuberance. He suggests therein that "American development is out of phase with the rest of the world. America, disenchanted with its own Marxists venture of the thirties, is learning the language of conservatism, and in finding itself ever more removed from the Asian and European worlds." I do not know if Feuer would still adhere to this severe judgment, but I am certain that he would continue to implore us "to re-learn the meaning of Marxism". Feuer (1959) well appreciated that as "freedom is reborn in Eastern Europe and Asia, it will speak in the Marxist idiom and try to disenthrall the universal humanist bearing of Marx's ideas from their Stalinist perversion".

What I should like to do in this essay is show how this process of perversion and enthrallment across generations has worked its way out as a special variety of the American social idiom. Specifically, I want to show how the phenomenon of left-wing fascism has become one way of accomplishing what Feuer terms a process of acknowledgment of that

S. Hook, W.L. O'Neill and R. O'Toole (eds.), Philosophy, History and Social Action, 245–266.

tremendous segment of reality which the Marxist philosophy has come closest to grasping (Feuer, 1975). To study the totalitarian phenomenon seriously means to get beyond dogmatism into actual ideological permutations and combinations which still retain a lively sense of politics. Let us proceed now to the task at hand.

The first purpose of this essay is to examine the social foundations of totalitarian doctrines in American life and the connection (or lack thereof) with their European counterparts. Essentially, my claim is that there is a powerful crossover between left and right, between communism and fascism, on American shores. This crossover is blurred by the exaggerated claims to uniqueness on both sides, and often obscured by the different demographic backgrounds of those adhering to left and right forms of totalitarianism. To provide some basis for this serious claim, I should like to examine recent American social history, in particular, those key historical figures who represented left–right linkages – sometimes knowingly, at other times inadvertently – as a function of the political context of American life at its peripheries. In so doing, I am seeking to explore the roots of political marginality in American political culture. For it is precisely the totalitarian characteristics of both extreme right and extreme left that have not only irrevocably fused them, but have doomed them with respect to any conquest of political power. I hasten to introduce a caveat: this is an exploratory statement. Much more work will need to be done to prove conclusively the point of view herein articulated. But it at least holds open the prospects for resolving long-standing debates not only about why America has had no 'socialism' but, no less important, why it has resisted 'fascism' at the same time.

The ideological roots of conservatism are partially located in its need to reconcile two conventionally opposed attitudes to life: the heroic and the moral. The former is most often associated with self-expression, the other with self-denial. The renunciation of joy in favor of destiny takes on a peculiarly Christian cast in twentieth-century conservatism. Writing of Oswald Mosley, his brilliant biographer, Robert Skidelsky, points out that "Fascism's fault, and by implication his own, as he came to see it, was too much paganism and not enough Christianity." But the seeming inability to fuse the higher men with political purpose meant that "Christian love could become the method only when the higher men are not underneath but on top? Till then it would be necessary to say with *Zarathrustra* that 'love hath an end' " (Skidelsky, 1975). Thus spoke Sir

Oswald. We are not dealing with either a foolish position or an incompetent individual. Indeed, Mosley in his journeys through socialism, labourism, fascism, coupled with a Europe purged of American capitalism and Russian communism alike, well articulates a neo-conservatism not far removed from its current expressions on both sides of the Atlantic. The question may be asked: Where do those of the Jewish persuasion fit into such a model ideology? The answer is that clearly, they do not. And to believe that some sort of commitment to a naked anti-communism could possibly paper over a sometimes manifest, often latent anti-Semitism, was at the core of Mosley's naivété. It distinguished a right and left in an age when the latter still gave lip-service to religious and ethnic freedom.

A frequently struck theme of conservatism has been the defense of Western values and Western civilization as a whole. A marvelous, and wholly prototypical example, is Charles A. Lindbergh, whose extraordinary technical exploits throughout his life were performed in the name of a rather ordinary ideological context in which "power to be ultimately successful, must be backed by morality; just as morality must be backed by power". As his biographer, Kenneth S. Davis (1959), succinctly notes: "the morality he referred to was synonymous with Christian ideals." The advocacy of a world organization guided by Christian ethical principles was a far cry from isolationism or a secular socialism. It was far removed from the Jewish persuasion as well.

The horrors expressed by Lindbergh of a 'ruthless dictatorial system' developed in Soviet Russia, in which millions were 'denied justice' and 'unknown numbers' labored as slaves, and by which 'a record of bloodshed and oppression never equaled has been made' anticipates precisely the position outlined by Solzhenitsyn in *The Gulag Archipelago* – and moreover the solution advocated. We cannot ignore the dangerous imbalance of so many conservatives in their rather tepid response to Nazism and the Holocaust in contrast to their impassioned response to communism and Babi Yar; this very absence of 'even-handedness' is what makes an alliance between conservative ideology and Jewish theology a curious alliance at best, and a misanthropic association at worst (Horowitz, 1984).

An elusive form of conservatism is that tradition linking its philosophical tenets of absolute morality and political probity with a critique of capitalism. As Sheldon Marcus (1973), the biographer of Charles Edward Coughlin, noted: "Regardless of prior ideological commitment on the left or on the right, they were attracted by Coughlin's slashing attacks against the concentration of power in the hands of big business and the federal government's passivity in the face of crisis." The prevalence of historical short-sightedness being what it is, we tend to view left-wing fascism as a recent phenomenon. But the roots are deep in American populism.

Coughlin regarded Marxism as the "parent of both Nazism and communism", left and right versions of the "same bird of prey". He was an early advocate of political evenhandedness: "We regard every organization against war and fascism as a menace until the officials of such organizations incorporate the war against communism in their program." His anti-Semitism, while late in developing, was a cornerstone of *social justice*, and probably the centerpiece of the paper to his followers. Nor should we forget his anti-war stance, his belief in an America "self-sustained and self-contained". Not unlike the anti-war advocates of today, he urged us to "be ready for the defense of our country and homes from alien aggressions, but never again be the aggressor".

Historians of the American Right have increasingly become aware that their formerly held neat distinctions between extremist politics were convenient but largely chimerical. In point of fact, fascist doctrines in America were often tinged by populist critiques of big business and oligarchical concentration, while communist doctrines were just as often permeated by elitist leadership principles and contempt for broad sectors of working people who failed to transcend "trade union consciousness" and more into "vanguard positions". Nor were these differences exclusively between left and right, they were also between European fascism's idea of unity and national purpose. State control of the economies were equal if not greater in Hitler's Germany as in Stalin's Russia.

Emphasizing the work of Huey Long and Charles Coughlin, Alan Brinkley (1982) points out that they resonated to a different set of concerns from their presumed European mentors.

> At the center of their message was a commitment to a major shift in the locus of economic power in America, not to the state, but to small community institutions and to individual citizens. Long and Coughlin contemplated a far more fundamental assault upon the 'plutocrats' and 'financial despots' than the European fascists ever attempted . . . Long and Coughlin envisioned a far different and more limited role for the government than the fascists proposed. Its purpose would not be, as in Germany and Italy, to subordinate

individual economic interests to the central goals of the nation. It would, rather, liberate individuals from the tyranny of the plutocrats, restore a small-scale, decentralized capitalism that would increase, not restrict, economic independence.

What we have then is a picture of the American right more in tune with what is now referred to as the 'new conservatism' than identifiable with the main currents of classical European fascism or communism.

In his work on the Protestant far right in America from the depression to the cold war, Leo P. Ribuffo makes essentially the same point, underscoring the concerns of far right and far left in operational terms. He points out that "far right activists and their opponents were influenced by the fear of deception that characterized the whole inter-war period. "When the United States gets fascism," Huey Long was supposed to have said, "it will call it anti-fascism." The remark was apocryphal, but the underlying premise that fascism would triumph in disguise, was shared by socialists and centrist liberals, Communist theoreticians and the editors of *Life*." Ribuffo (1983) makes the further point that "fear of the far right undermined the left's commitment to civil liberties". Indeed, this is the essential thesis of his book. What I find most interesting is this new awareness that extremism, totalitarianism as such, is a key aspect of American social history. The forms of such extremism may at times be right-wing at other times left-wing, but the core message seems unchanging; only the sources of positive goods, i.e. Germany or Russia, might be quite different. We are thus in a position to take this story up to the present. The period covering 1933–1953 – that is, from the emergence of Hitler to the death of Stalin (if not their 'isms') – is well represented in the literature. The more challenging task is to examine the current epochs – both as a function of this ambiguous totalitarian legacy and on its own terms – as a response to current social and political events.

In taking such figures as Mosley, Lindbergh, and Coughlin as exemplars of conservatism, I certainly do not mean to imply, even remotely, that a tradition as rich, deep, and marginal as this can be reduced to such quasi-political types. Rather, they typify a certain configuration of ideological thought that sought to transcend major struggles of the century: between communism and capitalism, totalitarianism and egalitarianism, intervention in and isolation from international conflicts. But the lynchpin in these efforts at synthesis, were appeals to absolute, dogmatic value systems, that, whatever their labels – Americanism or nativism being the clearest illustration – appealed to the organizing power of Christian doctrine. Again, while it would represent little more than calumny to identify an authentic Christian mission with such rabid political theologies, the appeal of Christianity was a serviceable tool in the exclusion of Jews and other non-believers and/or infidels.

This core aspect of traditional conservative doctrine, or 'left-wing fascism' is something the neo-conservatism of Podhoretz and Kristol, among others, seeks to ignore at its own peril. But even if they do not buy or sell the bundle of goods offered by the old conservatives, even if they expressly and emphatically repudiate its anti-Semitic core, they are left with parts of an argument that forms a *leitmotif* with the older conservatism – a defense of Western civilization and a perception of that civilization being directly and primarily threatened by world communism. The issues thus quickly move to the empirical, that is, can such an anti-communism sum up the troubles and tribulations of Western democratic societies?

Fascism has recently been called "the system of ruins" (Gray, 1983). In its 'pure' Hitlerian form, this was certainly the case, but in its 'neo' form, in its linkage of bringing an end to the social division of labor with the manifest rebellion against industrial-commercial society, fascism with its left posture, with its humanistic twinge, has taken on renewed meaning. No writer of our time has expressed these new linkages more ably than Ernst Nolte. It is worth quoting him fully (Nolte, 1983):

> Fascism can be directly compared with Marxism of the Soviet nature only in its radical form, in respect of its inner solidarity and its appeal to comrades of like mind in all countries; Italian fascism, in its phase as a development dictatorship, and more than ever the Croatian Ustase and the Rumanian Iron Guard were in fact, on the contrary, more like many of today's 'national liberation movements' than like late National Socialism . . .; there is nothing more grotesque than a "theory of Fascism" which denounces capitalism with much sincere indignation as the root of Fascism, at the same time overlooking that the theory identifies itself with conditions which show all the formal characteristics of Fascism. It is not astonishing that the liberal capitalist system produces Fascism under certain circumstances, but it is astonishing that in the great majority of cases Fascism has not succeeded in gaining power in spite of certain circumstances. The explanation can only lie in the fact that this social system with its peculiar lack of conception, its deep-rooted divergencies, its inborn tendency to self-criticism, its separations of economic, political and spiritual power obviously offers strong resistance to a transformation to fascist solidarity, and is aware that the deliverance which is promised would at the same time be loss of self. Thus capitalism is indeed the soil of Fascism, but the plant only grows to imposing strength if an exhorbitant dose of Marxist fertilizer is added to the soil.

What we have then is not so much a new phenomenon – left fascism has been with us for some time – but rather a new soil of a new continent on which the dry bones of inherited theory are being nurtured.

This is not to say that all distinctions between left and right were or are chimeric in character. Indeed, in the Utopian environment of the 1930s, sharp differences in attitudes between the extremes were manifest in everything from the ethnic and racial composition of the two groups to attitudes on a wide variety of measures to resolve the 'social problem'. As the decade wore on, pro-German versus pro-Russian beliefs became increasingly prevalent; especially insofar as these two nations embodied fundamental value orientations toward race, religion, labor, capital, and the confluence of ethical postures embodied in each nation-system. But with the Nazi–Soviet non-aggression pact of 1939, a brief period of cross pollination and cross-fertilization came into being which reduced many older ideological distinctions to rhetoric. As a result, a set of common negative sentiments toward American democracy replaced (displaced might be a better term) older left–right animus. The democratic essence became the core enemy.

It was these sentiments that led to a left-fascist coalescence in the late 1970s and early 1980s. The reason for emphasizing Father Coughlin and Huey Long as 'voices of protest' is to indicate that a tradition of right communism already existed; but it was nourished on infertile soil in America and hence was doomed to fail. The articulation of anti-big-business sentiments was often muted by more immediate concerns of social justice and emotional appeals of anti-Semitism and racism. But this was also characteristic of West European national socialism prior to the Second World War, and East European communism after that great war. The argument herein presented is not reducible to some simple-minded aphorism about extremes meeting; rather, it is an effort to understand the totalitarian penumbra which contains unto itself a variety of strategies, sentiments and slogans – all bent toward the termination of democratic societies.

Epiphenomena such as left-wing fascism or right-wing communism emerge as extreme forms of Utopian visions, and end as counter-Utopian onslaughts. If Utopianism in its essence contains an impatience with things and systems as they exist, extremist versions carry such impatience to a demand for the extirpation of evil, the rooting out of parasitical forms of social existence, a new instauration of moral virtue in the name of racial purity, class consciousness, or historical infallibility. Thus, it is that right-wing communism or left-wing fascism are identifiable with specific periods and movements. But these are not mechanistic decennial trends. They emerge as tactical efforts to deny democratic processes their due in the name of biology, history, or simply 'the people'.

I should now like to carry the discussion up to the present period; indicating that what first appeared on the scene of twentieth-century history as tragedy, does not reappear in our age as comedy, but quite the contrary, only as much deeper tragedy.

Vladimir Lenin issued in 1920 a stunning pamphlet on left-wing communism. Several decades later, a left-wing fascism has materialized, and the concerns of this essay are how, like its predecessor, it deserves to be characterized as 'an infantile disorder'. It was easy for Lenin to recombine elements in political society to forge new theories, yet it seems terribly painful for his followers to do likewise. For Lenin, the Bolshevik struggle was two-sided, chiefly against opportunism and social chauvinism, but also against petty bourgeois revolutionaries operating on anarchist premises. Lenin's own words, in *'Left-Wing' Communism: An Infantile Disorder (1920)*, are equally appropriate to the conditions of the 1980s:

> The petty bourgeois, 'driven to frenzy' by the horrors of capitalism, is a social phenomenon which is characteristic of all capitalist countries. The instability of such revolutionariness, its barrenness, its liability to become swiftly transformed into submission, apathy, fantasy, and even a frenzied infatuation with one or another bourgeois 'fad' – all this is a matter of common knowledge. But a theoretical, abstract recognition of these truths does not at all free revolutionary parties from old mistakes, which always crop up at unexpected moments, in a somewhat new form, in hitherto unknown vestments of surroundings, in peculiar – more or less peculiar – circumstances.

Fascism is not simply a political condition but is brought about by rooted psychological dislocations; these, linked to larger concerns, exercise an independent dynamic. The very term *infantile disorder* sharply focuses on the subjective qualities of fascism. Even a politically oriented analyst such as Leon Trotsky speaks, in *The Struggle Against Fascism in Germany* (1971), of the cycle of fascism as "yearning for change . . . extreme confusion . . . exhaustion of the proletariat . . . growing confusion and indifference . . . despair . . . collective neurosis . . . readiness to believe in miracles . . . readiness for violent measures." While these characteristics are invariably linked to a social class, a constant fusion–fission effect characterizes the momentum toward fascism. These terms also describe religious-political movements like Sun Myung Moon's Unification church and political-religious movements like Lyndon LaRouche's National Caucus of Labor Committees (NCLC), whose name was changed to the United States Labor Party (USLP).

It would be foolish, even dangerous, to suppose that a set of psychological variables or political positions determines disposition toward fascism. However, appeals to authority, tradition, or the mystique of nation, blood, or race are necessary preconditions. Ideological denunciation of appeals to evidence, discourse, rationality, individual conscience, decision reversals, or consensus for specific policies are also characteristic. Rejection of these elements of psychic or intellectual conditioning is a key factor in determining a propensity toward fascism. If the success or failure rate of fascism has to do with economic dysfunctions and political systems, preconditions for fascism have to do with social psychology, the mass psychology of a people. Inroads of cults into American life provide somber evidence of a propensity toward fascism.

We have so taken for granted distinctions among left, right, and center that it has become difficult to perceive new combinations of these categories. New practical political integrations disquieting the liberal imagination are hard to absorb. If Lenin was correct in his criticism of left-wing European communism for its exaggerated emphasis on purity at the expense of victory and on vanguard putschism at the expense of mass participation, similar phenomena of a different ideological persuasion are taking place in America. The purpose of this essay is to suggest the character of this recombination of political categories: how it functions in American life not simply to alter the nature of marginal politics but also to affect mainline political decision-making.

While my analysis is largely confined to the United States conditions, the state of affairs I call left-wing fascism is an international phenomenon. The following examples are illustrative. Massimiliano Fanchini was arrested in connection with the Bologna bombing. He first draw attention as a part of a Palestine Solidarity committee, which he helped organize with another fascist, Franco Fredo, who was jailed for killing sixteen people in Milan in 1969. Claudio Mutti, known as the 'Nazi academic' because of his post at the University of Parma, founded the Italian–Libyan Friendship Society and helped publish speeches by the Ayatollah Khomeini in Italian. The supposedly leftist Baader-Meinhof gang, which earlier only lectured the Palestine Liberation Organization on the need for armed struggle, bought its first load of small arms from the neo-Nazi Bavarian underground. Christopher Hitchens (1980), foreign editor of the *New Statesman*, noted in a recent article:

> There is a small and squalid area in which nihilists of left and right meet and intersect. There is a cross-fertilization, especially in Italy. Fascists often borrow demagogic leftist titles. One of the agreements facilitating this incest is a hatred of Israel; the other one is a hatred of democracy and a mutual conviction that Fascist/Communist takeover will only hasten a Fascist/Communist victory.

CHAOTIC ANCESTRY

Like all movements, left-wing fascism has a somewhat chaotic ancestry. Foremost is what might be called the later Frankfurt school, which emphasized in an uneasy mix the early Marx and the late Hegel and which was most frequently, if not necessarily properly, identified with the works of Adorno. The characteristics of the Frankfurt school derive from Adorno's strong differentiation between mass culture and elite culture, and his concern with the massification of society in general. For the first time in the history of Marxism, Adorno addressed a strong attack on mass culture. The obscurantist-elitist aspects of Adorno's, *later*, post-Second World War work does not refer to the democratic socialist analysis of the Frankfurt school offered by Franz Neumann, Max Horkheimer, and Herbert Marcuse, among others. Nor does it even refer to Adorno's own efforts at developing a political psychology while he was working in America on the 'Authoritarian Personality' project. But to deny the anti-popular and teutonic characteristics of Adorno's later works, worshipfully introduced in English by British and American scholars who should have known better, is to deny the obvious – and the dangerous.

Whether it be popular music or popular art, there is a clear notion in Adorno's work that mass culture is tasteless, banal, and regressive. His assumption is that such culture evolves in some abstract sense through commercialization of social classes and the existence of a worthless society. This critique is pointless since the emotional assault is on the masses for having such a culture. In many respects, Adorno thus sets the stage for a culture of left-wing fascism, which represents an attack on the popular organs of society for being what they are and a corresponding elitist demand that they be otherwise, that is, purified.

The Adorno line of reasoning, its critical negativism, assumes that what people believe is wrong and that what they ought to believe, as designed by some narrow elite stratum of the cultural apparatus, is essentially right. With Adorno, the theory of vanguard politics is carried over into the theory of vanguard culture. The cultural apparatus is blamed for the elevation of mass culture into high culture. Attention given the so-called Frankfurt school in present racial circles derives not from its origins in anti-totalitarian and anti-Nazi activities but rather from its elitist outcomes: attacks and assaults on masses and their culture. The pre-war Frankfurt school, with its emphasis on rationality as the basis of revolution, was corrupted in the hands of Adorno. It became a doctrine of rationality as cultural traditionalism. In this way, Adorno became central to the thinking of the *avant-garde* of left-wing fascism.

A second element of this belief characteristic of left-wing fascism is Marxism as praxis – or without Marx, and sometimes without Lenin. This results in what might be called 'praxis theory' or 'action theory'. It does away with the need for either an economic base to revolution, essential for Marx, or the political base of organization held essential by Lenin. What remains is a residual sentiment favoring revolutionary mysticism. The vaguely anarchical assumption is that the sum total of what one really needs is an action group, some kind of organized group or *foco*, usually clandestine, to create sufficient chaos or destruction of the state and society in selected periods of the capitalist economy. The combination of economic chaos and political protest will in itself somehow produce revolutionary action. Multinational terrorism will move into a breach, presumably magically offsetting the multinational economy.

This is often called the Cuban model of revolution, inspired by the works of Regis Debray. The transposition of a model from a small island like Cuba, with its special conditions of single-crop socialism, is quite difficult. What was originally a theory for social change in Cuba becomes enlarged into a universal theory of change; one is left with a 'theory' of the vital force. The theory of the putsch, the clandestine conspiratorial small group capable of seizing power at the proper moment, is common to fascism and until recently was not associated with Marxism. The theory of the *foco*, or revolutionary focus, first reduces Marxism by stripping away its sense of economic forces of oppression, then by stripping away its emphasis on the political sources of organization, and finally by stripping away its mass base. One is left with a theory of small group conspiracy, or terrorism in the name of Marxism, rather than a theory of Marxism as a source of social change and revolutionary action involving broad masses.

A third vital pivot is nationalism, in which the demand for revolutionary change is lodged in patriotic claims of the total system, a demand for the moment, having nothing to do with history and antecedents. Such a nationalist approach insists on spontaneity and is not necessarily linked with historical forces or recurring patterns. At this level left-wing fascism is fused to a theory of anarcho-syndicalism. The nationalist pivot involves not so much a doctrine of liberation as a doctrine of activity uninhibited by the need for social analysis. It is predicated on a notion of will and action at the correct moment to preserve the nation against its real or presumed enemies.

Other elements in the nationalist tradition fit into this left-wing fascist model: that every ordinary individual craves order over chaos; that one does not need a special theory of society to achieve revolutionary action; that individual economic origins are less important than social roles. These elements from Pareto, Sorel, and Mosca are incorporated into a left-wing fascist interpretation of the world, in which psychological mass contagion replaces social history as the interpretation of human events.

The sources of left-wing fascism are not abstract. Those who are enamored with appraisals of American society that seek immediate gratification and relief from ailments have become innovative in organizational form no less than ideological norms. Seeking ways to effect social change regardless of scientific or social base is the key to left-wing fascism. The effort to enlarge the cult into a state religion, the attempt to impose order and leadership on a society that seems purposively leaderless and fragmented, is characteristic of fascism – right or left, religious or secular.

In left-wing fascism we are dealing not so much with notions of traditional political involvement or traditional minor political means and ends. This involves inspiring others while servicing the needs of the actors, whether one is talking about special groups of nationalists or extreme self-styled radical groups seeking direct confrontation with other radical groups. The very act of confronting the enemy replaces any sense of organization or systems analysis. Action determines and defines one's place in the hierarchy of a political movement. In left-wing fascism the guerilla movement replaces the clandestine *bund* as the organizational vehicle. But its impact is not simply to be dismissed because of its barricade orientation. The latter provides the basis for the militarization of politics, its decivilianization, a central precondition to the fascist seizure of power.

ELITISM AND POPULISM

The main political source of left-wing fascism is its strange denial of America and the democratic system, together with the assertion of socialism as an abstraction. Left-wing fascists have the unique capacity to examine socialism without comment on the activities of the Soviet Union. They talk about the United States rather than about the loss of democracy. There is an inversion: the search for socialism becomes close to an abstract Utopian ideal, but when it comes to a discussion of democracy, discourse is critically and severely linked to the United States as a nation-state. The rhetoric constantly shifts. So-called enemies are unambiguously identified as the United States and its allies. When dealing with its own allies, however, left-wing fascism turns socialism into a generalized hypothesis rather than concrete forms of socialism as they are expressed throughout the world. What Adorno (1978) appropriately called the "collectivization and institutionalization of the spell" becomes the new fascist norm.

Left-wing fascism accepts socialism as an abstraction but rejects socialist practice and reality, and hence critique as a source of democratic renewal. The history of fascism in the United States mirrors that of Europe. Socialism, far from being dropped, becomes incorporated into the national dream, into a dramaturgy for redemption, for a higher civilization that will link nationhood and socialism in a move forward. This combination of words, *national and social*, generates a new volatility. These two words together can arouse stronger and more active participation than either of the concepts taken separately.

The weakness of traditional right-wing organizations is that they asserted the primacy and value of Americanism as nationalism apart from socialist values. The weakness of traditional forms of leftism is that they have asserted socialism over and against American or national values. The potential strength of left-wing fascism, such as that practised by the NCLC, is its unique combination, its ability to see how these concepts of Americanism and socialism can operate together as a mobilizing device in the development of a new fascist social order. The unique characteristic of left-wing fascism is its capacity, like its European antecedent, to combine very different ideological strains, traditional right-wing and traditional left-wing behavior, and come up with a political formula that, if it has not yet generated a mass base, has at least the potential for mass appeal.

The contents of left-wing fascism are heavily based on an elitist vision of the world. At every level of society it juxtaposes its minoritarianism against majoritarianism. It may take libertarian or authoritarian forms, but it always defends its leadership vision over any populist vision. Some examples are the hip versus the square, the gay versus the straight, the individualistic free soul versus the family-oriented slave, those who believe in the cult of direct action versus the fools who participate in the political process, those who practice non-violence over those who assert willfulness and violence as measures of human strength and courage, those who have strong affiliations with cults and cultism over and against the traditional non-believer (a marked departure from the anti-theological vision of most forms of leftist and socialist behavior), those who argue the case for deviance over and against mainline participation in the working class or in segments of class society, those who choose underground organizations in preference to established voluntary organizations, and ultimately, those who choose some type of deracinated behavior over and against class behavior and participation.

Historically, communists, like fascists, have had an uncomfortable attraction to both elitism and populism. The theory of vanguards acting in the name of the true interests of the masses presupposes a higher science of society (or in the case of fascism, a biology of society) than that susceptible of being reached by ordinary citizens. The superstructure of science, like culture generally, becomes a realm in which elites act in the name of the public. What happens to the notion of the people determining their own history in their own way? Here populism, or pseudo-populism, steps in to fuse formerly antagonistic trends. In some mysterious, inexplicable manner, these mass forces must be shaped or molded. Under communism, in sharp contrast to fascism, the stratification elements in the national culture are deemed unique or uniquely worth salvaging. But in the anti-ideological climate of the 'new world', people (class) and fold (race) blend, becoming the raw materials for fashioning the new society.

Left-wing fascism does not so much overcome this dilemma of elitism and populism as it seeks to harness both under the rubric of a movement. Having its roots in the 1960s, left-wing fascism views the loose movement, the focus, the force, as expanding upon the fascist elan and the communist vanguard. It permits a theory of politics without the encumbrance of parties. It allows, even encourages, a culture of elitism and crackpot technocism, as in the LaRouche emphasis on computer technology as a general ideology, while extolling the virtues of a presumed inarticulate mass suffering under inscrutable false consciousness.[2] The mystification and debasement of language displaces the search for clarity of expression and analysis, enabling a miniscule elite to harness the everyday discontent of ordinary living to a grand mission. Left-wing fascism becomes a theory of fault, locating the question of personal failure everywhere and always in an imperial conspiracy of wealth, power, or status (Kogan, 1968).

Fascism requires a focal point of hatred to unify behind. Thus, when fascists advocate anti-Semitism, they are simply using a tactic. It becomes not simply a tactic of fascism, nor is it opposed by communism. It becomes a modality of affixing the climate of a post-Nazi holocaust, a post-Stalinist Gulag, and the monopoly of petroleum wealth by forces historically antagonistic to Jewish ambitions. The new left-wing fascist segments, weak within the nation, can draw great strength from 'world forces' deemed favorable to its cause. The unitary character of anti-Semitism draws fascist and communist elements together in a new social climate. Anti-Semitism is the essential motor of left-wing fascism. The grand illusion of seeing communism and fascism as polarized opposites, the latter being evil with a few redeeming virtues, the former being good with a few historical blemishes, is the sort of liberal collapse that reduces analysis to nostalgia – an abiding faith in the unique mission of a communist left that has long lost its universal claims to a higher society. This catalog of polarities, this litany of beliefs, adds up to a lifestyle of left-wing fascism. Isolating any of these reified frameworks may lead to the conclusion that the dangers are less than catastrophic. But in this panoply of beliefs and practices, one finds the social sources of left-wing fascist participation and belief.

Even in the formation of the new left in the 1960s, the roots of a left-wing fascist formulation were in evidence. Now, in a more pronounced form, what has evolved is a strong shift from a class, party, or movement concept characteristic of the sixties toward cultism (Horowitz, 1968). We have cults not only in the strictly religious sense but in the political sense as well – marginal movements gaining small numbers of adherents but having a profound impact on the edges of society. Like the Nazi movement of the early 1920s, these left-wing fascist movements of the 1980s, such as the LaRouche organization and various socialist parties, are considered too small and inconsequential to have any impact on the body politic. But the danger to the society as a whole is that as the active element in the political process shrinks, this fringe becomes increasingly important. They do have sufficient numbers for a seizure of power once one takes into account that they do rely for victory on technology and not numbers. They rely on organization, swift movement, willfulness, and the ability to seize the critical moment. In this sense, left-wing fascist movements are not unlike the Nazi movement of the early twenties; although weak, marginal, and leaderless, they are in fact very much part of a social scene marked by powerful economic dislocations and putschist tendencies in segmented political processes (Harvey *et al.*, 1977).

Once left-wing fascism is seen in an authoritarian effort to destroy the legitimacy of the established system – a series of diminutions in voting participation, party affiliation, and faith in parliamentary systems and the achievement of social goals in an honorable and honest manner – then the potential of left-wing fascism becomes manifest. This also represents a decline in traditional socialist fallback positions of mass action, mass participation, and ultimately mass revolution. That collapse of trust in the popular sectors corresponds to the collapse in party sectors. What might be called the bourgeois or political parliamentary pivot on one hand and the popular or revolutionary pivot on the other are both viewed by left-wing fascism as a snare and a delusion, a mechanism for postponing the social revolution that is going to provide the cures to all ailments and remove all the temptations of ordinary people. Left-wing fascism ultimately represents the collapse of bourgeois and proletarian politics alike (Abel, 1980). It is not only the end of ideology in the traditional sense, or an end to participation in the political process, but an end of ideology even in the socialist sense of adherence to revolutionary processes that ultimately promise organized change and social justice.

END OF IDEOLOGY

Left-wing fascism assaults both mass and class notions of legitimacy, both Jeffersonian and Leninist visions of the world, both the rational discourse and the popular participation models. Left-wing fascism is that unique rejection of both elements and the incorporation of nationalism from the bourgeois ideology and utopianism from the proletarian ideology. These rejections and absorptions define the four-part paradigm of left-wing fascism: for nationalism in general; for socialism in general; against parliamentarianism in particular; against organized political parties in particular.

Left-wing fascism is much more than a political psychology. It develops direct-action frameworks in terms of specific modes of insurgency techniques and connects them to the tradition of fascism and extreme nationalism. There is a strong element of racism and anti-Semitism in this movement, cleverly rendered by a 'dialectical' pitting of Jews and blacks in archetypical terms. There is a further belief that the black movement must be subordinate to the class structure of American life and that blacks who see their own national destinies apart from this new movement are suspect. This *leitmotiv* of disdain toward successful blacks remains muted. The anti-Semitic modality is overt and made manifest first by its fashionable currency in the Soviet Union and the Middle East (Gottlieb, 1979). Historically, fascism has had a strong component of anti-Semitism: conviction of the need to liquidate the Jew as a political and economic entity and ultimately even as a biological entity. This easy glide from anti-Israeli to anti-Semitic visions has become part of the international left-wing rhetoric of our day. To move one large step further to left-wing fascism by utilizing anti-Semitism as a pivot becomes relatively simple, especially in the context of policy ambiguity concerning the legitimate claims of contending forces for national homelands.

Common wisdom has it that the most virulent forms of fascism in the twentieth century took anti-Semitic overtones. Less known but equally plain is that in light of Soviet politics from the end of the Second World War in 1945 until the present, anti-Semitism has been a *leitmotiv* of the Soviet system. There is no need here to argue whether anti-Semitism is at the center or periphery of Soviety orthodoxy. That such a controversy can even exist indicates the breadth and depth of anti-Semitism as a potent force in current affairs: it is the point at which the fascist and communist hemispheres are joined. Anti-Semitism is the cement providing a crossover from right to left in terms of both ideology and personnel.

The anti-bourgeois character of both fascism and communism in an American context has now extended to an assertion of anti-Semitism as a prototype of that anti-bourgeois sentiment. Probably for the first time, Jews rather than the customary Italians and Turks are now being blamed for drug traffic, with the consequent coming together of strange bedfellows. Again, we are led back to our prototypical organization, Lyndon LaRouche's group and its adherents. Despite the most manifest forms of racist appeal, LaRouche managed a united front with Wallace Muhammad, who in turn took the Black Muslims away from a black emphasis to an Islamic identification.

The effectiveness of the NCLC is seen most clearly in electoral activity and in the organization's success in building single-issue alliances with forces as diverse as the ultra-right-wing and anti-Semitic Liberty Lobby, the Black Muslims, and conservative-oriented Teamster union officials. . . . LaRouche and Black Muslim leader Wallace Muhammad formed an 'Anti-Drug Coalition' which has spread to at least eight cities. The coalition is based on LaRouche's theory that Jews are responsible for the drug traffic. The coalition's activites include mass rallies in ghetto churches; intensive and effective lobbying for stronger narcotic laws; and seminars in inner-city high schools. Wallace Muhammad has repeatedly refused to break off this alliance despite appeals from Jewish organizations and responsible Black leaders. The coalition has attracted an amazing range of clergymen, businessmen, mayors, law enforcement officers, state legislators, Masonic leaders, and trade union officials (Bellent *et al.*, 1977).

This indicates the emergence of a left-wing fascism that has learned to use the techniques of right-wing fascism with impunity (Beichman, 1981). It has also learned to appreciate the mass character of appeals to anti-Semitism. Historically, the problem of the American left has been its narrow socio-economic base, intellectual self-isolation, and above all isolation from the mainstream of workers. It perceives the working class as ready to be tapped, but only if the tactics are appropriate to the current internal situation – one in which Jews are perceived as isolated from working-class networks, gathered in the professional and middle strata of the population, and ideologically and organizationally distanced from their traditional Democratic party moorings. In such historical circumstances, left-wing fascist elements have seen this as an ideal opportunity to seize a political initiative and link up with social segments of the population never before tapped.

The ambiguity of the present situation harbors the sort of pseudo-populism that can easily accommodate to fascist and socialist ideals, both of which make appeals to totalitarianism in its most advanced, virulent forms. Left-fascism, unlike national socialism, no longer stands for a set of rural values over and against urban corruption, or mass sentiments over and against elitist manipulation. While masquerading under populist slogans, left-fascism is highly urban and elitist. The effectiveness of such marginal movements depends heavily on the state of American national interests: whether they can be sufficiently polarized to prevent concerted policy-making or sufficiently galvanized to reduce such left-wing fascist varieties of populism to manageable and non-lethal proportions.

Another serious element in left-wing fascism is its political mysticism, in which the cult of the group displaces individual conscience. Socialism becomes devoid of concrete practice or specific content. Socialism as negative utopianism becomes the order of the day. Real socialist practices are simply disregarded or at times privately rebuked. Like satanic lodges, new groups emerge with the strong feeling that Stalinism has been an oft misunderstood phenomenon that deserves to be supported once again. Just as there are cults of Hitler, there are now cults of Stalin – small bands of people convinced that history has assessed these leaders wrongly and that the source of strength of any future movement will involve a re-evaluation of these former political figures.

What we have described remains a nascent movement, an ideology and organization in the making. We are not dealing with finished ideological products of large-scale political movements capable of threatening established structures. Nor are they necessarily a threat to classical left-wing politics. Left-wing fascism does, however, provide an answer to a question plaguing our century: In what form will fascism come to America? What will be its ideology? What will be its social message?

American fascism could provide a focus with a series of left-wing components: monitarianism in the form of libertarianism; fundamentalism in the form of nationalism; a defense of socialist theory with denial of socialist practice; an assertion of nationalism and Americanism as values, with a denial of mass participation and mass belief systems; elitism as vanguard populism; a mobilization ideology in place of a mobilized population. These tendencies remain nascent; these fusions still remain to be crystallized in political practice. As we make attempts in the 1980s to overcome severe economic dislocations and a breakdown of organic union manifested in a hesitant attitude toward patriotism, and insofar as we exhibit a system with no apparent public commitment and only a series of specific egoistic requirements presumably underwritten by a doctrine of rugged individualism, while small groups make cynical determinations for large-scale policies and structures, to this extent we can expect to find left-wing fascism a powerful component in the political practice of the remainder of the twentieth century.

The twentieth century is polarized into diametrically opposed secular faiths. This dichotomy has taken hold because a century of war and genocide has given expression to competing messianic visions. After class annihilation comes classlessness, after racial annihilation, a triumphal master race. Subordination of the person to the collective is the common denominator. New totalitarian combinations and permutations are dangerous because they move beyond earlier hostility into a shared antagonism toward democratic processes as such. Concepts of evidence and rules of experience give way to historicism and intuition. Comfort with a world of tentative and reversible choices gives way to demands for absolute certainty. In such a climate, the emergence of left-wing fascism is presaged by a rebirth of ideological fanaticism. If the forms of totalitarianism have become simplified, so too has the character of the struggle to resist such trends. This awareness offers the greatest potential for democratic survival against totalitarian temptation.

The United States has been singularly prone to look benignly upon the forces of left-wing fascism because it so adroitly managed to escape the real thing, namely, European fascism on one side and Soviet totalitarianism on the other. As a people, Americans are thus more ready to assume the best, not only in people but in extremist propaganda systems. As a result, the process of intellectual cauterization against such extremisms was also an overseas, 'foreign' import. The migration phenomenon, with messages delivered by the victims of totalitarian systems, has become the essential source of resistance to the series of insidious and insipid banalities that go under the label of left-wing fascism. The work of Arendt (1966), Halevy (1966), Lichtheim (1973), Reich (1979), and Talmon (1981), however different in intellectual disciplines and cultural backgrounds, spoke the identical message on one key fact: the integral nature of the totalitarian experience and the fusion of left and right sentiments in converting such sentiments into systems.

The experiences of the Second World War – when conflict displaced fusion and nationalist ambitions pre-empted totalitarian tendencies –

had the effect of crystallizing differences between regimes, nations, and cultures. However, with the removal of the European war-time experience in space and time has also come the dismantling of traditional barriers between extremes of left and right. Their capacity to come together in a set of manifest hatreds for minorities, masses, and ultimately for democracy does not signify the end of ideological disputations characteristic of the century, but only a clearer appreciation that such distinctions among the enemies of democracy weaken the chances for the continuing survival, even slight expansion of new lands and nations, of societies in which every individual counts as one: no more and no less.

Rutgers University

NOTES

[1] Lewis S. Feuer, see in particular 'The Social Roots of Einstein's Theory of Relativity' in *Einstein and the Generations of Science*. New Brunswick and London: Transaction Books (1982), pp. 4–89, which provides an astonishing sense of his versatility in perceiving real disciplines in a fresh way.

[2] See Carol White, 'The Rieman-LaRouche Model: Breakthrough in Thermodynamics', *Fusion* **3**, 10 (Aug. 1980), 57–66; and Steven Bardwell and Uwe Parpart, 'Economics Becomes a Science', *Fusion* **2**, 9 (July 1970), 32–50.

REFERENCES

Abel, L. 1980. 'Our First Serious Fascist?', *Dissent* **27**, 4, Whole no. 121 (Fall), 430–436.

Adorno, T. W. 1978. 'On the Fetish Character in Music and the Regression of Listening' and 'Commitment' in *The Essential Frankfurt School Reader*, Andrew Arato and Eike Gebhardt (eds). New York: Urizen Books, pp. 270–318.

Arendt, H. 1966. *The Origins of Totalitarianism*, new edn. New York: Harcourt, Brace and World.

Beichman, A. 1981. 'The Myth of American Fascism' in *The Heritage Lectures*, no. 7. Washington, D.C.: Heritage Foundation, pp. 1–22.

Bellent, R., Gluss, S., Gordon, E., Kahn, H., and Ryter, M., 1977. *National Caucus of Labor Committees: Brownshirts of the Seventies*. Arlington, Va.: Terrorist Information Project.

Brinkley, A. 1982. *Voices of Protest: Huey Long, Father Coughlin, and the Great Depression*. New York: Random House, Alfred A. Knopf, pp. 281–282.

Davis, K. S. 1959. *The Hero: Charles A. Lindbergh and the American Dream*. Garden City, New York: Doubleday & Company, pp. 418–429.

Feuer, L. S. 1959. 'Introduction', *Marx and Engels: Basic Writings on Politics and Philosophy*. Garden City, New York: Doubleday Anchor, pp. xx–xxi.

Feuer, L. S. 1975. *Ideology and the Ideologists*. New York: Harper & Row, pp. 120–124.

Feuer, L. S. 1979. 'Cultural Detente at a Scientific Congress: The Secret Report of the Czech Sociologists' (ed. and introduced by L. S. Feuer). *Orbis*. **23** (Spring), 115–127.

Gottlieb, R. S. 1979. 'The Dialectics of National Identity: Left-Wing Anti-Semitism and the Arab-Israel Conflict', *Socialist Review* **9**, 5, Whole no. 47 (Sept./Oct.), 19–52.

Gray, J. 1983. 'The System of Ruins', *The Times Literary Supplement* (Dec. 30), Whole No. 4213, pp. 1459–1461.

Halevy, E. 1966. *The Era of Tyrannies*. New York: New York University Press.

Hitchens, C. 1980. 'Eurofascism: The Wave of the Past', *New Statesman* **231**, 18, 567–570.

Horowitz, I. L. 'Preface to the Paperback Edition of Radicalism and the Revolt Against Reason - Then and Now' In *Radicalism and the Revolt Against Reason*. Carbondale and Edwardsville: Southern Illinois University Press.

Horowitz, I. L. 1984. *Winners and Losers: Social and Political Polarities in America*. Durham, North Carolina: Duke University Press, pp. 179–191.

Kahn, H., *et al*., 1977. 'NCLS/U.S. Labor Party: Political Chameleon to Right-Wing Spy', *The Public Eye* **1**, 1 (Fall), 6–22.

Kogan, N. 1968. 'Fascism as a Political System' in *The Nature of Fascism*, S. J. Woolf (ed.). New York: Random House, pp. 11–18.

Lenin, V. I. 1920. *Left-Wing Communism, An Infantile Disorder: A Popular Essay in Marxian Strategy and Tactics*. New York: International Publishers (1940), pp. 10–11.

Lichtheim, G. 1973. *Collected Essays*. New York: Viking Press.

Marcus, S. 1973. *Father Coughlin: The Tumultuous Life of the Priest of the Little Flower*. Boston: Little, Brown & Co.

Nolte, E. 1983. *Marxism, Fascism, Cold War*. Assen, Netherlands: Von Gorcum, pp. 348.

Reich, W. 1979. *The Mass Psychology of Fascism*. New York: Farrar, Straus and Giroux.

Ribuffo, L. P. 1983. *The Old Christian Right: The Protestant Far Right from the Great Depression to the Cold War*. Philadelphia: Temple University Press, pp. xiii, 179, 220 *et passim*.

Skidelsky, R. 1975. *Oswald Mosley*. New York: Holt, Rinehard & Winston, pp. 465–480.

Talmon, J. L. 1981. *The Myth of the Nation and the Vision of Revolution*. Berkeley and Los Angeles: University of California Press.

Trotsky, L. 1937. *The Struggle against Fascism in Germany*, with an introduction by Ernest Mandel. New York: Pathfinder Press (1971), pp. 437–443.

MILTON R. KONVITZ

THE NATURE OF BRONSON ALCOTT

I

Bronson Alcott went forth from the city to converse in the West. His wife distrusted the venture: In 1852, for sixteen such conversations, he received the gross sum of $164; she saw not whence would come the bread for herself and the children. Neither did he, with the eyes of sense, but he knew that a purpose like his must yield bread for the hungry and clothes for the naked, and he would not wait for the arithmetic of this matter.

After one of his conversational tours in the West he came home late at night, waking the family by the bell. Louisa later remembered that Mother and the four little women embraced the half-frozen wanderer who came in hungry, tired, cold, disappointed, but smiling bravely and as serene as ever. They fed and warmed him, and brooded over him, longing to learn if he had made any money. But no one asked him until May said, after he had told all the pleasant things, "Well, did people pay you?" Then, with a queer look, he opened his pocketbook and showed one dollar, saying with a smile that made their eyes fill, "Only that. My overcoat was stolen and I had to buy a shawl. Many promises were not kept, and traveling is costly; but I have opened a way, and another year shall do better." Louisa never forgot how beautifully Mother answered him, though the dear hopeful soul had built much on his success: "I call that doing *very well*."

He could not bring himself to work for gain, except by teaching children, conversing in parlors, or splitting wood for neighbors at $1 per day. His schools for children did not last long. For he was a serious teacher when everyone else belittled the vocation. He treated the children as intelligent persons, as small ladies and gentlemen. He tried to make them comfortable, put backs on their chairs, gave them individual seats and blackboards, for which he paid out of his own meager funds. He abandoned the birch rod. He made the children's minds move from the known to the unknown, from the concrete to the abstract. He purchased from his own funds books for both pupils and parents. He

S. Hook, W.L. O'Neill and R. O'Toole (eds.), Philosophy, History and Social Action, 267–274.

arranged for evening classes and parent–teacher social gatherings. He tried to read the Gospels to the children as though they were a wonderful children's book written by a Bunyan. He encouraged the pupils to keep daily journals of their thoughts and feelings. He did not try so much to teach his pupils as to make it possible for them to teach themselves. His purpose was to set them free – free even from himself. To objectify their wrongs, he had the children inflict on him that punishment which they said they deserved. Then (this was in 1839), he admitted a little black child. The parents of the other pupils demanded that he dismiss this child. Of course he refused, and he was left with five pupils – a white child, the little black child, and three of his own.

The event did not shake his spirit unduly. He left that he was serving mankind as greatly in humble services – in setting trees and teaching his own children – as others did in the noisier and seemingly more widely useful sphere of public activity. Had he only a hundred and more little ones to meet daily, then would his cup have been filled to overflowing for the pleasure it would have given him. It was all one: persuading a child or man to uncover his own gifts, or persuading peas and cucumbers to grow in accordance with the laws of their own nature.

Reform itself needs reforming which plucks up the man in rooting out his sins. Heaven's kingdom comes not by violence but by persuasion. And persuasion, by teaching children or conversing with their parents, is always possible: for whenever two persons have conference with each other, an invisible, though implied person is ever prior to the concert, and to this third person deference is always given as to a common arbiter.

This ever-present possibility of agreement led Alcott to look for it, not for differences or controversies; for the person rather than the individual. He never sneered or ridiculed but sought for a point of easy connection. He should have kept an inn on the world's highway, said Thoreau. Rats and mice made their nests in him, he added; and a thought, he said, floated as serenely and was as much at home in his mind as a duck pluming herself on a far inland lake. The good Alcott, Carlyle called him, all bent on saving the world by a return to acorns and the golden age; a venerable Don Quixote, whom nobody could laugh at without loving. For many years he made plans for a Bible of the Nations, so that petty sectarian differences among men might be removed; but until the age of agreement was reached he preferred to belong to his Church of One Member. Meantime, too, he belonged to other groups of which he was the one member.

Two and a half years before Thoreau chose to go to jail and give an example of his theory of civil disobedience, Alcott refused to pay the town tax and chose to go to jail. It was his protest against what the federal government was doing to Mexico, or was refusing to do for the slaves and the soul of America.

The question of slavery was more in the mind of this non-political animal than in that of most professional moralizers. He refused to vote in any state or national election until 1860 (when he was sixty-one years of age). In that year he went to the polls to vote for honest Abe Lincoln and against slavery. He did all he could to avoid wearing on his person any cotton garment, because cotton was a product of slave labor. Nor would he, a later John Woolman, eat molasses for the same reason. He would not eat rice, this strict vegetarian, because it came from countries where serfdom prevailed.

He was a strict vegetarian from the age of thirty-six to his death at eighty-nine, a mere fifty-three years. He ate no meats, no eggs; to him, to eat these things was to invade the rights of animals. It was all a matter of consistency, of making the deed the end-product of the thought. When he went to market to shop for his wife and children, who clung to the fleshpots, he found the butchers and their wares an offense. Death yawned at him as he walked up and down in the abode of skulls. Murder and blood were written on the stalls. Cruelty stared at him from the butcher's face. He trod amidst carcasses. He was in the presence of the slain. The death-set eyes of the beasts peered at him and accused him of belonging to the race of murderers. Quartered, disemboweled creatures suspended on hooks pleaded with him. He felt himself dispossessed of the divinity. Walking among the stalls he thought of himself as a replenisher of graveyards, or as an animal prowling amidst unclean spirits and voracious demons for his prey.

Was this, perhaps, an overplus of conscience (which can be quite as disastrous as an excess of passion or of intellect, converting talents into demons)? But no one who knew Alcott thought so. The law did not mount him; he mounted the law and rode onward to victory. No Garrison was he, no demonic or satanic character. He was just the good Alcott, true to his own genius: a Diogenes with a lantern, but looking only for his own soul.

II

What Plato or Pythagoras meant to earlier generations, Alcott thought, was a fit subject for antiquarians. The only significant question to him was what they meant in his own day, to people living in Concord or Wolcott or Boston. The contemporaneity of spirit was his only concern. Plato, or Pythagoras, or Jesus, or the author of *Pilgrim's Progress*, became Bronson Alcott in so far as he thought their thoughts. Every Dedicated Mind has the obligation to recognize his kinship with the Infinite Spirit which makes itself known in manifold ways, through the work, thought, and life of affinitive souls. One must seek them out, or hold himself ever ready to recognize and receive them. Emerson said it was amazing how easy it was for his dearest friend Alcott to walk directly, in any strange library, to the books that belonged to him by ties of spirit though theretofore unknown to him. His search for 'Minds' was like the search of the soul for its other half, its partner, in the Platonic allegory.

Man, he said, must bring the ages into his own day. He must behold eternity in his own lifetime. If a man were to summon his soul into his presence would he not find a greater than Jesus or Paul standing before him? Wheresover we may lodge, there is only a thin casement between us and infinity. And how easy it would be to penetrate to the other side if only we would see first things first.

And the first thing is man's spiritual reality. The outward show of things which fills our minds and hours, even the stars in the firmament overhead, must surrender to ideas; only in ideas can one find their meaning and significance; mechanism must surrender to intelligence. There is, of course, an outside world; but events in the external world are so complex, so far removed from the simple powers on which they depend, that they become perplexing and of dubious meaning unless seen in the light of our spiritual sense. When thus seen, external events are of relatively small import; their insignificance stands exposed. Only the self is significant.

We study natural science to advance the knowledge of self, and of the self's reality. We throw ourselves outward upon nature that we may better look upon ourselves. We see the world with our eyes, but it is the pair of eyes behind the eyes that see and circumscribe the world outside. Nothing can show itself in the exterior that has not a prior being and shaping within. We must set out from the wide ground of the spirit. All else is manifestation of spirit.

War, cruelty, ugliness, man's beastliness, are manifestations of man's spirit, even as are peace, compassion, beauty, man's love. Body, the world, is spirit at its circumference. Except man be converted from devotion to the outward, and his vision be turned inward to the life of the spirit – aye, to the life of the spirit as it reveals itself in the consciousness of the little child – he cannot apprehend the true life of humanity. Man can bridge and belt the globe, tunnel and grade the earth and ocean, skate or steem around the poles, mine and map the particles all round, so that the planet becomes for the first time a small estate, with title thereto vested in him, become the possessor of cosmic domains; yet if this Adam bear a Cain in his loins, his mastery of the brute elements and his power to take the hemispheres tightly between his fingers and thumb and make them spin like a child's top will lead to destruction commensurate with his power. For conquest of nature without conquest of self – passage beyond one frontier of matter to another, without pasage beyond the real frontier that lies within us – means but the transmission of the inheritance from Adam to Cain.

It is not enough, however, to recognize the primacy of spirit yet to live our lives blindly, subject to brute force. Practice must be an unvarying comment upon principles; the correspondence must be visible and connected. One must realize and embody his idea as fully as his time shall suffer. Thoughts must not be detached from life; they must manifest themselves in words and deeds. The life of man must be the life of his spirit.

To make the primacy of spirit concrete means to save oneself from idleness, from tame and servile engagements, from compliance with the vulgar aims and pursuits of one's age; to lift oneself above its low maxims. It means to refuse to dine at the tables of the fashionable, the voluptuous, the opulent. A Dedicated Mind has the obligation to live a life which shall serve as a rebuke to all such meanness of spirit. He must earn his livelihood only by engaging in such activities as spring from his deepest nature. He must do nothing, to get money, which is not worth doing on its own account. Nothing is in better taste than poverty. Poverty alone permits one to be a whole person. The poor man alone can refuse to give his hand to a rich merchant.

Human life is a very simple matter. Breath, bread, health, a hearthstone, a fountain, fruits, a few garden seeds and room to plant them in, a wife and children, a man friend and a woman friend, conversation, neighbors, and a task lifelong, given from within: these are contentment and a great estate. On these gifts follow all others, all graces dance attendance, all beauties, all beatitudes that mortals can desire or know.

For one whose poverty is dedication, waiting is repose. It is the night of the soul without which it doth no miracles. For repose is faith perfected from watching, and all man's spirit acts in this state.

III

This man Alcott, harsh toward himself but gracious and gentle to all others, had many ways of winning happiness. Though the poorest man in town, he felt to be far happier than the bankers and merchants and freeholders. There were many times when he heard, ringing within him, all the bells of joy.

For one thing, he counted himself the richest of all men in possessing those friends of his: Emerson, Thoreau, Hawthorne, Channing. He esteemed them as the victories of his life. They were his country and his countrymen. It was a great day when he could write in his journal (one of the fifty volumes he wrote in his years): "All day with Emerson and Ideas, the rain pouring outside", or when he could write: "Eat strawberries and cream at Emerson's with Hawthorne, Thoreau, Sanborn, Hunt the artist . . . Come home with Hawthorne at half past 10." To Alcott these friendships were great creative events, products of the spirit as is music or painting. And he treated his friends as we do paintings and statuary, surveying them from the perspective of an affectionate idealism.

A garden, too, was to him a work of art. A garden represented to this Cowper in New England an example of the partnership between man and nature. He preferred the garden to the forest because the garden is an expression of man's spirit through materials supplied by nature. In its orderliness, purposiveness, it expresses man's conquest of himself as well as of the material forces. By mingling his mind with nature, and transforming the landscape into a thing conforming with his own nature, man generates the homestead and opens the country to civilization and the other arts. Gardening is the blending of man's genius with natural substance; it is an intermingling of mind with matter and a conversion of the earth into man-stuff through the hands and the mind. In gardening man humanizes nature, subdues it to his designs, makes it part of his pattern of life. Alcott criticized the colleges of his day for teaching agriculture (which is the cultivation of the soil from necessity and in bondage of man's wants), while neglecting gardening (which is the cultivation of the soil by choice, in freedom). Nature in the raw, which Thoreau praised, meant to Alcott the supremacy of mechanism over intelligence, necessity over freedom, force over grace, material over art. He looked for the genius of the home and household, the orchard and garden.

He loved the homely apple – for virtues other than those that nourish merely. Apples refresh the spirit by their taste and perfume. They are good for virtue and wisdom. A dish of the choice kinds standing in his study perfumed his composition and rejoiced his temper. While others might snatch wildness from the woods, he turned to orchards for subtlety of thought, for strong sense, for grace of diction, for ideas. Though so temperate that he would not drink tea or coffee, he could not resist the juice of the apple, even when improved by the process of time.

Apparently the only wild thing he really liked was the wild strawberry, which grew in luxuriance across the way in his neighbor's field. He loved these strawberries plucked while the dews were on them. It was a special treat to eat them from a bowl filled to the brim with cream. Strawberries and cream! One is not married without this tribute of summer. What shall one offer bride and bridegroom from the fruits of the seasons like this? A youth and a girl better choose June as their nuptial month and the country of Alcott's neighbor.

Friends and gardens and orchards. Other works of art there were too: one's home and a New England small town. Alcott filled a hundred pages of his journal with comments on his Orchard House. He learned the history of his house from town records, he respected and loved its character and antiquity, he made the alterations in the spirit of Thomas Fuller: "He that alters an old house is tied as a translator to the original." Home is the product of artistic impulse; it bears the image of the family; it imparts to the landscape qualities that remind a man of himself. It is the habitation of man: in each detail its human quality is apparent; it influences and molds, even as it reflects, the mind and heart. And the town is more than a mere cluster of houses, each occupied by persons whose hunger for human love, companionship, and neighborliness is never satisfied. It is good to visit a city like Boston, or

Brooklyn where Walt Whitman lived; but it is good, too, to leave the city. The small town, product of human genius, is the prime manifestation of neighborliness. It is man subdued, humanized, subjected to intelligence, sympathy, spirit. For it the world was expressly prepared; in it creation at last succeeds. It is a place whose daily life is the stuff of which our dreams are made, whose existence enhances the beauty and ampleness of nature herself.

In Bronson Alcott, who hoped that he would find heaven a place full of children and who gave thanks to the Friend of families and the Giver of good wives, the prayer of Socrates was realized perhaps more fully than in any friend of his: Pan and the other gods who haunted America gave him beauty in the inward soul; his outward and inward man were at one. He reckoned the wise to be the wealthy, he had such a quantity of gold as a temperate man, and he only, could bear and carry.

Indeed, both Socrates and Alcott were offspring of Phoenix, the teacher of Achilles, who taught his pupil "to be both a speaker of words and a doer of deeds". Alcott strove to fulfill, like Achilles, the true harmony of action and mind; to reach what Solon called Gnomosyne, which is both true insight and the will to put it into action.

In his Journal for 1857, Emerson wrote: "Last night in the conversation Alcott appeared to great advantage, and I saw again, as often before, his singular superiority. As pure intellect I have never seen his equal. . . ."[1]

Cornell University

NOTE

[1] Since I wanted to convey in this essay a sense of Alcott's presence and character, much of the language in the essay is traceable directly to Alcott's journals. See Odell Shepard, *Pedlar's Progress* (1937) and *The Journals of Bronson Alcott* (1938); also Madelon Bedell, *The Alcotts* (1980) and F. C. Dahlstrand, *Amos Bronson Alcott* (1982).

PAUL KURTZ

IS MARXISM A RELIGION?

I

Lewis Feuer occupies a special place in American pholosophical thought: first, as a sympathetic Marxist scholar and later as one of the most severe critics of Marxist philosophy.* For a long period of time he stood virtually alone – or with a few others – attempting to keep alive an interest in Marxist studies, at a time during the McCarthy period in America of the 1950s when all around him were fearful of discussing Marx for fear of contamination. I remember well his reading a paper on Marxism at a meeting at the American Philosophical Association, a lone voice attempting to keep open the spirit of free inquiry. Surely, Marx was the center of intellectual thought elsewhere having captured large sections of the intelligensia; and surely his American colleagues might at least attempt to wrestle with the Marxist conceptual framework.

The same Lewis Feuer later became an astute critic of Marx and Marxism. Following the odyssey of other former Marxists, he came back from his long journey familiar with the pitfalls and snares of Marxist dialectics and its frightful pragmatic consequences, and he was willing to openly submit these to rigorous criticism. Later when it was not fashionable for intellectuals or students to question the categories of left-wing rhetoric, he was willing to do so. Thus his disciples of the new left in the late sixties and early seventies considered Feuer to be a 'neoconservative' or 'reactionary', for he was questioning their revered principles, especially at the beginning of the student movement in Berkeley. His famous piece in *Harpers* magazine on the new left identified him as their foe.

Feuer has been his own man, ready and able to follow the dictates of his intellectual pursuits and often in opposition to the reigning intellectual fashions and their virtual hold on the intellectuals.

S. Hook, W.L. O'Neill and R. O'Toole (eds.), Philosophy, History and Social Action, 275–295.

II

One question that any serious student can raise is: Why has Marxism had such a profound impact on the intellectuals of our time? One interpretation that Marxists disdain is that Marxism is not simply a sociological or philosophical or scientific interpretation of human history, justifiable solely on intellectual grounds to its supporters, but that it was converted into a religion. And it is its religious and moral appeal that explains a good deal of its influence.

How do we explain this perplexing religious phenomenon? Jesus was hardly correct when he claimed that the Kingdom of God was imminent; nor were his disciples when they proclaimed him the Christ, Son of God. The advent of Christianity was not the fulfillment of God's Word upon Earth. It was, however, a self-fulfilling prophecy; for Jesus prophesied a new birth of faith on earth, and his wish became father to that fact for countless generations of Western civilization. The growth of Christianity no more vindicates the truth of its message than did the latter-day growth of Islam verify Mohammedanism. Historically, both religions satisfied deep-felt needs within the human psyche and, once institutionalized, achieved a power and effect far greater than their original impulses. Ideas by themselves are not causative until they are implemented in action; but once they are codified and regulated they take on a life of their own and have effects far greater than originally intended.

Is the growth and power of Marxism to be explained in Marxist terms – that Marx discovered the laws of history in the dialectical process, that he accurately foresaw the development of contradictions between the forces and relationships of production and the eventual replacement of one set of social relations by another? Marx predicted the emergence of proletarian consciousness ready to wrest control from the dominant capitalist class. Is the growth of Marxist consciousness the confirmation of his theory? Or, on the contrary, was Marxist ideology itself the basis for a self-fulfilling prophecy? And is its power to be explained in part by the fact that Lenin, like Paul, discovered that merely to promulgate an idea is not enough, that one needs to reinforce it by a legion of disciplined church or party cadres who are dedicated to its truth and unquestionably committed to materializing it and defending it against all counter efforts?

There is no doubt that Marx, like Jesus, was a heroic genius, a man possessed of brilliant insight and capable of profundity; and his works attest to the substantial contribution he has made to philosophy, sociology, and economics. But the power of Marxism cannot be explained solely by his theories; for these were at least partially limited by his nineteenth-century experience, and they have been superseded by the considerable development of the social sciences. The power of Marxism must therefore be located to a considerable degree in its religious impulse and its moral protest.

The theories of Marxism have at least a double function. On the one hand, Marx provided an analysis of society and an explanation of historical change. According to Engels, Marx's writings were unique because of "two great discoveries, the materialistic conception of history and the revelation of the secret of capitalistic production through surplus value" (Engels, 1892). The Marxist theory may thus be considered a scientific-philosophical account of the basic causes of historical change and of the role that certain determining structural factors exert in society – and in recent epochs in capitalist society. Marx's speculations are insightful and provocative and provide a useful tool for the sociological interpretation of history and society. But they are not omnicompetent to deal with all social changes; they must be modified and supplemented by non-Marxist categories and concepts if we are to understand and explain the dynamics of contemporary life. In any case, Marx, like Aristotle, Machiavelli, Hobbes, Hegel, Freud, and Dewey, made profound contributions to our understanding of man and society, and these remain as part of our intellectual resources. Even if Marx were the Newton of social science, however – and this is questionable – that would still not explain his tremendous appeal.

I submit that the explanation of his popularity lies in part in another function of Marxist language and philosophy: its moral and religious appeal. For Marx, like Jesus, provides a response to man's existential situation. He offers a set of normative ideals and prescriptions for how we shall live and in what direction we should work; and his message has been transformed by his disciples into a new faith, a new hope, a new promise. As mankind discards one religion, long since debunked and intellectually untenable, it seems all too prone to cleave to another, more sophisticated and subtle, yet clothed with a religious mantle. That Marx has had to suffer the excesses of his misinterpreters is most unfortunate. On which side of the barricades would he have been in a contest with Lenin, Stalin, or Mao? It is not difficult to say; surely he would be in opposition to the anti-humanism of many of his devotees.

I do not mean by these observations to minimize Marx's valuable perceptions of man and society, to deride his moral insights, or to deny the need for continuous social change in the world. Indeed, Marx stands as one of the most important ethical philosophers; he recognized the imperative necessity for social reconstruction. Perhaps it is not Marx but the perversion of Marx that is the seedbed for a new religiosity. The sons of great men are usually lesser men than their fathers; the disciples of geniuses lesser men than their teachers: Christ should not be blamed for Christianity, nor Marx for Marxism. Most likely both would disclaim paternity.

The appeal of Marx is based on his moral opposition to poverty and exploitation, oppression and ignorance, and his vision of a more perfect community of men. As a critic of hypocrisy, he would no doubt be a foe of the crimes committed in his name; he expressed profound outrage at what he considered to be the stupidities and inequities in civilization. Mankind, he believed, had progressed far enough in wisdom and power to correct these deficiencies. This was his essential message.

But there have been other critics of social injustices, others who have dreamed of ideal societies in which life was beneficent, peaceful, creative, and productive. History is full of moral indictments and Utopian visions. Wherein did Marx differ? Marxism offered no mere promise of immortal life on earth; it offered a concrete set of proposals in *praxis* to realize the goal. In answer to Fourier, Saint-Simon, Robert Owen, and other early socialists, his was a 'scientific' and 'realistic' socialism, not merely visionary or Utopian.

Yet, in saying this, Marxism disposed itself to a religious transformation. Given the insecure and often tragic character of human existence, Marxism offered a new ideal in which the limitations and failures of life, based on a corrupt social system, could be overcome; and it held the promise of a new, transformed life in which justice could prevail. It became a secular religion of hope, aimed at alleviating fear and focusing on a better tomorrow.

This moral-religious sense was grounded in a doctrine of dialectical progress which, like theism, was enshrined in a kind of inevitability. The Marxist prophecy was not mere wishful thinking; it was rooted in the nature of reality – not in static being, but in dynamic flux and process, the essence of history. The true believer in communism would suffer oppression by capitalist society, but he would be vindicated in the end by the victory of his ideal; thus his reward lay in the future. Accordingly, a religious hunger rooted itself in the weakest aspect of Marx, his philosophical-metaphysical dialectic of history, and this was used not simply an empirical tool of analysis – it could be useful as that – but as a dogma and creed. Although Marxism was predicated on atheism and science and rejected theistic nonsense, it nevertheless became for man a fulfillment of functions similar to those that underlay the discarded symbols of theism. This development in Marxism was especially pronounced in its Leninist–Stalinist version. The ideals of socialism degenerated in the Soviet system and other communist dictatorships into a form of state capitalism, which has led to a new kind of authoritarianism even more terrible than the authoritarian church of theism. It has imposed a rigid orthodoxy on philosophy and science and the arts. It has increasingly raised the specter of repression. In the name of an earlier call for emancipation and liberation of humanity there has emerged a clever new state religion more regimented than any of its predecessors.

It is important to clarify the sense in which a movement can be said to be religious. The term *religious* may be used to point to a belief in God as the "divine and hidden source of being." If we were to accept this definition in the strict sense, it would apply only to *theistic* beliefs, excluding a whole series of institutional beliefs, attitudes, and practices that are non-theistic, yet have religious characteristics and functions. There are, however, many religions that are non-theistic; hence a definition should not be so narrow as to exclude them: Buddhism, Confucianism, Epicureanism, Stoicism, for example; theism in any strict interpretation does not easily apply to them. In short, in defining a belief, we should refer to its experiential or behavioral components. *All* religious systems, in my view, are expressions of human behavior, and it is the kind of experience and the way of life they express that is important.

A *religious* experience or way of life involves a least these components:

(a) It offers a solution to the problem of the *meaning* of human life.
(b) It requires *commitment*. This means that there is some moral dedication or devotion, some binding character in a set of ideal beliefs that stimulate motivation and action.
(c) A religious commitment, as distinct from other kinds, is a *basic* or *fundamental* commitment to those ideals that function as first principles.

(d) Moreover, it differs from mere commitment to a philosophical outlook, which is rationally grounded, or from a commitment to basic postulates or axioms that guide scientific inquiry, because it is at base *infused or charged with emotion and passion*. It appeals to the whole person, going beyond thought and involving feeling and action.

(e) To be religious also implies an element of *faith*, an affirmation that results in one's staking his life on a certain style or course of action. Most religionists and theologians recognize the role of faith in religious belief.

To avoid confusion over this last element, we must make an important distinction between at least two meanings of the word *faith*. First, faith may be a belief that something is true of the world, that is to say, 'real', even though it has not been tested or proved. Thus the belief in God as an ultimate reality is a belief in an alleged propositional truth: The believer has faith that God, or X, or some other entity, really exists, though this cannot be fully demonstrated. Because of this leap beyond the evidence, this type of faith is only a reinforcement of a psychological belief-state. When someone asks a believer "Why do you believe in God, or in X?" he may reply, "Because I have faith," that is, he believes because he believes. There are several non-theistic religions, as well as theistic ones, that contain this kind of propositional assertion; for example, certain types of belief in the dialectic or in evolutionary progress are of this sort, functioning similarly to the God-belief.

On the other hand, there is a second sense of faith that is not propositional, but prescriptive and directive. One may say that he believes in some moral ideals that do not now fully exist in any perfect form. What one affirms by saying that he believes in them is that he hopes that they will come into being or prevail. Here there is faith in an *ideal* rather than in idea. This kind of faith is *normative*.

Now, some forms of Marxism try to give focus to life, and in doing so they fulfill all the criteria of religion. They offer (a) meaning and purpose to human existence, and they involve (b) commitment to certain (c) basic principles and ideals that (d) go beyond reason, express passion, and (e) involve a degree of faith, not simply in an idea but in an ideal. Marxism evokes commitment to a hope of a better tomorrow, not through an immortal soul but through a Utopian society.

But more, much like some powerful, moving, even fanatic religion, Marxism in some of its manifestations has seized and captured the imagination and dedication of its devotees. This is the primary secret of its appeal: not doctrine, but a passionate commitment to faith. I am not saying that religious faith is always bad; given the human condition, some religious commitment – moral and affirmative in character – may be necessary for many people. There are, however, good and bad, constructive and destructive religions. Some create a more serious problem than do others in their excessive leap of faith beyond the evidence. Some are based in fantasy and mythology to such a great degree that they distort most of the positive elements of life.

III

There is a great irony in the development of Marxism. Marx is one of history's great thinkers. Heir to the humanism of Greece and Rome, the Renaissance and the Enlightenment, he expresses the most eloquent values of the free mind. Indeed, his most lasting contribution may be not so much his theoretical explanations of society and history – instructive as these may be – but his contributions to moral philosophy.

Was he a humanist? If so, in what sense?

First, many consider Marx a humanist because he rejected theistic religion and defended atheism. As a materialist, Marx could not propose a spiritualized view of nature. Mind or consciousness does not have any independent reality, but is only a reflection of matter or body. This view enabled Marx to condemn religion for obstructing human development. "Religion", he maintained, "is the sign of the oppressed creature, the sentiment of a heartless world, and the soul of soulless conditions"; or, in his famous phrase, it is "the opium of the people". The abolition of religion and its promise of an illusory happiness is a necessary condition of the demand for "real happiness" (Dunaskaya, 1973).

Marx shared with other humanist thinkers a distaste for supernaturalism and defended a naturalistic world-view. But philosophical or abstract atheism is incomplete, he admonished in his early work *Economic and Philosophic Manuscripts*. Communism begins with atheism; but atheism itself is abstract; "communism is at once real and directed on *action*".[1] "Communism, as fully developed naturalism, equals humanism, and as fully developed humanism equals naturalism." It is, says Marx, "the *genuine* resolution of the conflict between man and nature and between man and man". It is "the riddle of history solved".[2] One cannot simply negate God, as the atheist avers; one must then postulate the existence of man. Communism is the next step, "hence the *actual* phrase necessary for the next stage of historical development in the process of human emancipation and rehabilitation".[3]

Marx's view that merely rejecting God is hardly sufficient is indisputable: one must take still other steps forward. Although he thought the next step was his own theory of communism, we may question also whether this is adequate or sufficient. Even to move in that direction leaves many problems of life unsettled. The core of his humanism can best be seen in his discussion of the relations between his proposed communism and humanistic values. It is clear that Marx was not interested in communism – that is, the abolition of private property and the destruction of capitalism – for its own sake, but rather for what it achieved. Communism as such is not the end of human development: man is. Marx wanted man to be free, spontaneous, creative, to realize his potentialities. Instead, he found him repressed, thwarted, deflected from satisfying his real needs and capacities. This deadening of human freedom, he thought, occurred because of the existence of economic classes; destroy and transform them by means of communism and man might at last be emancipated. Freedom and autonomy are the ultimate ends, but they cannot be achieved within the structure of society as we know it. Marx insisted in the *Communist Manifesto* that "the freedom of the individual is the basis of the freedom of all"; and in *Das Kapital* that "the development of human power, which is its own end", is the genuine "realm of freedom".[4] *The Grundrisse* (Marx, 1857) also shows the underlying continuity of Marx's thought. Humanistic concepts thus may be found throughout his work and were not merely a phase of his idealistic youth.

It is the division of labor, according to Marx, that fragments life and compels humans to focus their energies primarily on one form of activity. It is the existence of private property and the fact that their labor is appropriated by someone else that deprives the workers of their humanity. The worker becomes a slave of the objects that he produces. But more, his labor, controlled by others for profit, itself becomes a commodity. "What then constitutes the alienation of labor?" Marx asked.[5] It is the fact that labor is made "external to the worker, that is, it does not belong to his essential being". In his labor, man the worker "does not affirm himself but denies himself", does not feel content, does not freely develop his energy, but mortifies his body and mind. The worker thus "feels outside his work, and in his work feels outside himself". His labor is coerced, not voluntary. It does not satisfy a basic need, but is merely a means to satisfy needs external to it. The worker is thus estranged from both the product of his labor and the act of production. Since he depends upon nature for his life, he is also estranged from nature and from his species' life; his nature as a human being becomes alien to him. As a consequence he is estranged from others, who stand to him in a special kind of relationship. The worker is under the yoke of another, who, because of the existence of private property, can control him and consider him as a commodity. The only thing a serf or proletarian has to sell is his labor power, and the lord or capitalist purchases it at a low price, debases human values, and forces the worker into a state of subsistence in which his deepest qualities cannot be fulfilled. Money and exchange value comprise the final debasement. Man must sell himself for minimal wage, barely enough to keep his family alive.

The indictment continues: Because of profit, rent, and interest, the owner extracts the real source of value, the labor expended, from the worker and retains it for his own use, as in the case of capitalism, for capitalist accumulation. The Surplus Theory of Value on one hand is a description of capital formation in capitalist economies and on the other involves a moral indictment. 'Real value' is created by labor, thought Marx, but under an unjust system of social relations it is never fully returned to those who have created it. Marx shows a deep sympathy for the plight of the working man. Marxist humanism not only exalts freedom but equality and fraternity as ends to be achieved in a just society. Marxist humanists are impressed by the moving ethical character of Marx's analysis.

This is no place to enter into a critical analysis of the Labor Theory or the Surplus Theory of Value under capitalism. The question I wish to raise here is whether *all* human alienation may be traced to estranged labor. Surely modern man is estranged and his labor is fragmented, de-personalized, often devoid of significance; he functions as part of the machine factory and organization system. Moreover, income is disproportionately and unjustly distributed in many societies in which capitalist production prevails.

But even if we were to solve the problem of estranged labor and redress unfair wage rates, other forms of alienation might ensue. Alienation is pluralistic and cannot be given a simplified explanation; there are various kinds of alienation based on ethnic, religious, and cultural differences; racism; biological, psychological, and personality traits; the relation of man to nature; and so on. Not all of these may be attributed to the alienation of labor.

Can communism alone solve even the problem of alienated labor? It has not been solved in communist economies; a new bureaucracy and new class have emerged and the worker is again estranged from his work in large industrial organizations. The worker in present-day Poland eloquently demonstrates this fact. If Marxism is offered as a solution to the problem of alienated labor and cannot solve it, then Marx's recommendations for social and economic reconstruction may be thrown into question.

Leaving these considerations aside for the moment, however, we must look at Marx's general theory of human nature, including his observations on labor. According to Marx, man is first a producer, and it is the social conditions under which he labors and produces that define his being. "For the socialist man, the *entire so-called history of the world*", says Marx, "is nothing but the creation of man through human labor."[6] Accordingly, if we can remedy that condition, we can resolve the main problem of social existence. Communism is the means by which alienation is to be solved and human nature fulfilled; the materialistic condition of a society determines its philosophical, ideological, scientific, aesthetic, moral, and religious expressions – elements that reflect the underlying conditions governing the mode of production. Expand the forces of production and reconstruct the relationships of production and you can engender a creative burst of man's aesthetic, moral, and intellectual talents. It is precisely this materialistic interpretation of history – which Engels thought was Marx's major contribution – that is in question. Are the forces and relationships of production, "in the last analysis," "ultimate" and "decisive" in determining human history, as Engels thought? Or can other elements in the superstructure determine the mode of production? There is a direct relationship in history between the technological level of production and the way of life of a people. A slave society is based upon slave power; an agricultural society, upon animal power and agricultural techniques; a fishing community of course builds its life about the sea; the steam engine, fossil fuels, and atomic energy have led to dramatic leaps in technology and in the way communities live and work.

The great recent changes in society can be traced to technological advances related to scientific discovery and invention – intellectual and ideological elements that intervene in the forces of production. 'Consciousness' may have a more significant role than elements in the so-called material base. The productive ability of a people does not depend simply upon the relationships of production and whether or not they are fetters on the forces of production; rather it depends upon a constellation of values, moral and philosophical concepts. History cannot be understood as a product of one factor, as Engels recognized, because other factors intervene and modify the social structure. Although the economic interpretation of history may be, within limits, an analytic guide, it is not a total theory of society; nor does it serve, by itself, as a sufficient ground for normative morality or social policy.

What is unique in Mark's theory is that, unlike previous theories, it does not hold that one can save man from alienation merely by reforming him or making him rational, or by espousing ideals of justice and virtue. Emancipation is a *social* problem. It is only by a fundamental change in the social system, not by a conversion of the individual, that the human condition can be ameliorated. One cannot have a just man in an unjust society. It is *every* man, in social terms, not each man, that must be the starting point. For Marx, man is a social animal; the essence of him is tied up with his social relations. Given this conception, the problem of alienation can only have a social solution; it is the *structure* that must be changed, and the relationships that men bear to one another; the existing social classes must be destroyed if mankind is to recover its humanity.

In this basic Marxian premise there is both insight and limitation. While one must agree that many human problems have a social solution, one need not say that *all* do. There are individual biologically and psychologically based determinants of human behavior. Although a change in the structure of society is often essential, that it is sufficient is questionable. A new social system may bring in its wake even more terrible forms of oppression. There are questions for which we today have no easy solution, particularly in the light of the Marxists' efforts to remake economic and social structures. That they have succeeded in ameliorating the life of the average working man or improving his economic and cultural standard of living, in comparison with other economic systems in the world, is doubtful. Mere seizure of the instruments of production is no panacea. To transfer power without humanizing it makes little advance in the human condition.

An essential indication of Marx's ethical humanism is the central role that he entrusts to reason. This faith in reason at least is continuous with the humanism of classical Greece. For Marx, the history of mankind has heretofore been governed by blind and uncontrolled forces, of which the participants have been unaware. Moral reconstruction can occur only with social reconstruction; and this can only be achieved by a rational understanding of the causes of our alienation and a reordering of society in terms of them. In short, Marx insofar as he expresses a profound trust in reason and science, as he understood them, is a direct heir of the Enlightenment.

He uses reason to attack illusion, not only the mythology of religious illusion, but the pomposity and narrowness of moral shibboleths. He shows that religions and moral ideals often mask underlying economic and class interests. Values are relative. To glorify them as abstract and eternal is to miss their historical role in a social system and their consequent limitations. As Marx wished to develop the social sciences, he relied heavily on scientific modes of explanation. It is true that his science is limited by its Hegelian rationalistic and metaphysical components. Nevertheless, it was 'scientific socialism' that Marx talked about, based, in principle at least, upon the use of the scientific method. He envisioned the use of applied reason, functioning as an 'art' or as 'practical wisdom' in classical terms or as a 'policy science' in recent terms. The crucial point is that Marx thought the solution to human estrangement could come about in part by a self-conscious awareness of its existence and in part by a determination to use intelligence in order to redirect social forces into new and meaningful channels. He had confidence in man's ability to order and plan his destiny and to rebuild it by use of the natural principles discovered by science. Although man is a laboring animal, he is also capable of reason; and reason becomes the instrument, once embodied in *praxis*, of social transformation. Marx condemns the mere abstract contemplative use of reason; he lauds its concrete and applications.

Marxism, then, is an expression of ethical and humanistic courage. Unlike theism, it does not have a low estimate of the capacities of man. On the contrary, Marx had no illusion about reality. For man, life itself is the chief good, and it can be lived badly or well. All men seek a significant and enriched existence. If they cannot achieve the good life, it is because they are dominated by irrational and unjust social forces. Release them from these forces and provide positive conditions for the satisfaction of their needs, and the opportunity for development and the good life in a just society can be achieved. Marx had little doubt about the perfectibility of man; he thought his imperfections could be remedied, his tribulations alleviated, his anxieties quieted; if only man would face his condition and turn his efforts toward justice. Thus Marx's philosophy is life-affirming and optimistic; it has confidence in positive human powers; and it believes that, given proper conditions, they will eventually prevail. This focus is humanistic.

IV

The betrayal by latter-day Marxists of Marxism's essential morality and the undermining of its ameliorative view of man is perplexing to serious students of Marx. What we are confronted with is the existence of anti-humanistic elements side by side with humanism. The tendencies existed implicitly not only in Marx's theories but in his complex personality.

Lewis Feuer (1969) in *Marx and the Intellectuals* draws a highly uncomplimentary portrait of Marx the man. According to Feuer, at the same time that Marx was willing to sacrifice himself for humanity, he had an utter contempt for human beings. "Moreover, he seemed to exalt struggle and conflict." We should not commit the genetic fallacy and condemn a theory simply because of the personality of its formulator. Nevertheless, as Feuer has pointed out, a psychoanalytic critique of personality does help us at least to understand a man's theory. In Marx's writings we do find a strange mixture of humanism and anti-humanism. And we may ask, did Marx fully appreciate the subtleties of the human person, and does Marxism as a theory give sufficient weight to these components? Or was Marx so committed to his brilliant insight into the nature of man – the economic determinants overlooked by others – that he excluded other aspects of life? In particular, a central conflict emerges in his writings: an intolerance of intolerance, a hatred of injustice, an impatience with the errors and foibles of ordinary human beings. Thus Marx himself, and his personality, in one sense is responsible for the perversions that his theories generated.

The strange relationship between means and ends is the fundamental paradox that besets Marx and the Marxists. In Marx's view, Utopian socialists failed to create a just society because they did not attend to the practical kinds of action. But can we use any means to achieve humanistic ends? What about the tendency to hatred and intolerance? What of the use of violence and revolution? If Marx was a humanist in ideal ends, he was not a humanist in his moral principles. Engels made clear his belief in the extreme relativism of all moral considerations when he said:

> We . . . reject every attempt to impose on us any moral dogma whatsoever as an eternal, ultimate, and forever immutable moral law on the pretext that the moral world too has its permanent principles which transcend history and the differences between nations. We maintain on the contrary that all former moral theories are the product, in the last analysis, of the economic stage which society has reached at that particular epoch. And as society has hitherto moved in class antagonisms, morality was always a class morality. . . . A really human morality which transcends class antagonisms and their legacies in thought becomes possible only at a stage of society which has not only overcome class contradictions but has even forgotten them in practical life. . . .[7]

On the contrary, one may argue, without adhering to a doctrine of an absolute moral law, that there should be an independent role for an ethical humanism of principles; and certain principles can transcend the material base of a society. Truth, honesty, sympathy, beneficence, and kindness are humanistic virtues that are not time-bound even though they must be given a contextual interpretation. In any social system, if men are to live and work together in some sort of cooperative endeavor, then certain ethical principles ought to be respected.

Underlying the Marxist approach, however, is a basic, violent intolerance of alternative ideals and points of view concerning what is 'progressive' and what is 'reactionary' in social movements. One might ask: How can a new world of cooperative relationships be built on the dead corpses of hatred? Tolerance should be a cardinal principle of any moral idealism that claims to be progressive. This does not mean that we should tolerate unjust conditions, or that we should not seek, when necessary, to radically change attitudes and behavior. Yet the kind of self-righteous moralism that pervades Marxist literature is much like the impatient severity that classical theism – especially during the Inquisition – directed against opposing points of view. One cannot help feeling Marx's overpowering sneer at those with whom he disagrees; rather than maintaining indulgence toward contrary opinions, Marx is angry with the hypocrisies and stupidities of his opponents. Much of his attitude is predicated upon the notion that the major division in society is the conflict of classes; hence, hatred and struggle against the rulers is inevitable. But can a new world be built on bitterness? Marx did not blame the individual member of the bourgeoisie, only the system that made him what he was. But intolerance of a system inevitably spills over to those who occupy its seats of power, and also to those who compromise with it.

Democratic anarchism is naive; neither a socialist nor a capitalist society can function without elites. The need is to make certain that membership in an elite is based upon ability. Membership should be open to all, with equality of opportunity and education for people to rise to the levels of their capability. An elite should not be dominated by self-interest, nor selected by considerations of power or wealth. It must be innovative if it is to contribute to the common good. There is an elite under capitalism, increasingly the corporate managers rather than owners; and there is an elite under state socialism, the bureaucrats who are members of the Party. Marx's hatred for the elite, which he defends under the guise of class warfare, may do more to limit progress than to enchance it. The problem for workers in all societies is to get elites to be receptive to criticism, to hold them to account, to make them more responsive, to humanize their use of power.

The overriding problem that has split the world socialist movement concerns the means of bringing socialism about. This problem split the social democrats from the Leninist–Stalinists, the socialist parties from the communist parties, and was at the center of debate within the New Left. Marx recognized that to espouse a moral idealism or Utopian vision is not enough, that preachments are no substitute for action. He thought we needed an appraisal of the locus of power and the ways to destroy or re-order it. No class, he said, ever willingly gives up its power. If mankind is to achieve communism, it can only be by means of battle. The Marxist must be a revolutionary. He must be willing to smash the old social system, wrest control from the hands of the masters, and seize the means of production. Violence may be the only means available to force the issue. According to Lewis Feuer, Marx's socialist program is built upon a strategy of aggression and hatred, not love or compromise. "To reject violence", according to Engels (1870), was the "parson's mode of thought – lifeless, insipid, and impotent". In defending the use of hatred in Germany, Engels said: "In our country it is hatred rather than love that is needed, at least in the immediate future – and more than anything else a shedding of the last remnants of German idealism."[8]

It is true that there are portions of Marx and Engels that admit the use of peaceful democratic means to achieve socialism. Yet this is only one aspect of a revolutionary ideology that extolled the use of violence, emphasized struggle, and ruled out limitation by ethical principles as a kind of 'modern mythology'. All of this has been the seedbed for Leninist–Stalinist perversions of Marxism, which continues to be weighed with an ambivalent and contradictory moral position: It rejects modern society because of the hatred and injustice, cruelty and oppression in it; it focuses on an ideal state of moral perfection; nevertheless, it is itself ready to use hatred and violence to achieve the good as it conceives it. How ideal ends can be reconciled with a morality of evil means is the unresolved dilemma of Marxism.

The retort usually made to these doubts is that unless there is an uncompromising struggle against capitalism, socialism will not come into being. To which the democrat retorts: But at what price? Is the struggle worthwhile if in the process of achieving one's ends one betrays and brutalizes them by terrible means? To answer affirmatively is to invoke anti-humanism, defended as revolutionary in the name of humanism: an anti-humanistic humanism.

One of the sources of this intolerance may be traced to a first metaphysical principle of Marxism, and the degree of conviction with which it is held. I am referring here to the dialectical materialism to which both Marx and Engels were committed. If one is convinced that there are laws in history and that one has discovered them, then those in opposition to the thrust of history are seen as morally wrong; one's cause is morally right and just.

But what can be said of this dialectic, which is perhaps the most important yet the weakest axiom of the Marxist *Weltanschauung*? Positivists and empiricists have attacked it on at least two grounds: first, because it is an unverified thesis, and second and more devastatingly, because it is non-falsifiable. The highly general assumption that history moves by dialectical laws – thesis, antithesis, and synthesis, and the contradiction between opposing forces – has not been confirmed by the social sciences or historical inquiry. The dialectical laws of Hegel may be questioned on epistemological grounds as too formal and abstract to allow for concrete interpretation; and the more specific economic or sociological interpretation of history – that is, Marx's special application of the dialectic to human affairs – though immensely creative, is also so general that it has not been adequately tested; moreover, there appear to be many exceptions to the operation of the 'laws'. The dialectic lacks sufficient empirical meaning; yet it is offered to explain facts of social and historical existence.

According to Engels:

> The materialist conception of history starts from the proposition that the production of the means to support life – and, next to production, the exchange of things produced – is the basis of all social structure; that in *every* society that has appeared in history, the manner in which wealth is distributed and society divided into classes or orders is dependent upon what is produced, how it is produced, and how the products are exchanged. From this point of view the final causes of all social changes and political revolutions are to be sought not in men's brains, not in man's better insight into eternal truth and justice, but in changes in the modes of production and exchange.[9]

The theory promises too much; it is all-encompassing in its effort to reduce the diversity of social phenomena to a basic set of final and ultimate causes that are supposed to apply to *every* society, however important the economic factors may be in social systems undergoing change. Marx's theory in the last analysis is a quasi-philosophical-poetic theory more than a scientific hypothesis. It resists confirmation and is accordingly, as Karl Popper pointed out, not only unconfirmed but non-falsifiable. How would we go about testing its assertions? Perhaps by reference to specific cases, by examples and illustration of the theory. There are some striking examples, but these surely cannot confirm a total theory. Perhaps what is meant is that theory is tested by *verstehen*, by an intuitive method of apprehension. This is appealed to by some; but not everyone has the same *verstehen* or intuition. Some say that Marx's theory can be tested by experience and by its concrete predictions, but there are serious difficulties here. There is an intimate connection in human affairs between ideas and the events they are supposed to explain; often the ideals are self-creative, mobilizing the very forces needed to confirm them. The truth of the sociological interpretation of history in part is created by espousal of the theory. The ideas, rather than explaining human action, create and mold it. Thus Marxism in part is true because many of those who study Marxism believe in it, and behave accordingly!

The Marxist theory, then, like many others in history, is originative of the very factors that it claims operate in history. This is especially the case when a theory offered as an explanation is actually a prescription. The sociological theory of history and the language in which it is stated is first and foremost normative, directive, expressive. A religious component intervenes in a theory when men no longer consider it dispassionately – to be weighed and tested like all other critical scientific hypotheses – but emotively, requiring above all allegiance. If this is the case, then the theory is non-falsifiable as a descriptive scientific explanation, since it functions primarily imperatively and persuasively. Marxism is, if it is anything, a moral and religious declaration of intent and promise.

But there are a great number of predicted events that did not occur as expected in the century after Marx's death. Socialism did not occur by a revolution of the proletariat in advanced industrial societies, but rather in backward, semi-feudal, and underdeveloped societies, such as Russia, China, and Cuba. The proletariat in the advanced industrial countries did not develop full class consciousness, as expected, but were often the most 'reactionary' elements of the society. In the affluent capitalist societies, the worker was not reduced to a subsistence level; instead his condition improved immeasurably. Capitalist countries did not limit production by monopoly restrictions, but expanded markets by increasing wages and reducing prices; they did not bog down in economic collapse, but were often restored by compensatory fiscal policies adopted by governments accepting Keynesian recommendations. Socialization of the means of production did not lead to reduction of alienation or a marked improvement in the standard of living or freedom for the workers beyond what was accomplished in capitalist countries, nor did it destroy the class system, but only replaced an old bureaucratic class with a new one. These are only some of the empirical factors that confront Marxist theorists.

In retort, many Marxist scholars talk about capitalism's *impending* collapse in a gigantic economic crash, or of the endemic permanent lowering of the standard of living of the workers, or of the future growth of proletarian consciousness. Some of this may eventuate – history is more often contingent than determinate or fixed – indeed, the Marxist may help to bring it about by a self-fulfilling prophecy. But verification is always in the future; hence the theory is non-falsifiable and is modified *ad hoc* so as to account for discrepancies by rationalization. The test for a Marxist, more often than not, is a coherence test; all parts of his system of concepts and explanations must fit together. But under what conditions, if any, will it ever be said that the theory is false? Apparently none, for the true believer. His theory cannot be verified, even in principle.

In the last analysis the theory functions, like the God idea, as a *Weltanschauung*, giving coherence and unity to a point of view and moral guidance and direction to a way of life. While scientific humanism involves commitment to a critical rational mode of inquiry as tested by experience, Marxism in its first principle – the dialectic – abandons critical, experimental, and practical reasons and substitutes a vision of moral poetry and mythology for the real world.

V

Another crucial question that may be asked of any theory that claims to be concerned with man is what happens to the individual. Marx claims to speak for the individual, to rescue him from a repressive social system so that he may lead a free and creative existence. But the focus of Marx, as of Hegel, is not the individual but the social structure and historical process. Are there not some problems of life that have only an individual solution, not a larger social or historical one? Do not my cares, my values, my tastes and talents have some primacy? What guarantee do we have that the individual will not be smothered by the social relations that, according to Marx, define him? I am posing here a primarily moral question, concerning the ends of Marxism, not simply its means. It involves a somewhat different conception of human nature than Marx's.

We must ask whether Marx fully appreciates the uniqueness of individuality, as John Stuart Mill does, for example, in *On Liberty*. Does he really believe in the free creative personality who pursues his own lifestyle, who satisfies his own needs and desires as he wishes, who realizes his destiny on his own terms? It must be granted that no man is an island unto himself, that the problems of the individual often have social solutions, that individuality is fully realized only in social contexts. Yet should there not be reserved to individual personalities some measure of privacy? Marx talks about the freedom to fulfill one's nature, but does he appreciate the liberty to be left alone? May I not make my own mistakes and suffer for them, or discover my own truths and live for them? Why should others decide what is good for me? Why can't I decide for myself, even if I am in error?

Here liberalism and Marxism diverge. For the humanism of the liberal or libertarian humanist, unlike that of the Marxist-humanist, focuses on the individual rather than on the social good. This emphasis on the dignity and value of the individual cannot be dismissed as a mere reflection of bourgeois free enterprise. It transcends each and every social system; indeed, a social system can be evaluated by whether or not it preserves or enhances the rights of free agents.

This difference of ideal ends and values between Marxist-humanism and liberal humanism must not be underestimated; it has immense practical significance. Liberal humanism cherishes civil liberties, intellectual, aesthetic, moral, and political liberties, dissent and heresy. Marxism at times dismisses them or is willing to compromise them in the revolutionary struggle. The virtues of the commune are overvalued and the virtues of individuality undervalued. The tragedy of modern communist societies is that they have abandoned these fundamental human rights – as the Soviet Union, Poland, Hungary, China, and Cuba illustrate. Moreover, we may ask whether these rights were ever central in communist theory. Marx said he believed in them and defended the extension of democracy as a progressive program. Within Marxist literature, however, there are priorities, and the liberty of individualism is not as high on the scale as are social justice, equality, and fraternity. But can there ever be a just or humane society in which individual freedom is not respected as a *first principle*? Is not socialism *without* democracy as a necessary and uncompromisable principle a contradiction? Can the individual be sacrificed to 'progress', the revolution, the party, or mankind? Do these not exist for the individual?

On balance, we may say that humanist ideas and values are present in Marx. Perhaps it is the fate of all moral visionaries to be betrayed by their disciples. Marx's ideas have become official doctrine, sanctified by the Party-Church, proclaimed and enforced by a priesthood, taken as gospel orthodoxy. The cruel hoax of history is that the great heretic has been invoked to suppress heresy, the dissenter to crush dissenters, the deviationist to crush revisionism. Marxism has become for many an anti-humanistic religion.

State University of New York at Buffalo

NOTES

* This article was adapted from "The Religion of Marxism" in *Free Inquiry*. (Spring 1983) and is published with permission.

[1] Edited by Dirk J. Struik (New York: International Publishers, 1964) p. 136.
[2] *Ibid.*, p. 135.
[3] *Ibid.*, p. 146.
[4] *Das Kapital*, Kerr edition, vol. 3, pp. 944–54.
[5] *Economic and Philosophic Manuscripts, op. cit.*, p. 110.
[6] *Ibid.*, p. 145
[7] Engels in *Anti-Duhring* (1878, pp. 103–07) quoted in Howard Selsam and Harry Martel, *Reader in Marxist Philosophy* (New York: International Publishers, 1983), p. 252
[8] From Karl Marx and Friedrich Engels, *Selected Correspondence*, Moscow edn, 1953, pp. 367–368.
[9] Friedrich Engels, *Socialism*: *Utopian and Scientific, op. cit.*

REFERENCES

Dunaskaya, Raya. 1973. 'Marxism and Humanism' in *The Humanist Alternative*, Paul Kurtz (ed.). London: Pemberton.

Engels, Friedrich. 1892. *Socialism*: *Utopian and Scientific*, English translation by Edward Aveling. New York: International Publishing, 1935.

Engels, Friedrich. 1870. *Herr Eugen Duhring's Revolution in Science*, transl. Emil Burns. New York: International Publishers, 1939 pp. 209–210.

Feuer, L. S. 1969. *Marx and the Intellectuals*. New York: Doubleday.

Marx, Karl. 1857. *The Grundrisse*, ed. and transl. by David McLellan. New York: Harper and Row. (1971).

THELMA Z. LAVINE

JUDAISM IN THE CULTURE OF MODERNISM

Since the early 1960s there have been persistent discussions in articles and books presenting the idea that we have entered upon a new kind of mentality, or cultural style – the Culture of Modernism – and that we have been living with this kind of mentality without awareness of it, until now when we are beginning to recognize and objectify it. Are we indeed experiencing, participating in, and producing a new kind of culture, a kind of mentality which is being called Modernism?[1]

I. THE CONCEPT OF MODERNISM

What is this new or newly discovered mentality? What is Modernism?

The concept of modernism as a cultural style, or mentality seems to have appeared during the early 1960s in two different contexts: in literary criticism and in historical studies of the rise of German anti-Semitism from the middle of the nineteenth century. Among the literary critics, the influential literary critic, Lionel Trilling, seems to have been the first to identify modernism: modern literature he said, exhibits "hostility to civilization" and expresses specifically the hostility of intellectuals to the bourgeoisie. Similarly, Irving Howe identifies modernism in literature "as a revolt against [the] prevalent style" which rejects the Wisdom of the Ages and especially the morality and liberal politics of the bourgeoisie. Its outlook, he says, is totally subjective.

But in opposition to Trilling and Howe, who define modernism as hatred of the modern world, the historian Peter Gay views modernism as having a larger significance than that of literary rage against the bourgeoisie and industrialized technological society. Modernism, he argues, is a type of cultural style or mentality or angle of vision, like the Renaissance, or the Enlightenment, or Puritanism, or Victorianism, and like these other mentalities, modernism pervades not only literature and the arts, but is a way of viewing the world, and understanding human nature and social life. In this view of modernism as a cultural style, a wide sweeping mentality, Gay is surely on the right track. But

S. Hook, W.L. O'Neill and R. O'Toole (eds.), Philosophy, History and Social Action, 297–311.

Gay does not go on to examine modernism as a cultural style. Nor does he go beyond Trilling and Howe in explaining the rage and contempt which the modernist mentality feels for the industrialized modern world and especially for the middle classes.

At the same time that the literary critics were discovering the concept of modernism and modernity, these concepts begin to appear among historians. American historians such as Fritz Stern and George Mosse, studying the nineteenth-century roots of Nazi anti-Semitism and looking beyond economic and political sources, began to use the concept of modernism to explain German psychology with regard to the Jews: the Germans feared the approach of modernity, the developing industrialization, mechanization, urbanization as disrupting their world, and they hated the Jew as "the very symbol of modernity", the very symbol of everything that many Germans felt was a threat to their world.

What is Modernism?

Modernism is indeed, I suggest, a type of mentality, a cultural style but it exists not in isolation but in the form of counter-mentalities or oppositional mentalities which constitute it and only when we see this do we grasp the powerful tensions of the modern era from the mid-nineteenth-century German fear of modernity to twentieth-century American intellectual radicals' contemptuous alienation from bourgeois modernity, to the significance of modernism for our lives today. In Lewis Feuer's words (Feuer, 1969) concerning American intellectuals:

> They were alienated as an elite because they felt they were entitled to rule, to supplant the business elite, the legal elite, the military elite, even the democratically chosen elite.

Modernism as a mentality or cultural style, a cluster of cognitive, moral, and methodological claims concerning human nature, knowledge, morals, politics and history, has been in historical development all over the Western world since the eighteenth century. Modernism begins as Enlightenment Modernism; this is the mentality which has produced the modern world. From it come the features of what we now call modernity: the powerful forces of science and technology; industrialization; the continuous 'modernization' of agriculture, mining, machine production, transportation, communication; the rise of cities and of nation-states; the spread of literacy; the achievement of democracy and individualism, which brought the vast waves of immigration to the free world, especially to America. These powerful forces of Enlightenment Modernism were only in their infancy in the early nineteenth century when the counter movement to Enlightenment Modernism chose to attack it. Springing up in the context of literary and philosophical Romanticism, that counter movement may be identified as Romantic Modernism.

Each of the two mentalities of modernism may be presented briefly as a Weberian ideal type. The key to Enlightenment Modernism as it developed from the seventeenth century in England, and in eighteenth-century France and America is the importance it gives to human reason, its primacy, dominance, and power in human nature, knowledge and public affairs. Now that human reason has been freed through centuries of struggle from the controls and coercions of the religious absolutism of the church and the political absolutism of the state, it has the power to enlighten. Substantive reason yields true intuitions, self-evident truths concerning human nature, including the universality of rationality, the inalienability of natural rights, and equality under natural and positive law. Instrumental or scientific reason yields scientific knowledge and technology. Together, substantive reason and scientific reason yield a natural law of inevitable rational progress, in which expanding scientific knowledge is put into practice as technology, and natural rights truths are put into practice to democratize social institutions for the benefit of all mankind.

Enlightenment Modernism as a mentality not only enlightens through the truths of substantive and scientific reason; it promises liberation for the human individual through these powers of reason. In the political culture liberation proceeds by using the newly freed powers of reason to delegitimate factual falsity, superstition, and prejudice, especially in the irrational myths and dogmas of the authoritarian church and state, and in the exploitative beliefs, values and traditions of the feudal past. In the social realm Enlightenment Modernism liberates by mass public education, which both enlightens and delegitimates by exposing falsity and by educating to the national truths of science and democracy. Liberation proceeds by demystifying. Its goal is a world of truth, in which reason is the ground of science, morality, and politics, and progress is the law of history.

In the realm of science, liberation proceeds by power of science and technology to know and to reconstruct the world through physics, engineering, medicine, new methods of agriculture, mining, industrial production, transportation, communication. These were the means of liberation from disease, hunger, plagues, excessive physical toil, isolation and ignorance. Scientific reason was the source of these modes of modernization, the continuous process of reconstruction of the world through scientific knowledge, technology and their products; this was the transformation of the world with which Marx credited the bourgeoisie. Scientific reason is the source of the continuous process of rationalization, secularization, bureaucratization which Weber regarded as the characteristic intellectual and sociological tendencies of the modern world.

No other scholar has matched the erudition and moving eloquence with which Lewis Feuer has detailed, especially in *The Scientific Intellectual*, the liberating force of Enlightenment Modernism (Feuer, 1963):

> The scientists of the seventeenth century swept away the miserable universe of death, famine, and the torture of human beings in the name of God. They took a world that had been peopled with demons and devils, and that superstition had thronged with unseen terror at every side. They cleansed it with clear words and plain experiment . . . They taught men . . . to take pride in their human status, and to dare to change the world into one which would answer more fully to their desires.

By the beginning of the nineteenth century the views of Romantic Modernism had emerged and soon crystallized into oppositional categories, into a passionate counter-mentality, fighting off ferociously the growing domination of Enlightenment Modernism in Europe. Romantic Modernism, as the second of the two mentalities of modernism, may now be presented as an ideal type. Romantic Modernism rejects all of the fundamental claims of Enlightenment Modernism: it rejects the primacy of instrumental reason and the claims of scientism, rejects a world of fact alone, it rejects the doctrine of political liberalism, it rejects the delegitimating analytical strategies by which Enlightenment Modernism destroys cherished beliefs and traditions.

In profound opposition to the primacy of reason held by Enlightenment Modernism, Romantic Modernism claims that not reason, but the human spirit, the striving of the personal and the collective will, have primacy in human affairs. It is not cold, external, objective fact which is significant for human life, but the spiritual world of the private self and the dynamics of its inner life, its conflicts and longings for wholeness, its feelings of alienation and defiance, and its yearning for community with other human beings, and for communion with nature. In opposition to Enlightenment's celebration of the brotherhood of man, Romantic Modernism shrugs off universal mankind and celebrates concrete historical groups, racial, ethnic and national in their tribal or organic unity and their deeply rooted historic traditions. Romantic Modernism is, however, torn between nostalgia and contemptuous disregard for the past, either of the group or of the self. In opposition to the Enlightenment Modernism view that truth is discovered by instrumental or intuitive reason, Romantic Modernism insists that truth is not discovered at all, but is created by the poet ("Poetry is truth" – Novalis) or is embedded in the traditions of a people (Rousseau). And thus, rather than honoring the scientist for discovering truths of nature, as Enlightenment Modernism does, for Romantic Modernism it is the artist who is to be honored for creating poetic-symbolic truth.

Finally, Romantic Modernism has its own counter-style of liberation. To Romantic Modernism, fighting against the growing power of Enlightenment Modernism, liberation is felt to be liberation from the hegemony of the Enlightenment mentality, from the oppressiveness of the bourgeois public world of spreading industrialization, urbanization, and governmental bureaucracy, and from the levelling effects of liberal democracy and of universal mass education. In opposition to this bourgeois public world, Romantic Modernism sponsors alienation, hatred, and defiance of this established world and its coercive structures; its concern is for the victims of the bourgeois world, for the oppressed, the poor, minorities, dissenters, rebels, and revolutionaries against the dominance of bourgeois Enlightenment Modernism. Regardless of merit, Romantic Modernists rush to embrace perceived victims, oppressed, dissenters. Regardless of merit, Romantic Modernists find no empathy but only disdain for the middle class.

Romantic Modernism is characteristically anti-bourgeois, anti-capitalist, anti-rationalist, anti-liberal. Romantic Modernism builds politically not upon the autonomous individual of Enlightenment democracy, with his inalienable rights and independence, but upon the collectivity, the group in some form. It moves in the direction of a politics of organic unity, collectivism of the left or right, toward some variant of socialism, communism or fascism.

For Romantic Modernism, freedom is not the possession of inalienable rights to life, liberty, and happiness, as it is for the Declaration of Independence. True freedom for the individual is the power to be authentically expressive of himself and thus creator of his own truth. Only the artist or the individual who is capable of identifying with, and expressing the group will (as Hegel and Marx have shown) can achieve this substantial freedom.

We have thus seen the structure of modernism, accelerating since the end of the eighteenth century in its Enlightenment and Romantic modes. Modernism is a mentality of two modes, two dialectically opposed cultural styles, Enlightenment and Romantic, each subverting, undermining the other; each overtly or tacitly contemptuous of the other; each delegitimates the other's conception of the human individual, truth, morality and politics, and the methodology for knowing them.

This is the newly discovered conflicted mentality of our time.

II. MODERNISM AND THE RISE OF GERMAN ANTI-SEMITISM

Let us turn now to the mentality of modernism in relation to the rise of anti-Semitism in Germany. A pioneering and insightful contribution to this problem remains the study by Fritz Stern, the *Politics of Cultural Despair*.[2] It is a socio-historical analysis of three men (Paul deLagarde, Julius Langbehn, and Moeller van den Bruck) whose lives span the period in Germany from the middle of the nineteenth century to the rise of Hitler's Third Reich. They were conservative social critics who wrote in despair as they witnessed the gradual disintegration of the institutions and culture of the old Germany, and their lives were spent in attacking the rise of modernity which they saw as the destructive force. In their despair over the decline of the old Germany their diagnosis of Germany's condition built upon the hatred of Enlightenment Modernism which had already been poignantly expressed by early German Romantic writers (Novalis, Friedrich Schlegel), proclaiming the ascendancy of spirit over Enlightenment reason, poetry over Enlightenment science, spiritual nature over the world-machine, irony, dialectic, and symbolic truth over the scientific laws.

They were also able to build upon the doctrine of the primacy of will which had been absorbed into the intellectual culture from Fichte and Hegel. These philosophic exemplars of Romantic Modernism established the German metaphysical and epistemological tradition of voluntarism, in which will has metaphysical primacy as infinite and finite will, and epistemological primacy as projecting, mastering, and re-appropriating the object. From this philosophic heritage these despairing critics appropriated the conception that the spirit or will of a people is a manifestation of the divine will; and the conception of the will as relating to objects through mastery, overcoming, and self-aggrandizing re-appropriation. Lewis Feuer has illuminated this aspect of Romantic Modernism by a typology of modes of perception (Feuer, 1963b):

> The sadistic mode of perception tends to see history as the field in which a particular nation or group realizes its supremacy; the Nazi philosophy, which raised to an extreme all the latent sadism in the German tradition, made history the vehicle for the expression of a racial soul, which could destroy other people to fulfill its agressive compulsions, and which could bend all reality to its purpose.

Modernity was the enemy. Modernity signified to these embittered cultural critics the growth of political liberalism, which they regarded as the foundation of modern society; they hated liberalism because it connoted the French Revolution and Napoleon's humiliation of Germany; and because it connoted the bourgeois world of representative government and political parties, rather than the strong leadership of a heroic figure symbolic of the people.

Modernity also signified capitalism and materialism, and they denounced the spiritual emptiness and the economic selfishness and exploitativeness of the commercial mass society which they saw Germany rapidly becoming. Modernity also signified the decline of Christian faith, without which they believed life would be unbearable. They were also in despair with regard to the domination of the natural sciences in the German universities, which were delegitimating the Christian faith and promoting scientism, the view of science as the sole source of truth. And finally, these critics attacked the newly emancipated Jew as the very symbol of this new and hated modernity; in attacking the Jews they were of course able to draw upon deeply rooted Christian and German anti-Semitism.

The Jews, having been newly emancipated in central Europe, seized upon the opportunities which Enlightenment liberalism now made available to them. They entered rapidly into the universities, professions, journalism, the arts, and business where they were rapidly perceived as aggressive, materialistic, unscrupulous, Godless interlopers, whose emancipation had been a mistake. This was the perception especially of the lower middle classes (small shopkeepers, tradesmen) who were being 'pushed to the wall' by the Industrial Revolution which had no place for them. The emancipated Jew was thus seen as the clever exploiter of the modernity which was insidiously dissolving German culture.

Lagarde, Langbehn, and Moeller become ideologists calling for a conservative revolution in which the will of a strong Fuehrer would take control of the nation's destiny, destroy the evil forces which were insidiously weakening the will of the German people, undermining their institutions, and displacing the ideals of German culture. The leader would inspire the German collective will and restore the mythic spiritual unity of a people destined to achieve hegemony over the world. The ideology prescribed a creative destruction: only by eradicating the evils of the present order could the breakthrough to the restoration of the German soul and culture be achieved and the national will will be exalted.

This ideology for a conservative revolution, developed with remarkable similarity by these three writers, expresses a fantasy rooted in cultural despair, which is virtually indistinguishable in form and content from the ideology of National Socialism for which it had prepared the way. The themes of the fantasy remain the same from Fichte in 1810 to Lagarde 1878 to Hitler in 1933. The only major different is the increasing grandiosity of the role assigned to the Jews. In the Nazi ideology Jews are no mere symbols of a hated modernity; they are elevated instead to the status of a powerful international conspiracy of industrial capitalism threatening the entire world, producing economic crises and social upheavals. Global industrial capitalism is personified as the Jew: it is the Jew who is the force behind both capitalism and socialism, manipulating both; it is the Jew who is responsible for the vulgarities of modern materialist culture in journalism, literature, and art; it is the Jew who is responsible for the decline of ancient group traditions and values. Understandably, in this fantasy, the extermination of this international Jewish conspiracy takes priority over all other goals.

The politics of cultural despair and its call for a conservative revolution develops in situations in which the traditional culture of a people is seen to be threatened by modernity. The revolution brought into being by Hitler's National Socialism is a paradigm of a type of revolution which arises in modernity, the conservative revolution in opposition to the changes taking place in consequence of the dominance of Enlightenment Modernism. The politics of cultural despair is a cultural and psychological phenomenon, conscious and unconscious. It draws upon the key themes of the mentality of Romantic Modernism: the primacy of spirit, the dialectic of personal and collective will; conflict, alienation, defiance and the longing for wholeness and unification; and the group will as the locus of truth.

The genesis of cultural despair is a function of many dimensions of objective conditions and human experience. Cultural despair in Germany intensified from the middle of the nineteenth century to the Nazi era in response to social disruptions in the wake of modernization; repeated economic disasters; the political catastrophes of defeat in two world wars; the narcissistic wounds and humiliations to the German collective will and its self-image of a people destined for world dominance; the cultural and philosophic tradition of the primacy of the aspiring, self-aggrandizing will, personal and collective, from Faust to Fichte, Hegel and Nietzsche; and the age-old Christian cultivation of hatred and contempt for the Jews as the killers of Christ who may therefore be justifiably subjected to any conceivable form of suffering and death.

Under the stress of the interaction of these elements as they came together in historic Germany, an ideology was developed, crystallized, and reiterated in diverse social strata, in which the threat of Enlightenment Modernism to the traditional German culture was to be overcome by the destruction of the Jews, the evil incarnation of Enlightenment Modernism, in order that the destiny of the German people for world domination would be fulfilled. This ideology is sometimes called the German ideology. We can more sharply identify it as an ideology expressing the fascist form of Romantic Modernism in its pathological hatred for the scientism, liberalism, universalism of Enlightenment Modernism. And the millions of murdered European Jews may be seen as the victims of Romantic Modernism carried to vicious excess in the Nazi ideology of hatred for Enlightenment Modernism. As Jackson Lears (1981) writes: "The war to exterminate the Jews marked the ultimate extension of one form of anti-modernism."

III. MODERNISM AND AMERICAN JEWS

The Destruction of the European Jews is the most overwhelming outcome of the clash between the two mentalities of modernism. If, however, we move outside the realm of the ideological fantasies of Nazi

Romantic Modernism and its horrors, what can we say are the contemporary implications, or outcomes, of the two mentalities of modernism, for the Jews and specifically for the Jews of America?

Enlightenment Modernism

Enlightenment Modernism is the dominant mentality of America. What does it imply for Jews?

First Outcome: Most important of all, for Jews, Enlightenment Modernism implies emancipation. Enlightenment Modernism emancipates, liberates; it includes Jews in the brotherhood and sisterhood of humankind, with the universal rights of all humanity. It provides political equality; it liberates by universalizing.

Second Outcome: Despite struggles and frustrations, Enlightenment Modernism has opened up the world to the Jews – in sciences, trade, business, industry, agriculture, technology, law, medicine, education.

Third Outcome: But Enlightenment Modernism not only emancipates, liberates, and provides access and opportunity, it also delegitimates. As we have seen, Enlightenment Modernism liberates from beliefs which do not meet the tests of reason and science. It strips away the legitimacy, the truth value of all such beliefs, which it calls irrational superstitions. The world of Enlightenment Modernism is a disenchanted world. The first victim of this delegitimation is religious belief. Enlightenment Modernism delegitimates Judaism along with all other religious beliefs.

Other effects of delegitimation: The delegitimation of orthodox Judaism has led some Jews to invest their ethical idealism, withdrawn from Judaism, in political movements such as forms of socialism or Marxism; thus they find themselves within the mentality of Romantic Modernism and at odds with the dominant Enlightenment Modernism mentality of America. The delegitimation of Judaism has led other Jews into skepticism, rejecting the beliefs of Judaism and all other belief systems. According to Thorstein Veblen, this skepticism contributes to the prominence of Jews as creative innovators, solving existing problems and opening new horizons. And another conspicuous outcome of the delegitimation of religion is 'interfaith' marriage: if religious belief has no truth value, then it is no barrier to interfaith marriage. Usually in interfaith marriage faith has been delegitimated on both sides.

Fourth Outcome: Enlightenment Modernism not only emancipates, provides opportunity and delegitimates, it also disrupts. The bourgeosie world of democratic capitalism, the world of modernity, is a world of change, movement, process in response to expanding and shifting growth patterns. It disrupts family life; long gone is the community life of the idealized *stetl*. In the Diaspora, under the conditions of modernity brought about by Enlightenment Modernism, there is a further dispersion of Jews in the larger society.

Romantic Modernism

What are the outcomes, implications, consequences of Romantic Modernism for American Jews?

First Outcome: Just as emancipation is the crucial significance of Enlightenment Modernism for Jews, so the crucial significance of Romantic Modernism for Jews is that in its pathological form it provided the ideology of Nazi Romantic Modernism which culminated in the Holocaust.

Second Outcome: Whereas Enlightenment Modernism individuates, emancipates the individual as autonomous, Romantic Modernism unifies; it provides the strengths and gratifications of belonging, *Gemeinschaft*, group membership, group solidarity, the spiritual rootedness and shared aspirations of a people, a community. But the significance of this idealization of the group offered by Romantic Modernism is that it generated not only the blessings and gratifications of group inclusion, it also generated the terrors of group exclusion, and this became the role of the German Jews. Romantic Modernism in its idealization of the spiritually unified community generated a role for the Jews as outsider, as a needed personification of threatening and negative identity to the group. (Freud states, in *Civilization and Its Discontents*, that in mediaeval Christian communities celebrating Christ's commandment to love one another, the role of Jews was to be a needed outlet for hate.) Historically, the unification and self-idealizing which Romantic Modernism provided for European groups, generated the rise of nationalistic movements hostile to outsiders, frequently casting the Jew in the role of the devil.

Third Outcome: Romantic Modernism not only unifies and provides group inclusion and also exclusion – it also spiritualizes: in the field of theology Romantic Modernism invested Protestant Christianity with personal, spiritual intensity and the potency of self-realization. It idealized Catholic Christianity as a religion of love and beauty, and of mystical union. But Romantic Modernism not only spiritualizes, it despiritualizes. Romantic Modernism in the field of theology has treated Judaism as mere legalism, a religion of laws, and has turned Judaism into stone, into petrified legislation.

Fourth Outcome: Romantic Modernism cultivates insight into the self, in its conscious and unconscious aspects; sensitivity to group relations and to group ideologies and differences; and interpretation of expressiveness. And these areas are the concerns of philosophy, psychoanalysis, literature and literary criticism, and political ideology, to which many Jews have been drawn in America. Is it that the Jew is admirably suited for these tasks by his historical heritage of alienation? Does his continual sharpness of perception in these areas follow, as Veblen says, from his tendency to skeptical estrangement from both Judaism and from the Gentile host society? Are these areas magnets for ethical idealism displaced from religion? In any case, these areas of insight, sensitivity, and expressiveness play a valuable complementary role to Enlightenment Modernism: they provide the leavening of inwardness and subjectivity, a critical distance, sources of self-understanding for the individual and for the group, sources of insight and change.

IV. THE OUTLOOK FOR SYNTHESIS; THE FUTURE

Two further questions may be raised about the two modes of modernist mentality of our time: first, is it not possible to bring them together in a viable, strong, consistent, integration or synthesis? To answer that question most briefly, we turn to the magical three names of Marx, Freud, and Einstein. Their names are often invoked to show the extraordinary influence of Jews upon culture. They are often identified as innovators, and as 'modernists' in the sense that they reflect and produce the accelerated process of change in modern society. But have they in their theoretical products achieved a synthesis of the two kinds of mentalities, the two kinds of methodologies, the two angles of vision of modernism? A quick glance will show that the answer is, predictably, no. The hold on modern mentality of the two modes of modernism is too strong for any theorist, any innovator in theory or practice to escape them; but they are too radically oppositional for successful synthesis.

Marx: The writings of Marx we now see to be deeply divided between the writings of the Young Marx, stressing alienation, conflict, the self, subjectivity, metaphysical vision (expressive of Romantic Modernism); and the mature Marx of scientific socialism which had previously been regarded as his consistent, single viewpoint (expressive to a high degree of Enlightenment Modernism). Moreover, his writing is split, deeply divided between a Romantic Modernist attack upon all bourgeois values, and an affirmation of certain of these values. An enormous and growing literature examines these and other conflicts in Marx's thought.

Freud: The writings of Freud have been perceived to be deeply divided between his instinct theory, his so-called biologism and deterministic mechanisms and on the other hand, his explorations of the realms of the self: dreams, symbolism, myths, humor, the forgetting and slips of tongue of everyday life. A literature as enormous as the literature on Marx has grown up with regard to the dual aspect of Freud's thought. As with Marx, the debate continues, the conflict is unresolved.

Einstein: The writings of Einstein have long been known to exhibit a shift in thought from his theory of relativity of 1905 to his resistance after the First World War to quantum theory and to the principle of indeterminacy. We now owe to Lewis Feuer (1972) a careful analysis of the influence on Einstein of a radical political group and its theory of social relativity which led him to think in 1905 of relativity in the field of physics. But by the end of the First World War in Berlin he suffered feelings of depression and a loss of his early political revolutionary optimism, and his relativist mood disappeared as is evidenced by his famous anti-relativist statement that God "does not play dice with the universe". As he perceived that his relativity theory had been influenced by the radicalism of social relativity, so he now saw indeterminacy influenced by the nihilism of the age. In Einstein too we find the influence of both scientific Enlightenment Modernism and revolutionary Romantic Modernism – in unstable, unresolved relationship.

Finally, what of the future? Will one of the two competing modes of modernism win out, triumph over the other? Let us hear from two voices: one, the immensely influential Max Weber, who predicts the triumph of Enlightenment Modernism but at a terrible cost: rational controls, mechanization, bureaucratic structures are extending their power from the institutional world to the human soul; the modern world has become an 'iron cage'. But in opposition to Weber the American sociologist Daniel Bell insists that the triumph has gone to the other side, to Romantic Modernism. Bell points to left-wing politics; to the hatred of the bourgeois world which exists among almost all of the intellectuals of the twentieth century; and to the pervasive antinomian type of self which is liberated from any authority. These have won; Romantic Modernism has triumphed. But each of these pessimistic views is time bound, trapped in the limiting perspective of its own specific time and place. Perhaps it is more sober to say that the opposition between Enlightenment and Romantic Modernism which began with Rousseau in the eighteenth century will reappear in varying cultural and historical forms as far as our present horizon permits us to see.

As for American Jews, they are citizens of a modern nation founded upon Enlightenment Modernist principles. As an Enlightenment Modernist country, America has no Romantic philosophy of an idealized unifying past or of a future apocalyptic destiny of the American people which would exclude the Jews. Moreover, the Jews of America never experienced in America a subjugated status which had to be painfully and uncertainly overcome; there is therefore no 'Jewish question' in America. Jews live in America with the advantages and stresses of Enlightenment modernity; they are sensitive as well to the significant insights of Romantic Modernism. But they are aware that no other politics in the Diaspora guarantees, as does the American Declaration of Independence and the Constitution, the inalienable rights of its citizens and thus protects the Jews. The 'high culture' of Romantic Modernist Europe did not protect them. Radical politics (socialist or communist) has its own final solution for them. The crucial painful sensitivity for Jews in all parts of the world is to the pathological form of Romantic Modernism which culminated in the Holocaust; and the crucial hopeful commitment for American Jews is to America as a fortress of freedom, as an Enlightenment Modernist nation.

George Washington University

NOTES

[1] A version of this paper was presented to the National Scholar Series of the Washington Hebrew Congregation, Jan. 23, 1983. In both versions, this paper is a tribute to the magisterial contributions of Lewis Feuer to the sociology, psychology, and history of knowledge.

[2] Fritz Stern, *The Politics of Cultural Despair* (Berkeley and Los Angeles: University of California Press, 1961). See Ch. 2, 'The Idealism of Antimodernity.'

REFERENCES

Feuer, L. S. 1963a. *The Scientific Intellectual*, New York: Basic Books, p. 17.

Feuer, L. S. 1963b. *The Scientific Intellectual*. New York: Basic Books, pp. 257–258.

Feuer, L. S. 1969. *Marx and the Intellectuals*. New York: Doubleday, p. 2.

Feuer, L. S. 1972. *Einstein and the Generations of Science*, 2nd edn. New Brunswick and London: Transaction Books, 1982.

Jackson Lears, T. J. 1981. *No Place of Grace: Antimodernism and the Transformation of American Culture 1880–1920*. New York: Pantheon Books.

GEORGE S. N. LUCKYJ

PANTELEIMON KULISH:
A UKRAINIAN ROMANTIC CONSERVATIVE

The Ukrainian national revival dates from the early nineteenth century and is closely tied to Romanticism. The greatest Ukrainian romantic poet, Taras Shevchenko (1814–1861) was also the chief promoter of the national awakening of his country, which had lost the autonomy it had enjoyed under Cossack rule in the seventeenth century. The Cossack past fuelled the romantic imagination of Shevchenko and his contemporaries who, under the influence of German idealist philosophers (Schelling, Herder) and Russian and Polish Romanticism, attempted in their works to carve out a new national identity for the Ukraine. When this identity began to emerge in mid-century it owed its existence to these men of letters, primarily poets. Among them was Panteleimon Kulish (1817–1897), second in importance to Shevchenko. Although one of his first works was a long poem 'Ukraina' (1843), Kulish was more drawn to the historical novel and was strongly influenced by Walter Scott. His poem was written in Ukrainian, but his first historical novel, *Mykhailo Charnyshenko* (1843) was in Russian. Kulish gradually changed to Ukrainian in his prose, thus sustaining the efforts of Ukrainian writers to establish a new literary language and to stress their cultural separateness from Russia. Unlike Shevchenko, who was born a serf, Kulish came of an impoverished family descended from the Cossacks. His formal education was scanty in the sense that he never completed the *gymnasium* and only audited lectures at Kiev University, but, endowed with high native intelligence, he read widely in foreign languages and became steeped in the study of Ukrainian folklore and history. This was also the main preoccupation of the small circle of Kievan intellectuals led by Kulish, the historian Mykola Kostomarov and the poet Shevchenko. Shevchenko, who had gained his freedom from serfdom and who now lived and painted in St Petersburg, was the author of a small collection of poems *The Minstrel* (Kobzar, 1840) which became the cornerstone of modern Ukrainian literature. Soon, in 1845, Kulish left for St Petersburg.

A few months before his departure from Kiev, Kulish, Shevchenko, Kostomarov, and a few others formed a secret society – the Brother-

S. Hook, W.L. O'Neill and R. O'Toole (eds.), Philosophy, History and Social Action, 313–321.

hood of Sts Cyril and Methodius, dedicated to the enlightenment of the peasants and the creation of a pan-Slavic union of which Ukraine would be a constituent member. The 'brethren' were particularly dedicated to *narod* (the people), a word which in both Russian and Ukrainian could also mean *nation*. Modern Ukrainian nationalism is often traced to this origin and to this particular Kievan cycle. Yet, as is often the case, the application of the term 'nationalism' is an oversimplification. Within the Brotherhood there was a wide spectrum of views, though there was also general agreement as to the ultimate goals. While some (Shevchenko) were more radically inclined, Kulish held a conservative position. In 1846, while still in St Petersburg, he even published a small 'pocket-book' for landowners, in which he opposed the abolition of serfdom and pleaded for a more humanitarian treatment of serfs by the landlords (Hladky, 1846). This was written while his friend Shevchenko was fulminating in his poems against the tsar and the entire tsarist system of government. Kulish's conservatism is usually explained by Soviet critics as being due to his class origin, but there is more to it than that. He himself recognized that, at the age of 28, he was a conservative by temperament. In his autobiography, written in the third person, Kulish stressed his 'aristocratism' which showed itself in "the orderliness of his tidy person". He "loved order in things and time . . . searched for an equilibrium of heart and mind" and his ideal was "cold energy, silent yet unconquerable by either happiness or sorrow" (Kulish, 1868). These qualities and aspirations of the young Kulish were hardly those of a Romantic. His departure from that stereotype was, however, facilitated by the conservatism inherent in Ukrainian Romanticism which managed to combine the individualism and radicalism of Western European Romanticism with the specific task confronting it in the Ukraine. This task was, first of all, to rediscover and preserve the national way of life. Research in folklore and history was aimed at the conservation, not the demolition of old values. This research was encouraged by the official policy of the tsarist government which, in the 1830s, embarked on the development of *narodnost* (nationality or national spirit) as the third part of the official slogan "autocracy, orthodoxy, nationality".[1] The originator of, the spokesman for, this tsarist policy was the minister for public instruction, Count Uvarov, and although he "assigned a passive and secondary role to the people" (Thaden, 1964), the romantic nationalists and the Slavophiles used the official doctrine for the propagation of their ideas. There were also other supporters of what became known as 'official nationality' – the promoters of nationalistic Orthodoxy. All those groups, including the Romantics, were deeply conservative. The new policy encouraged Ukrainians to explore their own *narodnost* as well as to look for support among other Slavs, not necessarily Russians, but Czechs and Poles. Kulish himself became very friendly with the conservative Polish writer Michal Grabowski and other Polish intellectuals. It became clear to him that the Ukraine, caught in the middle between Russia and Poland, had to reach an understanding with both her neighbors rather than develop in isolation. This view, which he developed later and to which he held fast all his life, was not shared by many Ukrainian nationalists. Yet it was the hallmark of his conservatism, since, "conservatives glorified in variety. Everywhere men were different because everywhere they were steeped in particular national and local traditions and customs" (Dowler, 1982, p. 12).

The activities of the Kievan circle came to an abrupt end in the spring of 1847. Early that year Kulish and his newly-wed wife were on their way abroad. They planned to go for an extended tour of Western Europe and Shevchenko was to join them later. The Kulishes reached Warsaw in March. Everything looked very promising. Then, without any warning, Kulish was arrested during a reception in Warsaw. The other members of the Brotherhood of Sts Cyril and Methodius were swept into the police net at the same time and all of them were brought to trial *in camera* in St Petersburg. Informed of the existence of the Brotherhood by a student's denunciation, the Third Section (the secret police) charged the Ukrainians with illegal clandestine activity. The interrogation of the accused lasted over a month, after which stiff sentences were handed down to nearly all the members of the Brotherhood. The most severe punishment was meted out to Shevchenko, who was sentenced to serve for life as a private in the Orenburg Corps. After ten years of exile this sentence was abrogated and the poet was released, unforgiving and unchanged in his attitude of no compromise with the regime. Kulish was to be exiled to Tula for a period of three years, a sentence which was comparatively mild. Yet the arrest and the interrogation, during which he denied his participation in the Brotherhood, were a great trauma for him. Not only was his career (he had been promised a university appointment) halted, but he was convicted of disloyalty and crime. Although privately he never acknowledged his guilt, soon after the trial he began to write penitent letters of recantation, hoping to alleviate his sentence by submission. Shevchenko behaved more courageously and the harshness of the sentence produced no submission in him. Kulish, however, cringed before the authorities and yet had to remain in exile until December 1859, when he regained his freedom.

The bitter years of exile did not make him abandon the Ukrainian cause. After his liberation he slowly returned to it, but now his approach was dictated by caution and circumspection. Tsarist repression very often resulted in compromise rather than in continued defiance. It is the defiant ones, however, who were preserved for posterity. This is one of the reasons why Shevchenko's name is better remembered today than Kulish's. A prodigious and prolific writer, Kulish set out first of all to complete the work he had begun earlier. He published his two-volume collection of Ukrainian folk literature *Notes on Southern Rus* (*Zapiski o iuzhnoi Rusi*, 1856–57) and the novel *The Black Council* (*Chorna rada*, 1857) in Ukrainian. They alone secured him lasting fame in Ukrainian literature. In the epilogue to the novel, which he published in Russian in the Slavophile journal *Russian Conversation* (*Russkaia beseda*), he pleaded for the unity of Russia and the Ukraine, each with a different culture, but with a common historical origin (Kievan Rus). In future, the Ukrainian contribution would be in the field of culture, Russia's in politics. The call for this symbiosis was not popular among the Ukrainians, while the Russians were at no time likely to consider them as equal partners. It was regarded as the creed of a solitary eccentric and went unheeded by either side. Kulish, in the meantime, started to work on the first biography of Gogol, who was to him the embodiment of such a symbiosis, as a writer of Ukrainian origin contributing his masterpieces to Russia. At about the same time Kulish became acquainted with the Slavophile Sergei Aksakov, whom he visited on his estate. In March 1855 tsar Nicholas I died and many intellectuals greeted the ascendancy of the new tsar, Alexander II, with high hopes. Kulish was sympathetic to Aksakov's ideas about the peasantry being the backbone of Russia's renewal. But at the same time Kulish envisaged the task of the Ukrainian intelligentsia being, as he wrote to Aksakov, "to reach that time when the seeds they have sown will produce a rich harvest and may even feed the tribe [the Russians] which now ridicules them."[2] A fully developed Ukrainian culture, he believed, must rest on folk culture, but, unlike the Slavophiles who based culture on the Russian peasant commune (*obshchina*), Kulish wanted the free peasant homestead (*khutir*) to be the basic unit in the Ukraine. He himself had been born on a *khutir* and later in life he returned to this type of independent farming retreat. In 1861 Kulish published his *Letters from the Homestead* (*Lysty z khutora*), outlining his conservative philosophy. Village life, especially on a *khutir*, he regarded as far superior to life in the city. Moreover, the village is Ukrainian, the city – Russian, and by sending children to city schools the Ukrainians risked losing their national identity. He was skeptical of modern urban civilization, except for America, where "it seems to have taken a good turn".[3] This retreat to the *khutir* remained Kulish's credo even later in life, when he himself settled on one and was very aggravated by his rural existence.

Kulish's conservatism was put to the test by events which unfolded in Russia, Poland, and the Ukraine in the 1860s. In 1861 serfdom was abolished. Kulish welcomed this turn of events and attempted, unsuccessfully, to translate the text of the tsarist proclamation into Ukrainian. The language of his translation proved unacceptable to the authorities. These authorities were preparing to further tighten the screws on the development of Ukrainian culture, and in 1863 the Minister of the Interior, Valuev, issued a secret circular, banning Ukrainian Sunday schools and prohibiting many publications in Ukrainian. While most Ukrainian intellectuals actively tried to oppose these measures Kulish obtained a high official post in Warsaw and was indirectly instrumental in the tsarist repression of the Polish movement after the unsuccessful uprising in 1863. It seems that Kulish deliberately chose to collaborate with the repressive regime, fearing that any open opposition would only lead to more severe repression. Such was the logic of a conservative, who hoped that the authorities would in time seek a dialogue with the opposition. As far as Russian official policy was concerned, this hope was as vain in 1863 as it was in 1983.

In the Ukraine things got worse, not better. While the Ukrainian intelligentsia began to organize cultural and educational circles, called *hromady* (communities), the government began new attacks on all those who were either involved in them or who worked for Ukrainian culture independently. In 1876 tsar Alexander II signed a secret *ukaz*, banning virtually all publications in Ukrainian. Kulish was very disappointed, but he was by now too far away from the mainstream of Ukrainian intellectual life to be able to join the opposition. Instead, after abandoning government service in 1867, he travelled and wrote, concentrating his attention on the Cossacks. He had discerned in the Cossack movement some individuals who had tried during the Cossack wars against Poland in 1648 to enrol the help of wealthy magnates in their opposition to the abuses of Polish rule. Yet he came to the conclusion that the Cossack movement as a whole was both destructive and anarchic and he did not hesitate to condemn it *in toto* as the darkest period of Ukrainian history. Quite justifiably he linked the Cossack movement with the peasant *haidamak* rebellions of the eighteenth century, in which he saw only senseless bloodshed and violence. He scoffed at his contemporaries, populists and radicals alike, for seeking to glorify the Cossacks and emulate the *narod*. Peasants, according to Kulish, needed culture, not politics and to this task of bringing culture to the Ukraine, mainly through his translations of the Bible and Shakespeare, Kulish devoted the rest of his long life. His historical works, especially *The History of the Reunification of Rus* (*Istoriia vossoedineniia Rusi*, 1874–7) and *The Separation of Little Russia from Poland* (*Otpadenie Malorosii ot Polshi*, 1888–89), although unpopular with his compatriots, were the first attempts at Ukrainian revisionist writing from a conservative point of view. Rejected and deserted by his compatriots he retreated to the *khutir* to continue his scholarly work alone. Before doing this he did occasionally show a desperate desire to demonstrate his frustration with official policy. The last time he did so was in 1882 when, on a visit to Galicia (the part of the Ukraine under Austrian rule) he decided to give up his Russian citizenship and become an Austrian subject. After a while, however, he changed his mind and returned to Kiev. Although on the surface controlled and rational, Kulish was a very emotional individual and was known to his friends as 'fiery Kulish'. Unable to secure for himself a sastisfactory position in government service and constantly frustrated in his efforts to pursue a literary and scholarly career, he was saved solely by his own extraordinary capacity for work and by the companionship of his long-suffering wife.

Where does he stand in relation to the Russian conservative thinkers of the day? Some, like Sergei Aksakov, exerted, as we mentioned, a strong influence on him. His early career in St Petersburg was assisted by A. Pletnev, the rector of the university and a staunch conservative. Yet neither of these men had a profound influence on Kulish because they showed no sympathy for the Ukrainian cause. The only Russian with whom Kulish was very much attuned in his view of literature was the critic, Apollon Grigorev. While belonging to conservative rather than radical intellectual circles, Grigorev considered himself a descendant of Vissarion Belinsky, whose historical criticism of literature he supplanted with his own theory of organic criticism (Dowler, 1982, pp. 45ff). One of the tenets of this criticism was the view that "the artist was the full representative of the entire breadth of national life . . . and was the voice of the ideals of the whole people" (Dowler, 1982, p. 47). Moreover, art was inseparable from morality, while at the same time it expressed the national identity. This identity was contained not only in the life and more of the common people, as the Slavophiles believed, but in the life of the entire nation. Grigorev was cognizant of the birth of modern Ukrainian literature, wrote an article on Shevchenko (in which he also mentioned Kulish) and was very sympathetic to the Ukrainians. This alone would have endeared him to Kulish, since few Russian critics viewed Ukrainian literature favourably. But there was also Grigorev's view of national literature as an organism which appealed to Kulish. This organism, rooted in national history and culture, was often compared by Kulish to a tree. It must be cultivated and watered but there must be no interference with its growth. Any attempts at social or governmental reforms were regarded by Kulish as threats to cut the tree or its branches and he rejected these radical measures as inadmissible. This, surely, was the hallmark of a true conservative.

At the same time Kulish was hostile to Russian revolutionary thinkers, whose philosophy also had its roots in conservatism. On several occasions he crossed swords with the arch-reactionary journalist of the time, Mikhail Katkov (1818–1887), who was also a great enemy of the Ukrainian cause. The Russian intellectual scene was divided between the radicals and the ultra-conservatives, and Kulish, quite properly, belonged to the moderate conservatives and was therefore outside the Russian intellectual battles. Among the Ukrainians he represented the right wing, but even here he was not doctrinaire and had friends among the Ukrainian socialists. Late in his life he corresponded with two prominent Ukrainian socialists, Mykhailo Drahomanov and Mykhailo Pavlyk, and admired their cultural activities while disagreeing with their politics. He was aware that a nation had to unite all its human resources, regardless of any political affiliations.

There is no evidence that Kulish was familiar with the writing of Edmund Burke, the father of European conservatism. Another conservative thinker, Friedrich von Savigny, was close to Kulish's philosophy. Kulish translated into Russian part of Macaulay's *History of England* and admitted that Macaulay's Whig philosophy appealed to him. On another occasion, however, in a letter, he wrote that he was "not quite a

Tory and not quite a Whig". If conservatism stands for "protection of the social order – family, neighborhood, local community, and region foremost – from the ravishment of the centralized political state" (Nisbet, 1982), then Kulish may be called a conservative. Though never advocating political activism, through his publications he helped to create a sense of Ukrainian cultural identity which proved a great obstacle to tsarist authoritarianism. After 1882 Kulish retired to his *khutir* and lived the life of a recluse. He continued working tirelessly on two major scholarly projects: translations into Ukrainian of the Bible and of Shakespeare's works. With the help of others the Bible was published in Ukrainian posthumously, as were ten plays of Shakespeare in his translation. No mean achievement for one man, who was determined to build the nation's cultural foundations on traditional values and well-tested premises.

Kulish's reputation after the 1917 revolution underwent drastic changes. At first, during the liberal 1920s, he was highly regarded and his works were published and studied. A leading communist writer, Mykola Khvylovy, praised him as a "true European, a person who neared the Western European intellectual" (Khvylovy, 1925). Then, with the onset of Stalinism, Kulish was banned and labelled a 'bourgeois nationalist'. Today he is still virtually proscribed in the Soviet Ukraine. Some studies of Kulish have appeared in the Ukrainian Diaspora, where he is disliked for his pro-Russian views. The Soviet ban on Kulish is an attempt to eradicate from Ukrainian literary history a major figure whose views are unacceptable today chiefly because they were the views of a conservative.

University of Toronto

NOTES

[1] Cf. N. V. Riasanovsky, *Nicholas I and Official Nationality in Russia, 1825–55*. Berkeley, 1959.

[2] As quoted in G. Luckyj, *Panteleimon Kulish; A Sketch of His Life and Times* (New York, 1983), p. 66.

[3] *Ibid.*, p. 111.

REFERENCES

Dowler, Wayne 1982. *Dostoevsky, Grigor'ev and Native Soil Conservatism*. Toronto, p. 16.

Hladky [P. Kulish] 1846. *Karmannaia knizhka dlia pomeshchikov* St Petersburg.

Khvylovy, Mykola 1925. *Kamo hriadeshy*. Kharkiv, p. 43.

Kulish, P. 1868. 'Zhyzn Kulisha', *Pravda* (Lviv), pp. 285–286.

Nisbet, R. 1982. 'Conservatism', *Prejudices*. Cambridge, p.

Thaden, E. C. 1964. *Conservative Nationalism in Nineteenth Century Russia*. Seattle, p. 20.

CZESLAW MILOSZ

IDEA

(*English and Polish Versions – 1976*)

Afoot, on horseback, with bugles and baying hounds,
We looked down at last on the wilderness of the Idea,
Sulphur yellow, like an aspen forest in late fall
(If the memory of a previous life does not deceive me),
Though it was no wood, but a tangle of inorganic forms,
Chlorine vapor and mercury and iridescence of crystals.
I glanced at our company: bows, muskets,
A five-shot rifle, here and there a sling.
And their outfits! The latest fashions from the year one thousand
Or, for variety, top hats such as Kierkegaard,
The preacher, used to wear on his walks.
Not an imposing crew. Though, in fact, the Idea
Was dangerous to our kind no more, even in its lair.
To assault poor shepherds, farm hands, lumberjacks
Was its specialty since it had changed its habits.
And the youngsters above all. Torturing them with dreams
About justice on earth and the Island of the Sun.

Dmąc w rogi, prowadząc psy na smyczy, kto konno, kto pieszo,
Przybyliśmy na te zbocza skąd widoczna jest knieja Idei,
Siarczano-zółta, jak osinowe lasy późną jesienią
(Jezeli nie myli mnie pamięć ubiegłego zywota).
Choć nie był to las, tylko gęstwa kształtów nieorganichznych,
Z chlorowym wyziewem i rtęcią i tęczowan em kryształów.
Spojrzałem po naszej kompanii: łuki, rusznice,
Pięciostrzałowy sztucer, gdzieniegdzie proca.
A stroje! Ostatnia moda roku tysiącznego,
To znów cylindry jakie nosił na spacerach
Kaznodzieja Kierkegaard.
Niezbyt dobrane grono. Choć i tak Idea
Dla takich jak my nie straszna, nawet w jej legowisku.
Pasterzy ubogich napadać, rolników, smolarzy,

S. Hook, W.L. O'Neill and R. O'Toole (eds.), Philosophy, History and Social Action, 323–324.

To tak, odkąd zmieniła swoje obyczaje.
A nade wszystko młodzianków. Udręczając ich snami
O Wyspie Słońca, sprawiedliwości na ziemi.

University of California, Berkeley

ROGER O'TOOLE

ORGANIZATIONAL WEAPONS AND POLITICAL SECTS

Though written in the present tense, this essay is based upon sociological research conducted more than a decade ago. My preference for avoiding the past tense is rooted not merely in the greater sense of immediacy communicated thereby but in a personal perception of the curiously timeless quality of political sectarianism. Specific sects come and go, but, as I try to suggest in what follows, the essence of what Daniel Bell terms "these opaque, molecular worlds" appears to be remarkably perennial. The organizations and members described below have contemporary counterparts in any city in North America or Western Europe. While the precarious and precipitous paths surveyed by contemporary political sects may lead in new political directions, those who tread them make the same social and psychological journey as generations of their predecessors.

While hastening to offer the traditional solitary *mea culpa* regarding any defects of this essay, I am delighted to express my gratitude to Lewis Feuer, not only for his encouragement of my work on political sects, but for his constant stimulation of my curiosity concerning those murky areas on the borders of religion, science and politics.

Sociological analysis of social movements has not ignored their 'irrational' aspect, but as regards revolutionary political organizations it has emphasized their 'rational', 'instrumental' or 'goal-oriented' characteristics. Such a bias is evident in, for example, Smelser's treatment of 'value-oriented movements'.[1] Within such a context, the small, exclusive political organization which proclaims itself the vanguard of the revolution and which prides itself on an essentially martial approach to organization, strategy and tactics, appears to represent coldly calculating instrumental activity *par excellence*. The classic example of such an organization, and the model for countless imitations is the Bolshevik party forged by Lenin in the years preceding his seizure of power in Russia. Though his blueprint for a revolutionary machine has found many critics, among revolutionaries no less than others, its characterization as an 'organizational weapon' expresses, from the standpoint of the sociologist, a position long held by such opponents, as well as by Lenin's self-styled heirs.[2]

It is the contention of the present essay that what will be termed the 'political sect' – and defined as a small, distinct, exclusive political group in conflict with the wider society and composed of those who, entering its ranks on a voluntary basis, have been selectively chosen as guardians or carriers of some specific variant of political truth – is not *necessarily* a

S. Hook, W.L. O'Neill and R. O'Toole (eds.), Philosophy, History and Social Action, 325–348.

revolutionary organizational weapon and may indeed be something very different. It is suggested that while the exclusive form of the political sect may be conducive to instrumental efficiency, it also provides a setting for an entirely contrary orientation. Such an assertion may not surprise astute political activists but it may necessitate a changed perspective on the part of some sociologists.[3]

While it is certainly not the case that all political sects base their organization upon Leninist principles, they all claim to be in rational pursuit of specific goals and view their organization (together with their doctrine) as *instrumental* in this regard. This claim is regarded skeptically in the present paper, and an alternative appraisal of the nature of political sectarianism, based upon research into a number of Canadian Marxist organizations, is offered.[4]

THE LEAGUE FOR SOCIALIST ACTION

The LSA is a revolutionary organization fraternally affiliated to the Fourth International: the World Party of Revolution founded by Leon Trotsky.[5] Acting as a section of this organization, the LSA is organized as an exclusive Bolshevik vanguard, modelled on Lenin's conception of the revolutionary party, and stakes its claim to leadership of the Canadian working-class and of the Canadian revolution it anticipates. It accepts the basic Trotskyist view of the need for a world revolutionary party to overthrow both capitalism and Stalinist bureaucracy. Viewing reformist social democrats and trade-union bureaucrats as enemies, who, the LSA maintains, prop up a capitalist social system which Trotsky diagnosed nearly half a century ago as being in its death agony, it is committed to the analysis and strategy found in Trotsky's Transitional Programme.[6] The tiny but dedicated forces of Trotskyism in Canada have on occasion approached extinction and their existence has mirrored that of the international organization to which they adhere; most of the years of the Fourth International (since its foundation in 1938) have, by its own admission, been lean ones. The LSA therefore claims the support of at most two to three hundred people in the whole of Canada. Insignificant as these numbers may seem, they represent the largest Trotskyist formation in Canadian history, surpassing the previous peak of the movement in the 1930s and representing a period of growth beginning in the mid 1960s.

Its social composition is difficult to assess with accuracy. The LSA itself likes to present an image of balanced composition with an emphasis on a strong proletarian element and certainly it includes a wide range of individuals, from semi-retired veterans of over 40 years of Trotskyist struggle to young university freshmen and from genuine proletarians to the sons of prominent Canadian families. However, the present membership appears young and not decidedly proletarian, while individuals with university education, seem to be disproportionally represented. Predominantly a movement of the under-30s and, to a great extent, of the under-25s, its age composition matches that of the Fourth International generally.[7] Coser (1954) observes that:

> The political sect may be seen as consisting of those who have cut themselves off from the main body of society, [who] have formed a restricted and closed group which rejects the norms of the inclusive society and proclaims its adherence to a special set of rules of conduct . . .

In his view:

> (The sect) as distinct from the political party which aims at encompassing a high proportion of the mass of electors, consists of specifically qualified members, 'professional revolutionaries' certified Marxists or Alte Kaempfer. The party aim[s] at recruiting all men of goodwill, the sect aims at recruiting all men of goodwill, the sect aims at recruiting an elite of . . . politically qualified 'performers''.

The LSA is a political sect in as much as it lays an exclusive claim to truth and is organized as an exclusive organization which prizes membership as an honour to be granted only to the worthy. Its perception of its own nature is expressed succinctly in its emphasis upon the word 'vanguard' with its suggestions of a select, highly- trained, highly committed elite corps.

Members like to emphasize, particularly to youth, that they are a hard, steeled, revolutionary group whose members are subject to democratic-centralist discipline and who devote to it the whole of their lives. They present themselves as the most conscious and dedicated revolutionaries with tremendous stamina, capacity for work, and organizational and propagandist skills.[8] The invitation to join their organization is a call to participate in the work of a moral and intellectual elite which operates in semi-secrecy. It is the bestowal of the highest honour, and acceptance is seen as a voluntary act of total commitment for life to the LSA cause and organization.

Considered in terms of the fundamental option which Berger applies to religious organizations – "a choice between hanging on to or surrendering cognitive deviance" – the LSA insists that it is firmly "cognitively deviant", indeed, "cognitively defiant" (Berger, 1970), not only with regard to the wider society, but also other radical organizations. It believes that in its version of Marxism it has the key to scientific truth and is thus the only Canadian organization attuned to reality. It sees itself as an intellectual elite able to expose false views of the world, as a moral elite able to expose and combat evil, and as uniquely fitted to provide the leadership of the Canadian working class by virtue of the rich legacy and tradition which designates its members as the only legitimate Canadian heirs of Marx.

Its distinct insight into reality enables the LSA to scorn outsiders who stigmatize it as a tiny, ineffectual organization whose predictions remain unfulfilled and whose claims to historic destiny are groundless. In explaining the fallacy of such observations, LSA members "draw heavily if not exclusively on a definition of self reflected in [their] sect" (Lemert, 1951), reassert their grandiose claims and follow the advice of the founding father of North American Trotskyism:

> The surest way to lose one's fighting faith is to succumb to one's immediate environment; to see things only as they are and not as they are changing and must change; to see only what is before one's eyes and imagine that it is permanent (Cannon, 1954).

LSA members are therefore assured of the correctness of their analysis of events and of their ultimate triumph and vindication.

Along with its view of itself as an intellectual elite in possession of the key to truth, the LSA also sees itself as a moral elite charged with the task of righting historical wrongs and leading the Canadian proletariat in its struggle for emancipation. Spurning bourgeois morality, the LSA sees its members as truly moral in their dedication to truth and to assisting in the unfolding of history's plan. Regarding capitalist society as not only inefficient and wasteful but also as an evil cause of human suffering, LSA members are pledged to destroy it. Engaged in a moral mission, they view themselves as "the only really moral people" (Cannon, 1968). Coser asserts that "the sect by its exclusive structure creates a morality opposed to that of the rest of society".[9] The LSA confirms this proposition.

As the only upholders of this higher morality in Canada, LSA members experience the satisfaction of devotion to duty in the noble struggle for a better world. Such satisfaction is entirely legitimate, for Trotsky declared:

> Yes, our party takes each one of us wholly. But in return it gives to every one of us the highest happiness – the consciousness that one participates in the building of a better future, that one carries on his shoulders a particle of the fate of mankind and that one's life will not have been lived in vain.[10]

So satisfying for LSA members is involvement in their moral mission that they are quite candid in seeing themselves as an elite by virtue of it. A a leading member declared at a public meeting:

> We, as revolutionaries, know that the best of us are more completely human than most.

Members' pride in their status as a moral elite is, therefore, no less than their pride at being an intellectual one.

As an exclusive organization with a claim to the monopoly of both truth and morality, the LSA stands defiantly against the outside world. The attempts of members to proselytize in the wider society and to cooperate with other organizations meet the demands of Trotsky's Transitional Programme, but present a real danger to the sect. In encouraging its members to build bridges to the wider society, the LSA must ensure that these do not become mere exit routes. The member's belief in his sect's version of truth and his commitment to its organizations must be maintained. Concerned with retaining its integrity and identity, the LSA attempts to insulate its membership from the temptations of the world.

Kurt and Gladys Lang (1961) note:

> To the extent that the sectarian group monopolizes the loyalties of its following, it lowers self-consciousness in terms of the norms of the larger world by means of a group code which provides insurance against conventional values and wordly possessions. The group climate is thus conducive to the free expression of unconventional behavior.

In attempting to ensure that members of their exclusive elite remain uncontaminated by conventional values, the LSA attempts to monopolize the loyalties of its following by developing total involvement on the part of members. Lenin's insistence that revolutionaries be prepared to devote "not only their spare evenings but the whole of their lives"[11] to their organization is echoed by the LSA, which seeks to retain and intensify the commitment of its members by providing facilities for members to live a full life within the sect. It attempts to minimize the necessity for activities and relationships outside its boundaries (although neither of these is forbidden). LSA members are tempted rather than compelled to spend the whole of their lives within the sect. When they are not proselytizing, demonstrating, or improving themselves intellectually and politically at internal meetings, public forums or study classes, they have available not only informal companionship during day-to-day activities but also a full and varied social life. LSA members do not live by politics alone, but participate in a continuous program of films, dances, parties, poetry-readings, theatrical visits, and summer camps. With such opportunities for recreation, amusement and companionship, strong effective relationships are likely to develop among sect members. Wherever possible, the LSA tries to infuse its social activities with symbolic and practical content. Some work activities, such as leafletting, preparation of flags and placards for demonstrations, and poster-hanging, are incorporated into parties with food, dancing, and entertainment. Other recreational activities are given a symbolic celebratory theme (for example, Lenin's birthday or the anniversary of the Tet offensive) and traditional celebrations such as Hallowe'en are given a lighthearted political slant. There is certainly no lack of opportunity for mutual reinforcement of members in the world view of their sect. The rebukes, sneers and criticisms of hostile outsiders can be rejected with contempt, while members' individual doubts can be removed in earnest discussion with sympathetic comrades and friends.

By informally insulating its members socially as far as possible, the LSA tries to secure their total commitment. Faced with the hostility and ridicule of outsiders or tormented by doubts, the LSA member is able to re-charge his commitment in the refuge of the sect. Like the religious sectarian who wishes to retain cognitive deviance,

> he has only one effective remedy against the threat of cognitive collapse in the face of these pressures: he must huddle together with like-minded fellow deviants – and huddle very closely indeed. Only in a counter-community of considerable strength does cognitive deviance have a chance to maintain itself. The counter-community provides continuing therapy against the creeping doubt as to whether, after all, one may not be wrong and the majority right. To fulfil its function of providing social support for the deviant body of 'knowledge', the counter-community must provide a strong sense of solidarity among its members (a 'fellowship of the saints' in a world of rampant with devils) and it must be quite closed *vis-à-vis* the outside. . . . In sum it must be a kind of ghetto.[12]

Sect members are reluctant to consider the *satisfactions* of membership; for they wish to portray their groups as organizational weapons, mere means to an end (Cannon, 1962):

> The Leninist party must be, above all, a combat party intent on organizing the masses for effective action leading to the taking of power. . . . It cannot be a talking shop for aimless and endless debate. . . . Neither can it be an infirmary for the care and cure of sick souls or itself a model of the future socialist society.

Although LSA members vehemently deny that theirs is a therapeutic community they would nonetheless endorse Trotsky's observation that "for a revolutionary to give himself entirely to the party signifies finding himself".[13] In demanding of members "the whole of their lives", the sect appears likely to become a home as well as a professed revolutionary weapon. While it would be mistaken to view the LSA as "an infirmary for the care and cure of sick souls", it would be equally wrong to accept it at face-value as a purely utilitarian combat group. Consideration of how far the LSA is an expressive rather than an instrumental organization is required.

Despite their assertions of faith in their ultimate success, repeated avowals that all their activities are directed towards the goal of revolution, and self-portrayal as men and women of destiny, LSA members experience more immediate satisfactions than the calculations of their rate of progress towards socialism. Their sect offers the continually reinforced reward of membership in an exclusive intellectual and moral elite which enjoys a certain notoriety in both radical and political circles and the wider society. Younger members in particular seem to derive satisfaction from belonging to a select body with an established reputation for political subversion. In such a situation even hostility directed against them becomes a kind of compliment and source of pride to members. For the individual member, the goal of the socialist utopia may fade into the background as he engages in the busy daily round of activities and short-term projects. Not only can the LSA member become immersed in performing a variety of tasks with congenial companions, he can also become heavily involved in the group's rich social life. This mixture of enjoyable communal work and play, which also offers considerable emotional support for individuals, may not necessarily mean that the LSA final goal is reduced to nothing. It does suggest, however, that an analysis of this sect, which gave a portrayal of it purely as an instrumental 'organizational weapon' and which failed to indicate the importance of its expressive aspect, would be inadequate. In belonging to this small, exclusive community, members enjoy an experience which could not be replicated in the inclusive atmosphere of the mass political party. Irrespective of their attitude to the final goal, it is certain that many LSA members derive great satisfaction from the sectarian experience itself.

THE INTERNATIONALISTS

The Internationalists are Maoists in the style of the young people portrayed in Jean-Luc Godard's film *La Chinoise*. As members of a small organization founded in Canada and dedicated to the making of a communist revolution in accordance with the thought of Mao Tse-tung, they have adopted the stilted language of *Peking Review* in oral and printed discourse. Though the Internationalists place less emphasis on an identity as a Leninist vanguard party, in fact they are a far more exclusive and isolated sect than the LSA.[14] They claim to be an open and tolerant organization willing to welcome all sincere persons to their ranks, and eager to cooperate with any progressive group, yet despite frequent rallying calls "to the Canadian Working Class, Revolutionary sections of urban and rural petty bourgeoisie, genuinely peace loving and democratic people", the members of this sect do not practise what they preach.

Their avowed policy of building a "broad-front anti-imperialist movement" does not involve arrangements with other groups; for the true attitude of the sect toward them is expressed succinctly in the words of a member to his comrades:

> It's necessary for us to distinguish ourselves from all other groups by denouncing them!

Thus, the Internationalists express willingness to cooperate with all worthy organizations, but find no worthy allies. Regarding any organization which is not completely *for* them as being *against* them, they reject any possibility of genuine political alliances. For these is substituted a grand fraternal alliance of the sect's own creation, which has meaning only within the closed world of the sect itself.

Internationalist isolation from other organizations is accompanied by a policy of exclusivism with regard to membership. Again a sharp distinction must be drawn between the group's ideal portrayal of itself as tolerant, open and accessible, and its actual situation as an exclusive sect with clearly defined boundaries. It is evident that the Internationalists have never really practised what they preach in this regard; for even a sympathetic early observer noted:

> I reacted against the aggressive attitudes of members of the group . . . to individuals, myself being one, not directly associated with it . . . the tone of approach of the Internationalists served to alienate more than endear . . . they seemed to equate rejection of their criticisms as the mark of ignorance or satisfaction with the system outside the group; they did not seem to accept anyone to be in a transient stage, they seemed to make no allowances for people without the necessary intellect to evaluate their attitude to their immediate context. In a word, they seem intolerant, a closed circle. . . .[15]

Despite the comings and goings of visitors to its study groups and meetings, "the hard core of Internationalists remains intact and close",[16] a closed circle in total political isolation. Despite their protests to the contrary, they guard against any threat to their existence as a small, exclusive group, and potential sympathizers or recruits are discouraged rather than warmly welcomed. Recruitment is minimal, despite glowing official descriptions of expansion and progress. Although Mao Tse-tung Thought does not appeal to Canadians to the extent suggested by Internationalists, it is apparent that this sect's small numbers are due not only to this fact, but to a deliberate policy of the Internationalists themselves. Members of the sect appear determined to maintain a small, tightly disciplined, exclusive organization and thus "alienate rather than endear" outsiders attending Internationalist public gatherings and those with whom they have informal contact.

Public events sponsored by the Internationalists thus invariably become intolerant demonstrations of the sect's self-image as an exclusive elite. The sect's rules for mass-democratic discussion – "No investigation; no right to speak. No rhetoric or sophistry"[17] – allow its members to control and stifle discussion by sweeping condemnation of arguments and individuals they dislike. Sympathetic outsiders repeatedly exhibit shock and disbelief on being rebuffed, insulted, threatened, and reduced to silence at Internationalist public meetings. Outsiders, unaffiliated to other organizations, find themselves denounced as 'Trotskyites', 'fascists', 'reactionaries', and 'petty-bourgeois colonialists', and suffer the indignity of being shouted down.

A causal observer is therefore correct in commenting that at their public meetings, the Internationalists "usually end up talking to themselves".[18] Potential allies are treated with no more tolerance or encouragement in informal conversation. They are insulted and harangued, while their arguments are treated with indifference, hostility or contempt. Even their most inoffensive statements are tortuously subjected to the worst possible interpretation, as the Internationalists prove their skill in exposing reaction by magnifying disagreement and rejecting conciliation.

This organization's claim to feverish activity in spreading its message and winning recruits for a broad revolutionary movement is manifestly hollow. Despite a bookstore, public meetings and frequent conferences, Internationalist evangelizing is less an occasion for spreading the word than for demonstrating the existence of the sect as a unique intellectual and moral elite, totally distinct from all other organizations. Members are able to proclaim haughtily their status as conscious individuals and to confirm it by provoking persecution by 'agents of imperialism' present in their audience. It would be superficial to view such activity merely as a genuine, though misguided, attempt to gain support and recruits. Evidence that Internationalists are less concerned with imparting a political message than with asserting their own unique group identity is provided by their practice of spreading their good news in a stilted Chinese-style jargon which is incomprehensible to most outsiders and hardly calculated to win them to the cause. Indeed, at times, members of the sect show a distinct disinclination to engage in any public activity, and only hold meetings, unheralded by 'bourgeois publicity', at which admission is 'strictly by invitation only'. Thus the Internationalists engage essentially in pseudo-proselytization or pseudo-evangelization, while preserving their sect as an alternative to a mass movement rather than the basis of one.

Despite their outward humility and their denials that they see themselves as such, the Internationalists are in fact constituted as an intellectual and moral elite. While asserting that seeking truth is an ongoing process, members of the sect argue that they are, as changed people, the 'conscious' members of society. Having overcome 'anti-consciousness', they are able to expose the nature of "Imperialist cultural oppression", and in developing "a conscious understanding of Marxism–Leninism–Mao Tse-tung Thought" they are scientists subjecting history, society, and the universe to the only correct analysis possible in North America. The Internationalist, therefore, regards himself as a member of a select group possessing a monopoly of truth, of an intellectual elite with the only genuine perception of reality based upon facts. For the Internationalist a state of consciousness and accord with reality inevitably leads to the perception of moral truth. The repression of one's 'fascist being' involves the individual in a realization of the evil of egotism and of the moral imperative to serve people. As the only people who genuinely serve rather than mislead the masses, the members of the sect see themselves as comprising a moral elite. Blumer's characterization of the sect as a "select group of sacred souls"[19] fits the Internationalists perfectly. This intellectual and moral elite "regards the outsider as not participating in grace, as not belonging to the select, as not yet having the fortitude or capacity to adhere to revolutionary principles, [and] sees him as an exponent of a lower morality."[20] It seems mistaken in such circumstances to view the Internationalists' commitment as being maintained by involvement in a growing social movement. Those who remain within the sect might better be considered as gaining satisfaction from the sectarian experience itself, an experience actually very different from that portrayed in Internationalist descriptions of a grand new democratic alliance of oppressed and progressive people.[21]

A clue to the real rewards experienced by members of this sect may be found in the writings of "the founding member of the Internationalist [who] has consistently participated in all facets of the organization's development",[22] for referring to the fact that 'some Internationalists' derive a 'kind of security' from their 'new society', he notes:

> For many people, the Internationalists have become a new historical crib, a new perspective through which they can rationalize their position in nearly all circumstances, they can say that "I am an Internationalist, therefore I am a developing person by definition." . . . The malaise of the Internationalists can be thought of in this way: they reinforce their own personalities within the group, and derive satisfaction from being called rebels outside of it.[23]

The outside observer should accept and extend this perception, viewing the sect as an expressive rather than instrumental movement which provides its members with an alternative to the world rather than the means of changing it. Members of this sect appear to derive satisfaction from being part of a small, politically isolated, cognitive and moral elite, a closed circle with a certain notoriety in the outside world. Sharing a distinct sense of reality derived from facts in which "relevant things [are] relevant and irrelevant things are irrelevant",[24] the Internationalists accord themselves an elevated status further enhanced by its exclusive nature and its rarity. To its members the sect offers a closed environment in which they and the significance of their acts are completely transformed. It provides them (Bell, 1952) with

> the illusions of settling the fate of history, the mimetic combat on the plains of destiny, and the vicarious sense of power in demolishing opponents all [of which] provide a sure sense of gratification which makes the continuance of sectarian life desirable.

Within this "opaque molecular world",[25] young intellectuals attain cognitive certainty, a sense of status and moral worth, a feeling of belonging, and an awareness of a meaning in life.[26] Rather than viewing this sect as a means to an end, it is worthwhile attempting to understand it as an end in itself.

THE SOCIALIST LABOUR PARTY

The Socialist Labour Party of Canada is a small sect, virtually unknown to both the general public and to the members of other political bodies. If other political sects are outside the mainstream of Canadian political life, the SLP is even outside the mainstream of sectarian politics. Yet, it is not a recently formed group struggling to become established, but an organization with a long North American history. It is over 90 years old as an American political group; has had a separate Canadian organization for over 40 years; and indeed, has been dubbed 'The Grand Old Party'. Essentially a living fossil, a relic of a once-powerful political group, the SLP now barely manages to hold onto existence.[27]

While recognizing organizational predecessors from 1877, the SLP credits its real foundation in 1890 to Daniel De Leon, "the founder of the present day Socialist Labour Party, the discoverer of the Socialist Industrial Union programme, and the greatest of modern Socialists".[28] It regards itself still as a De Leonist party, and cannot be understood without reference to the towering figure of its ruler (from 1890 to 1914). An intellectual leading a party composed primarily of working men, a professor revered by his uneducated followers, De Leon exhibited dogmatism, intolerance and ruthlessness. Under his guidance the SLP became identified with the idea of Socialist Industrial Unionism – his own vision of the revolutionary road to the socialist Utopia – and took the steps which led to isolation and decline. By the dawn of the new century the SLP was already totally isolated from the mainstream of American labor, and although from the editorial office of the *Daily*

People, De Leon "continued setting the scene in immaculate fashion for the revolution that never came",[29] his death should have been accompanied by the demise of his organization. The fact that the epitaph of the SLP remains to be written marks this group as one of the most unusual and interesting in contemporary politics. Still drawn primarily from the ranks of blue-collar working men, North American members of the SLP now number only in the hundreds, and total Canadian membership can only be a few score.

Although the SLP is a tiny group clinging on to life, its members claim possession of a doctrine (Marxism–De Leonism) which is as precise, logical and unemotional as a mathematical theorem, and express confidence that their organization will eventually spearhead a revolution. Regarding Leninism as a Marxist heresy, and specifically condemning Lenin's conception of the revolutionary party, SLP members nonetheless view their own organization as a weapon, referring to it revealingly as "the head of the lance". Despite their denials, SLP members undoubtedly regard themselves as belonging to an exclusive cognitive and moral elite. They derive satisfaction from viewing themselves as exclusive guardians of an absolute scientific truth; they are 'cognitively deviant' and 'cognitively defiant' in maintaining a distinct and exclusive sense of reality:

> Only the Socialist Party has achieved and retained Marxist integrity. It has done this by unswerving adherence to science . . . The SLP teaches scientific Socialism, the only Socialism worthy of the name. As a result, the products of its education are men and women who understand the forces of history and the meaning and implications of the class struggle.[30]

Like religious sectarians, who claim to possess the grace which eludes the practitioners of more powerful established religions, SLP members declare themselves to be in possession of the truth which official science is incapable of recognizing and which intellectuals, with years of formal education, are unable to grasp. With the anti-intellectualism which has long characterized this sect,[31] members express their satisfaction in 'exposing' and 'laying bare' reality and in attaining knowledge denied to those hampered by formal learning.[32] SLP members derive satisfaction from being a *cognoscenti*. As guardians of truth they are an exclusive 'band of the chosen', with a status very different from that which they enjoy outside the sect. The possession of truth gives them fortitude and peace of mind, while exclusive possession provides them with the satisfaction of sharing secret knowledge.[33]

SLP members belong to a moral, as well as a cognitive elite; to an exclusive possession of truth they add an exclusive hold on morality. As a small, unique group charged with the historic moral task of showing mankind the path to redemption, they constitute a "select group of sacred souls"[34] and "[regard] the outsider as not participating in grace, as not belonging to the select, as not yet having the fortitude or capacity to adhere to revolutionary principles [and] as an exponent of a lower morality".[35] They regard De Leon's work as having moral as well as intellectual importance:

> It was not genius alone that De Leon brought to the movement for working-class emancipation. He brought also a high moral character and a purity of purpose that matched his intellectual gifts.[36]

The moral satisfaction of SLP activity is underlined by an official SLP publication:

> It were a pity that the knowledge of scientific Socialism be acquired and not put to use in the service of our class. Much is lost then. But it is not the movement alone which is the loser; the individual who deprives the movement of enlightened devotion loses also.[37]

SLP members deny that they regard themselves as an elite, and adopt an attitude of humility, portraying themselves as mere servants of mankind. Such a stance suggests the humility of the exalted, however, and the term, 'enlightened devotion', used within the SLP, sums up its members' conception of themselves as a cognitive and moral elite, serving their cause with both 'heart and brain'.[38]

SLP members acknowledge and even emphasize the fact that their group is small; but denying that smallness implies significance, they claim to regard it as a mark of honor rather than of failure. Arguing that quality rather than quantity is the only measure of the worth of a revolutionary group, members of the sect portray the smallness of the SLP as confirming rather than contradicting their conception of its historic role. It is identified with total dedication and with the preservation of doctrinal purity, primarily due to their concern with upholding rigorous standards of membership. Arguing that though many are called few are chosen, sect members emphasize their status as an elite. In their view, the SLP is small in size but great in knowledge, dedication and potential. It is a unique, exclusive organization composed of individuals carefully selected from the best elements in society.

The SLP views its political isolation in the same way: as a matter for pride rather than shame. Isolation, like smallness, is proclaimed as desirable, especially for the defence of doctrinal purity. Maintaining that "only the Socialist Labour Party has achieved and retained Marxist integrity",[39] the sect conspicuously declares its isolation from all "other parties calling themselves Socialist or Communist".[40] Declaring itself the only genuine revolutionary Marxist movement, the SLP vilifies other radicals, labelling them 'reformists' and 'fakers'.[41]

Appreciation of the fact that SLP members derive satisfaction from membership in a small, isolated and exclusive cognitive and moral elite leads to a consideration of the group's proselytizing activity.

Although the SLP clearly does not have vast numbers of workers seeking admission, its members view it, nonetheless, as a *select* organization which remains small mainly because of their insistence on a high quality of membership:

> In all revolutionary movements, as in the storming of fortresses, the thing depends upon the head of the column – upon that minority that is so intense in its convictions, so soundly based on its principles, so determined in its action, that it carries the masses with it, storms the breastworks and captures the fort. Such a head of the column must be our socialist organization to the whole column of the proletariat.[42]

Viewing themselves as the 'head of the column', members regard the SLP as an exclusive organization while at the same time, they claim to engage in an energetic campaign to win recruits and expand their organization "by conducting a ceaseless, vigorous programme of socialist education among the workers".[43] Underlining this professed emphasis on proselytization, the sect announces:

> We do not seek to develop mere bookworms or barren philosophers. Man of action are the need of the hour.[44]

The SLP is at pains to document its missionary activities in precise detail, and such entries as the following typically appear in its regular Canadian publication:

> One of the highlights . . . was a trip made to Oshawa, Ontario . . . under the direction of National Office. 3,235 leaflets and 54 back copies of the *Socialist Press Bulletin* and *Weekly People* were distributed on this occasion.[45]

> A Daniel De Leon Commemoration public meeting was held on December 14th. In honour of the occasion Comrade X delivered a thought-provoking address entitled 'The Issue is Survival' to a small, but attentive audience.[46]

The close observer, however, is aware that rational bookkeeping methods are never applied to the *results* of proselytization, and that "a small but attentive audience" denotes a handful of individuals, mostly sect members. He learns also that when "literature distribution continues to result in a good number of write-in enquires, personal visits and phone calls to Section Headquarters in search of further information",[47] a maximum of one or two responses is involved. SLP members thus appear, at first glance to be meeting disappointment and keeping their spirits up by transforming their failure into mild success.

Certainly, the sect disguises the failure of its proselytizing activities by emphasizing the energy and diligence with which they are carried out rather than their results, and by stressing the *potential* effect of such endeavours. Yet, a simple conception of SLP proselytization as a sincere attempt to spread the De Leonist word and to seek recruits to the sect seems ill-founded. Plausible though it is, a view of SLP members engaging in earnest proselytization and coming to terms with failure and disappointment by the construction of a rationalization which demonstrates their success is not supported by close analysis of the sect. Far from engaging in the energetic activity of which they boast, SLP members appear to be maintaining what Robin Williams (1961) terms a 'cultural fiction' by merely going through the motions of educational and propaganda activities and, indeed, aiming many such activities at targets which virtually guarantee their failure.

It may be suggested that participants in the sect's proselytization activity are not really disappointed by its failure, first, because they are aware that objectively it has little chance of success, and secondly, because it goes against their subjective inclinations. This point may be illustrated in the context of the main form of SLP proselytization: the anonymous distribution of leaflets presenting the basic ideas of the sect. Compared to the sophisticated canvassing and propaganda techniques used by many modern political groups, SLP leafleting seems traditionalistic, uneconomical, unproductive, and rather shame-faced. Unpromising as it might seem to leave bundles of such literature in the lobbies of apartment buildings or behind the windscreen wipers of cars parked at shopping centers,[48] the activity appears to be assured of failure by a calculated avoidance of its most strategic targets. Leaflets are *never* distributed in factories or outside factory gates, a situation which parallels the refusal of sect members to present their ideas in a trade-union setting.

Thus, SLP members appear to be paying lip-service to the goal of proselytization and their insistence that they do the best they can, given their numbers and resources, is unconvincing. The challenge of snatching recruits from the enemy's lair by directing a leafleting campaign toward factory workers, for example, seems to impress SLP members less than the possibility of encountering hostility toward their ideas. The nature of SLP leafleting is symbolized by an incident in which a leading member of the sect was asked which leaflets had been chosen for distribution on a recent leafleting campaign and by the reasons behind the choice. Rather than by replying in political terms by stating that, for example, a particular issue was important or timely and that a certain leaflet dealt with this issue, he merely stated:

> We decided we'd better get rid of this pile as they were beginning to yellow. They've been around for some time and we decided we'd better move them.[49]

The SLP members' lack of enthusiasm for whole-hearted proselytization also seems apparent in their attitude toward firebrands. Although the sect claims to seek 'men of action', its members express great hostility toward those who can be labelled with this well-established SLP epithet. One sect member expressed his sentiments in the following day:

> The sort of people we don't want in the movement are 'firebrands'. *They're* going to shape-up the movement, *they* are going to put it into action, and so on! A friend of mine was just like that but he burned himself out, that's what they do, they burn themselves out – and then drop out of politics because they get disillusioned. We generally tell these guys – "look, calm down, that's not the way we work in the SLP!" For us, building a revolution is slow, painstaking work – there's no big drama. Unless someone is willing to submit to the discipline of the party and drop all these ideas . . . we can tell him to get out. The party can do without him!

Consideration of SLP leafleting activity and of attitudes toward zealous activism leads the informed observer to view with skepticism the sect's claims to energetic involvement in spreading the word. Rather than the widespread, urgent activity portrayed by sect members, an outsider is likely to perceive half-hearted, ritualistic and by now traditional attempts at proselytization. SLP members appear to be acting out a role as revolutionary evangelists and their activities seem to constitute a pseudo-proselytization or pseudo-evangelization aimed, at best, at maintaining the sect at its present strength. Despite their avowed aim of carrying "the life-giving principles and programmes of Marxism–De Leonism, the hope of humanity, the beacon light"[50] to the working class, the activities of members of the sect seem less than heroic.

Although SLP members would, of course, reject such an interpretation, by stressing their conception of the party as a highly selective, exclusive and carefully isolated means to an end, their heavy emphasis upon the 'fellowship' to be found within its ranks indicates an explanation for their reluctant evangelism and suggests the primary reason for the sect's continued existence. The importance attributed by the SLP to the fellowship within its ranks is striking, and certainly, *esprit de corps*[51] can be important in maintaining members commitment to sect beliefs, as a leading SLP members suggested to a comrade:

> I've been in the party for twenty years, and I've held on to the party ideas – and God knows, its been a hard job at times to hang on to them with the pressures of work and life – and I've been tempted just to let it all go. . . . But I'd come here [to SLP Headquarters] and talk to the other SLP fellows, and do you know, that was a real help in meeting all the problems. . . .

However, although this statement illustrates the effect of group reinforcement on individual commitment, it is more interesting for its demonstration of a sect member's awareness that his relationship to SLP comrades helped him meet 'all the problems' which he identifies with 'the pressures of work and life'. This member credits the congenial companionship he has found within the sect with enabling him to overcome the difficulties of life outside it. Other SLP members have confirmed to this writer that the sect provides a warm supportive environment in which private troubles may be shared with sympathetic listeners:

> No matter what pressures get you down in work and life – and they get us all at times – you'll find that belonging to the SLP will make you able to live with those pressures better and to get over them.

SLP emphasis on fellowship is shown in published reports of activities; for accounts of public meetings and leaflet distribution are accompanied by items which appear odd as progress reports of an avowedly revolutionary organization:

> The section held its annual picnic on Sunday, August 2 . . . the weather was excellent and forty comrades and friends from Toronto, Hamilton and as far away as Buffalo, Detroit and Ohio enjoyed a fine day of SLP fellowship. Games were held for the women and children and plenty of corn on the cob was available.[52]

The Saturday night social which in the past was held in Comrade X's home had to be switched to Comrade Y's home. Mrs. X who always prepared a delicious smorgasbord dinner for this annual get-together was ill. However she was able to come to the affair and enjoy the evening's festivities with the rest of the comrades and symphatizers. We are grateful to Comrade Y and his wife . . . for pinch-hitting and coming through with home runs. I delivered an informal talk at the conclusion of the eating session and took up a collection which netted the impressive sum of $102. Needless to say, everyone was extremely pleased with the all-around success of this impromptu party.[53]

The attempt to infuse a serious purpose into such activities by an appeal for such funds appears to emphasize rather than deny their essential social nature. The best-attended sect activities are similar to those of such non-revolutionary voluntary associations as veteran and church groups, and as such provide a congenial, familial atmosphere in which individuals may experience a sense of belonging.[54]

SLP emphasis on fellowship is perhaps the cornerstone of the sect's existence, in terms of which members' lack of concern with changing the sectarian *status quo* may be most fully understood. SLP members gain satisfaction from belonging to a cognitive and moral elite, but this is sustained by their satisfaction with the sect as an oasis, a place of refuge from the pressures of the outside world. This sect is a warm, sheltered 'old soldiers' home',[55] an organization which is essentially expressive rather than instrumental, and in which membership, like virtue, is its own reward.

CONCLUSION

The implication of the findings of the above case-studies is that an automatic presumption of the instrumental nature of political sects is unwarranted, and that a uniform perception of them as 'organizational weapons' is most inappropriate. A sharp contrast appears between ideal and actual activity, a distinction frequently blurred even for sect members, and it is clear that, in certain cases, political sects may be classified as expressive rather than as instrumental organizations.[56] Thus while smallness, exclusivism and isolation may be important attributes of an organizational weapon, they may also provide a setting which is as conducive to retreat as to revolution. Revolutionary Marxists have long condemned Eduard Bernstein's famous maxim "the movement is everything, the final goal, nothing",[57] yet it may be suggested that this phrase sums up accurately and concisely the experience of many political sectarians. A sociological perspective which analyses the political sect as an end-in-itself as well as a means-to-an-end avoids an "over-rationalized conception of politics"[58] and contributes to a more sophisticated understanding of the phenomenon of the social movement.[59]

University of Toronto

NOTES

[1] See Neil J. Smelser, *Theory of Collective Behaviour*, New York: The Free Press, 1962. For discussion of the terms rational and irrational in this context, see Joseph Gusfield, *Symbolic Crusade* (Urbana, Illionois: University of Illinois Press, 1963), p. 179.

The distinction between expressive and instrumental social movements has a long history within the subdiscipline of Collective Behavior. Its origins may be seen in Scipio Sighele, *Psychologie des Sectes* (Paris: Giard et Brière, 1898); and in the work of Park and Blumer on expressive and acting crowds. See R. E. Park and E. W. Burgess, *Introduction to the Science of Sociology* (Chicago: University of Chicago Press, 1924), pp. 870–874; H. Blumer in A. M. Lee (ed.), *Principles of Sociology* (New York: Barnes and Noble, 1951), pp. 214–216. See critical discussions of the distinction in K. and G. Lang, *Collective Dynamics* (New York: Crowell, 1961), pp. 500–511; and in J. R. Gusfield, *op. cit.*, p. 179. See also the similar distinction in R. Turner and L. Killian, *Collective Behaviour* (Englewood Cliffs, N.J.: Prentice-Hall, 1957), p. 431.

[2] This is Selznick's term. See P. Selznick, *The Organizational Weapon* (Glencoe: The Free Press, 1952).

See also Lenin's pamphlet *What is to be Done*? (Many editions), Alfred G. Meyer, *Leninism*, (Frederick A. Praeger, 1957), and the classic Marxian attack on Leninism in Rosa Luxemburg, *The Russian Revolution and Leninism or Marxism*? (Ann Arbor: University of Michigan Press, 1961).

[3] The generally perjorative use of the terms 'sect' and 'sectarian' by political activists is often based upon an appraisal of their activities similar to that presented here. Sociologists who emphasize the importance of the sect as a 'nucleus' of a mass social movement (see Note 1 above) will be most affected by the perspective presented here.

[4] This research is reported more fully in the author's *The Precipitous Path: Studies in Political Sects* (Toronto: Peter Martin Associates, 1977). Material related to the thesis of the present paper may also be found in "Underground" Traditions in the Study of Sectarianism: Non-Religious Uses of the Concept "Sect" in *Journal for the Scientific Study of Religion* **15**, No. 2, 1976, pp. 145–156.

[5] It should be noted that a number of bodies claim to be the authentic Fourth International. The LSA is affiliated to the re-unified body of that name which dates, in its present form, from 1963, and may be considered the largest Fourth International.

[6] Leon Trotsky, *The Death Agony of Capitalism and the Tasks of the Fourth International* (New York: Pioneer Publishers, 1964). The Transitional Programme adopted by the Founding Conference of the Fourth International; first published, 1938.

[7] See 'SWP Convention Maps Fall Actions', *Intercontinental Press* **7**, No. 29 (Sept. 15, 1959), p. 808: "The great majority of the participants in the convention were young people ranging in age from their mid-teens to mid-twenties"; (The SWP is the U.S. Socialist Workers Party, an ally of the LSA within the Fourth International.) and *Intercontinental Press* **7**, No. 26 (July 14, 1969), p. 666: "A noticeable feature of the congress [Ninth Congress of the Fourth International] was the youthfulness of many of the delegations. They represented the most politically conscious sector of the new generation of rebel youth that is stirring the world today."

[8] LSA members pride themselves on their political prowess, making such statements as "One of our people is worth ten of anybody else." This same pride in the superior abilities of Trotskyists in political activity may be seen in J. P. Cannon's statement: "There is some peculiar physical quality about Trotskyists – I don't know what it is. Normally they have no greater physical endurance than other people, sometimes not as much. But I have noticed more than once that in political fights when it is a question of fighting for some political idea, Trotskyists can stay awake longer and speak longer and more frequently than people of any other political type." *The History of American Trotskyism: Report of a Participant* (New York: Pioneer Publishers, 1944), p. 213. (Cannon may be regarded as one of the founding fathers of American Trotskysism.)

[9] See Coser (1954), pp. 360–361. See also Leon Trotsky, 'The Moralists and Sycophants Against Marxism' in *Their Morals and Ours: Marxist versus Liberal Views on Morality* (New York: Merit Publishers, 1966), pp. 53–54.

[10] Leon Trotsky, in a speech celebrating the 10th anniversary of the foundation of the SWP. Quoted in J. P. Cannon, 'Trotsky on America', *International Socialist Review*, **21**, No. 4 (Fall 1960), 105.

[11] Lenin stated in the first edition of his paper *Iskra* (Dec. 1900): "We must train men and women who will devote to the revolution, not merely their spare evenings, but the whole of their lives." See 'The Urgent Tasks of Our Movement' in V. I. Lenin, *Selected Works*, 3 vols (Moscow: Progress Publishers, 1970), Vol. 1, pp. 114–118. The phrase has been made famous, or infamous, by Benjamin Gitlow in *The Whole of Their Lives* (Belmont, Mass: Western Islands Publishers, 1965). First published 1948.

[12] See Berger (1970), pp. 17–18.

[13] Leon Trotsky, as quoted by J. P. Cannon, *International Socialist Review* **21**, No. 4 (Fall 1960, 105).

[14] Although the Internationalists officially remained an 'open' organization of 'progressive' youth and students, they officially launched a full-scale Leninist party with the foundation of the Canadian Communist Movement (Marxist–Leninist) on March 23, 1969. Within months this organization had been restyled the Communist Party of Canada (Marxist–Leninist). For the official statement on the founding of CCM (M-L), see *Canadian Internationalist* **1** No. 2 (March 1969), pp. 1–2. Some idea of the heavy student and intellectual composition of the group may be gained from its inordinate concern with cultural rather than economic matters: "Our analysis shows that the principle (*sic*) aspect of our struggle is going to be on the cultural front. Our pinciple (*sic*) struggle is the defeat of U.S. imperialism and modern Soviet revisionism, and this means waging uncompromising and resolute struggles against imperialist ideology, art and literature, social and political organizations." *Mass Line*, **1**, No. 10 (Sept. 17, 1969) p. 9.

[15] Letter to *Words International* **1**, No. 2 (1967), 3.

[16] *Ibid.*, p. 4.

[17] These rules govern all Internationalist meetings. See, for example, the statement in *Canadian Student* **2**, Nos 2–3 (Feb.–March 1969), 23.
[18] University of Toronto Student 'Handbook '69' (Sept. 1969).
[19] Blumer, *op. cit.*, p. 216.
[20] See Coser (1954), p. 361. The Internationalists' main slogan is: "Seek Truth – Serve People". Their leader states: "To seek truth is to serve people . . ." (*Necessity for change*, p. 13) and "for the revolutionary, service to others is seeking the truth" (*Ibid.*, p. 8) See also Mao Tse-tung, 'Serving the People', in *Quotations from Chairman Mao Tse-Tung* (Peking: Foreign Languages Press, 1966), pp. 170–174.
[21] See Mao Tse-tung, *On New Democracy* (Peking: Foreign Languages Publishing House, 1964).
[22] See *Canadian Student* **1**, No. 1 (Sept. 1968), 18.
[23] See *Necessity for Change* (The Internationalists, 1967), p. 5.
[24] See *Necessity for Change*, pp. 7, 22.
[25] Bell's phrase; see Bell (1952), p. 222.
[26] This sect provides 'status' to 'up-and-outers' rather than 'down-and-outers'. See the discussion of M.R.A. in C. S. Braden, *These Also Believe* (New York: Macmillan, 1949) p. 409. Also on this topic see: B. R. Wilson, *Religious Sects* (New York: McGraw-Hill, 1970), pp. 31–32; H. Cantril, *The Psychology of Social Movements* (New York: John Wiley, 1941), pp. 41–45. On the communion of the sect, see the discussion of *esprit de corps* in H. Blumer, op. cit., pp. 214–215; and Eric Hoffer, *The True Believer* (New York: The New American Library, 1958), pp. 44–45. On the problem of meaning see Talcott Parsons' introduction to Max Weber, *The Sociology of Religion*, trans Fischoff (Boston: Beacon Press, 1963), pp. lvii–lxix. See also H. Cantril, *op. cit.*, pp. 53–77.
[27] Organizations which might also be called fossils in the same way as the SLP are remnants of the Industrial Workers of the World (IWW), and the Socialist Party of Great Britain (SPGB). On the modern IWW see Stewart Holbrook, 'Last of the Wobblies', *The American Mercury* **62** (April 1946), pp. 462–468, and on the SPGB see George Thayer, *The British Political Fringe: A Profile* (London: Anthony Blond, 1965), pp. 148–150; and Robert J. Alexander, 'Splinter Groups in American Radical Politics', *Social Research* **20** (1953), pp. 288–289. See also the description of the remnants of the Populist Party in Walter Lippmann, *A Preface to Politics* (New York and London: Mitchell Kennerly, 1913), p. 276.
[28] Eric Hass, *The Science of Socialism* (published by SLP), 1967, p. 5n. Discussions of De Leon may be found in most histories of the American socialist and labor movement, but see especially Don K. McKee, 'Daniel De Leon: A Reappraisal', *Labour History* **1**, No. 1 (Winter 1960), 264–297, for a recent critical discussion of De Leon's place in U.S. labor history. For SLP appraisals of De Leon, see *Daniel De Leon: The Man and his Work*, A Symposium (New York: National Executive Committee, Socialist Labour Party, 1934); and the following, published by the SLP 'New York Labour News Company': Arnold Petersen, *Daniel De Leon: Social Architect*, 2 vols; Arnold Petersen, *Daniel De Leon: Socialist, Pioneer, Editor*; and Olive M. Johnson, *Daniel de Leon: American Socialist Pathfinder*. SLP literature presents an invariably glowing picture of De Leon, but as McKee notes, "the eulogies to De Leon written by SLP historians are transparent overstatements. In particular, they greatly magnify his role as a theoretical innovator and exaggerate his ideological contribution to the philosophy of the IWW", *op. cit.*, p. 267.
[29] See Daniel Bell (1952), pp. 243, 246.

[30] *The Science of Socialism*, pp. 5–6. On maintenance of a separate sense of reality, see Coser (1954), pp. 366–367; and J. L. Simmons, 'On Maintaining Deviant Belief Systems: A Case Study', *Social Problems* **11** (1964), 250–256.

[31] Hostility to intellectuals, other than De Leon himself, is long established in the SLP. In 1905 De Leon presented "probably the ugliest portraiture of the 'Intellectuals' that an intellectual ever wrote". See L. S. Feuer, 'The Political Linguistics of "Intellectual": 1898–1918', *Survey*, No. 1 (Winter 1971), 165. Anti-intellectualism entrenched itself in the SLP between 1905 and 1914 when intellectuals left its ranks to join the new Socialist Party.

[32] Just as Eduard Bernstein noted of the communistic sects of seventeenth-century England that they exhibited ". . . a contempt for academic learning combined with a great interest in education". *Cromwell and Communism* (New York: Schocken Books, 1963), p. 235.

[33] See Michael Walzer, *The Revolution of the Saints* (New York: Atheneum Publishers, 1968), pp. 317–319, and G. Simmel 'The Secret and the Secret Society' in Kurt H. Wolff (ed.), *The Sociology of Georg Simmel* (New York: The Free Press, 1950), pp. 307–316.

[34] Blumer, *op. cit.*, p. 216.

[35] Coser (1954), p. 361. Bittner refers to the 'arrogant self-righteousness' exhibited by members of some radical organizations. See Egon Bittner, 'Radicalism and the Organization of Radical Movements', *American Sociological Review* **28**, 1963, 928–940.

[36] See Eric Hass's Introduction to Daniel De Leon, *Two Pages from Roman History*, p. 1.

[37] See *The Science of Socialism*, p. 59.

[38] Phrase used by Arnold Petersen, National Secretary, SLP of America from 1914 to 1969. Quoted in *Socialist Press Bulletin* (April 1969), p. 5.

[39] *The Science of Socialism*, p. 5. In *Socialism: Questions Most Frequently Asked and their Answers*, the SLP is said to have built ". . . the most solid, self-disciplined Marxist organization, which for scientific Socialist integrity, has no like in the world" (p. 60). (Pamphlet published by SLP, n.d.)

[40] Such a declaration appears prominently in all SLP publications.

[41] On reformism versus revolution, see Daniel De Leon, *Reform or Revolution* and *The Science of Socialism*, pp. 14–15.

[42] See Daniel De Leon, *Reform or Revolution*, pp. 22–23.

[43] See *Socialism: Questions Most Frequently Asked and their Answers*, p. 60.

[44] Arnold Petersen, National Secretary of the SLP of America, in a document presented to all new members. Quoted in *The Science of Socialism*, p. 59.

[45] See *Socialist Press Bulletin* (Aug. 1969), p. 5.

[46] See *Socialist Press Bulletin* (Jan. 1970), p. 6.

[47] See *Socialist Press Bulletin* (April 1969), p. 6.

[48] In terms of classic social science knowledge such activity would appear likely to prove unproductive. See Paul Lazarsfeld, Bernard Berelson, and Hazel Gaudet, *The People's Choice* (New York: Columbia University Press, 1948), in which the statement is made: "In the last analysis, more than anything else people can move other people" (p. 158).

[49] Urgent propaganda, yellowing at the edges, might almost be taken as symbolic of the SLP. George Thayer noted that the literature of the Socialist Party of Great Britain was similarly yellowing with age. See Thayer, *op. cit.*, p. 150.

[50] See *Socialist Press Bulletin* (June 1970), p. 5.

[51] Herbert Blumer's term. See Blumer, *op. cit.*, pp. 214–215.

[52] See *Socialist Press Bulletin* (Sept. 1970), p. 6.

[53] See *Socialist Press Bulletin* (Oct. 1969), p. 5.

[54] Familial in both a metaphorical and literal sense; for the sect seeks to involve members' families in its social gatherings. On the relations between sects and families, see Coser (1954), p. 362; Bittner, *op. cit.*, pp. 928–940; and Lemert (1951), pp. 220–221.
[55] Bell (1952), p. 243. Bell used this phrase to refer to the SLP of 1888, but its use is appropriate today.
[56] Blumer notes that expressive movements "do not seek to change the institutions of the social order or its objective character. The tension and unrest out of which they emerge are not focused upon some objective of social change which the movement seeks collectively to achieve. Instead they are released in some kind of expressive behaviour. . . ." See Blumer, *op. cit.*, p. 214.
[57] See Trotsky's reference to "the famous (essentially meaningless) formula of Eduard Bernstein" in Leon Trotsky, 'Centrism, Sectarianism and the Fourth International', reprinted in *Against the Stream: Two Articles by Leon Trotsky* (London: W.I.R Publications, 1965), p. 9; and Eduard Bernstein, *Evolutionary Socialism: A Criticism and Affirmation*, trans. Harvey (New York: Schocken Books, 1961). Bernstein refers to his proposition: "To me that which is generally called the ultimate aim of socialism is nothing, but the movement is everything" (p. 202). See also p. xxix.
[58] This phrase plays upon the title of Dennis H. Wrong's well-known article, 'The Oversocialized Conception of Man in Modern Sociology', *American Sociological Review* **26**, No. 2 (1961), 183–193.
[59] As well as contributing to the general study of sectarianism, such a perspective sheds light on the topic of process, development and 'careers' of social movements. It also has importance for the discussion of 'political-religious' or 'secular-religious' distinctions. In this context, see, for example: Margrit Eichler, 'Some Comments Concerning Murvar's "Messianism in Russia: Religious and Revolutionary" ', *Journal for the Scientific Study of Religion* **11** (June 1972), 187–191; N. J. Smelser, *op. cit.*, pp. 313–381; and Paul Wilkinson, *Social Movement* (London: Pall Mall Press, 1971), pp. 75–79.

REFERENCES

Bell, Daniel. 1952. 'Marxian Socialism in the United States' in Donald Drew Egbert and Stow Persons (eds), *Socialism and American Life*, 2 vols. Princeton, N.J.: Princeton University Press, 1952, Vol. 1, p. 222.

Berger, Peter L. 1970. *A Rumour of Angels: Modern Society and the Rediscovery of the Supernatural*. Garden City, N.Y.: Doubleday Anchor Books, pp. 17–19.

Cannon, J. P. 1954. 'Trade Unionists and Revolutionists', *Fourth International* (International Socialist Review), Spring, p. 60.

Cannon, J. P. 1962. 'The Vanguard Party and the World Revolution', *International Socialist Review* **28**, No. 5 (Sept.–Oct.), 26.

Cannon, J. P. 1968. *Letters from Prison*. New York: Merit Publishers, p. 93.

Coser, Lewis A. 1954. 'Sects and Sectarians', *Dissent* **1**, No. 4 (Autumn), pp. 360–361.

Lang, Kurt and Lang, Gladys. 1961. *Collective Dynamics*. New York: Crowell, p.202.

Lemert, Edwin. 1951. *Social Pathology*. New York: McGraw-Hill, pp. 225–226.

Williams, Robin M., Jr. 1961. *American Society*. New York: Alfred A. Knopf, p. 391.

RICHARD POPKIN

MILLENARIANISM IN ENGLAND, HOLLAND AND AMERICA: JEWISH-CHRISTIAN RELATIONS IN AMSTERDAM, LONDON AND NEWPORT, RHODE ISLAND

Jewish-Christian intellectual relations during the Middle Ages were often quite bitter, taking the form of disputations about whether the Messiah had come and denunciations of Judaism, and were often followed by grim consequences for the Jewish participants. In the late medieval period Christian interest arose concerning Jewish modes of interpreting Scripture. Starting with Nicolaus of Lyra, who was probably a convert, Christians started wading into the Jewish sources, the Mishna, the Talmud, Midrashim, Response, etc. As interest rose about the actual original text of Scripture, rabbis were consulted by figures such as Luther, by Papal authorities, and Henry VIII. Converted Jews became professors of Hebrew in England, France, The Lowlands, Germany, Italy and even New England.

Two developments, the Reformation and the expulsion of the Jews from Spain and Portugal, intensified the contacts, both positively and negatively. Some of the Reformers saw themselves as closer to the Jewish view, and saw their movement as leading very soon to the Millennium, starting with the conversion of the Jews. They blamed the Catholic falsification of the Christian message for delaying the conversion and other events that would lead to the Millennium. Early in the seventeenth century, divines especially in Holland and England, started interpreting the books of *Daniel* and *Revelation* as forecasts of history that was about to begin. They worked out a calculus for interpreting the texts which led to two formulas, that the Millennium would begin 1260 after the Fall of the Rome Empire and 1290 years after the profanation of the faith. These indicated that it would happen *circa* 1650–1670. In addition, wise theologians took note that, according to the best chronological evidence, 1656 years elapsed between the creation of the world and the Flood. Therefore some equally monumental event should occur 1656 years after the birth of Jesus. They decided the event must be the Conversion of the Jews.[1]

On the Jewish side strange developments occurred as a result of a

S. Hook, W.L. O'Neill and R. O'Toole (eds.), Philosophy, History and Social Action, 349–371.

Spanish campaign launched in 1391 to convert the Jews by force. Most Jews of Spain became converts. Those unconverted were expelled in 1492. Many went to Portugal, but were expelled from there in 1497. Those who remained, called *conversos*, were accused of being false Christians called *Marranos*. The Spanish Inquisition was established to police them, and to punish them severely if they were secretly keeping up any Jewish practices. As a result many Marranos fled when they could, and moved to safer areas including France, Italy, and the Lowlands. In Italy they could practice Judaism freely in ghettos. In France they formed secret Jewish colonies. However, the doctor of Marie de Medici, Elijah Montalto, insisted on publicly practicing and agressively defending his Judaism. He only agreed to come to Paris on condition that he did not have to work on Saturday, and that he be given police protection in practicing his religion. He polemicized against Christianity and sought to convince Marranos to become observant Jews.[2] To the court of the time, Marie de Medici had brought two Italian Jews, Concini and his wife Leonora Galigai, who ran the country after the murder of Henri IV. Doctor Montalto's secretary, Saul Levi Morteira, was also from Italy. When Montalto died in 1616, Morteira took his body for burial to The Netherlands because there was no Jewish cemetery in France. The next year Concini was assassinated by young Louis XIII, and the secret Jewish cell in the Louvre was exposed.[3]

However Morteira and others found real freedom to practice Judaism in Holland, a state of affairs that grew out of the Dutch rebellion from Spain. This official tolerance was formalized in a treaty suggested by Hugo Grotius. Thereafter Jewish synagogues and schools were established. Amsterdam, especially, became a Mecca for Marranos, who soon established a thriving Portuguese and Spanish Jewish community that played a vital part in the rapid expansion of Dutch overseas trade. Through their contacts, these Jews could provide an international banking service, and a basis for international trade.

The Jewish community of Amsterdam, unlike communities in Italy, and in most of Germany, was not ghettoized. From the beginning the Jewish area of the town was contiguous with the rest of the city, open in both directions. Rembrandt lived at the beginning of Joodenbreestrad, a block from the Portuguese synagogue. The Niewekerk was on the edge of the Jewish district, and the Church of Moses and Aaron a few doors from where Spinoza was born. The Jewish community was walled off for the first time when Hitler seized the city in 1940.

Dr Montalto's secretary, Morteira, rose to be the chief rabbi. He wrote polemics against Christianity. He organized Jewish schools to carry on both Spanish learning and study of the Jewish religion.[4] One of the first teachers in these schools was a young Marrano, Menasseh ben Israel, born in Madeira. His parents fled the Inquisition and he was born and raised in La Rochelle, France. At the age of eighteen he went to Amsterdam and became a Hebrew teacher. He then started printing Hebrew texts. The first to do so, he soon made Amsterdam the center of the Hebrew printing business. However, Menasseh's most important contribution – one that has hardly been recognized – was the series of works he published, in Latin and Spanish, describing fundamental Jewish doctrines. His account was read by Christian theologians all over Europe, and from him they learned what the Jews really expected to happen in the future. Theologians in England, France and Holland began consulting him by mail and in person. He became known as '*The* Jewish philosopher'. Rembrandt did his portrait, and illustrated one of his books. Christians, including the Queen of England, the wife of Charles I attended the synagogue in order to hear him.[5] He reported that more than half of those who attended services were Christians,[6] and the paintings of the services that exist seem to bear this out.[7]

Menasseh was a Messianist, expecting the Messiah in the near future. He was vague about making predictions. In 1644 a Portuguese Marrano came to Amsterdam from Brazil and reported that he had run across an Indian tribe that was practicing Judaism in the Andes mountains. Menasseh had the account recorded before a notary, and then told various Millenarians about it. Some in England soon published it as proof that the American Indians were Jews *and* that they were the Lost Tribes of Israel, who were to reappear at the end of days before the Millennium. They used this news, that the Indians were Jews, to raise money for the New England missions.[8] They convinced the cautious Menasseh that the discovery of the Jewish Indians was a real sign. This led him to write his most well-known work, *The Hope of Israel*, which was published in Hebrew, Latin, Spanish, Dutch and English and dedicated to the English parliament. The book indicated that the Messianic Age was near at hand, and that the political events taking place in England, plus the discovery of some part of some of the Lost Tribes, showed that the end was very near indeed.[9] The English translator added a postscript about the impending conversion of the Jews. Menasseh followed this work with a *Humble Address* to Oliver Cromwell, telling him that the culmination of history would occur when all the prophecies in the Old Testament were fulfilled. One of the most important outstanding prophecies was that in Daniel, saying that the Jews would be dispersed to the four corners of the world. Jews were widely scattered, but Menasseh pointed out, they were not allowed to live in England, having been expelled in 1290. If Cromwell would re-admit them, then great things might follow.[10]

Leading English Millenarians then came to Amsterdam to negotiate with Menasseh about the conditions under which (a) he would go to England to deal with Cromwell, and (b) the Jews would come to England.[11] These talks were getting nowhere when, in late 1654, Menasseh journeyed to Brussels to see Queen Christina of Sweden, who had just abdicated her throne. The usual accounts, trying to trivialize Menasseh's activities, say he went in the hope of collecting a bill she owed him for Hebrew books he had sent for her collection.[12] We don't know what transpired, but on returning to Amsterdam, Menasseh went to the home of Peter Serrarius, the dean of Amsterdam Millenarians. They had freely met over the years, discussing common interests. In the tolerant atmosphere of Amsterdam, Serrarius held a continuous open-house for all the different kinds of *avant-garde* theologians who came to the city. When Comenius was in Amsterdam he stayed with Serrarius; so did the English Millenarian leader, John Dury. The first Quakers in Holland used his house as a base, and various debates about theological views were held in his home.[13]. Into this atmosphere Menasseh entered and announced to the assembled group that he now knew that the coming of the Messiah was imminent. This touched off feverish activities among some of the Christian believers. One who was there, Paul Felgenhauer, a Czech Lutheran Millenarian, rushed home and wrote a work, dedicated to Menasseh, entitled *Good News for the Jews*.[14] Appended to it is a letter from Menasseh giving his reasons for declaring that the coming of the Messiah was imminent. This included letters he had received from English, German and Bohemian theologians who expected the Millennium any second, plus what Menasseh had just read in a book he apparently borrowed from Queen Christina, *The Recall of the Jews* by the French Millenarian, Isaac La Peyrère, who was the secretary of the Prince of Condé and, at the time, the next door neighbor of the ex-Queen.[15] La Peyrère had announced that the Jews would be recalled to France, converted and sent off to rebuild Jerusalem, there to prepare for the Messiah's arrival. When He came, He would rule the world with the King of France as his regent, and the converted Jews as His court.[16] Guillaume Postel had earlier advanced this view about the special role of the King of France in the Messianic Age.[17]) Although La Peyrère's work was published in 1643, Menasseh apparently had not seen it until he visited Christina.[18]

While his friends in Amsterdam and England were making their Christian preparations for the beginning of the Millennium in 1655, Menasseh made his Jewish preparations, deciding to leave for England to see Cromwell. Records of the Amsterdam Jewish community indicate that they did not want him to go, believing he could only cause them grief by dealing with the rulers of a rival state. Nonetheless, Menasseh appointed himself agent for the Jews of the world. He sent letters to three Jewish communities in Italy and Germany announcing his appointment and offering to negotiate for any and all of them. He set sail for England in September 1655. The next year and a half in both England and Holland were full of the greatest excitement and expectation of the fulfillment of the final prophecies of Scripture.[19]

In Holland, meanwhile, the Jewish community was carefully watched for indications of its imminent conversion. The astrological signs were calculated for clues. Geological and political events were scrutinized for their relationship to the prophecies in *Daniel* and *Revelation*. The recently formed English religious group, the Quakers, sent missionaries to Holland to convert the Jews since they realized that as there were hardly any Jews in England it wasn't worth the effort there. The Quakers attended the synagogue and argued with the Jews after the services[20] (We now know that they probably co-opted the recently excommunicated Baruch de Spinoza to translate some of their pamphlets into Hebrew.[21])

In England Menasseh ben Israel's arrival was greeted by various Millenarians as a crucial prelude to the conversion of the Jews. Leading figures like Henry Jersey, John Dury, Nathaniel Homes, Ralph Cudworth, Robert Boyle, Henry More and Henry Oldenburg, called on Menasseh, and some of them negotiated with him on behalf of the British government. They tried to find out why he did not become a Christian. They also tried to find out what special information he had about the imminent events to come. A top level government commission of politicians and theologians met for weeks. He acted as the agent for the Jews of the world (so described in Jessey's official account of the meetings). Oliver Cromwell met with him many times. Menasseh journeyed to Oxford and Cambridge, met their Hebraits, and looked over their Hebrew holdings. He sold some manuscripts and books to various scholars. He was the only really public Jew in England at the time.[22] The Quaker leader, Margaret Fell, wrote a pamphlet-letter to him urging him and all of his brethren to convert.[23]

We are also told that some other rabbis came with Menasseh. They were looking into the possibility that Oliver Cromwell was himself the Messiah. (After all, he had destroyed one of the kings of the earth.) They scoured the birth records to see if Cromwell was from the House of David, but were sadly disappointed.[24] A rather strange, though important, Welsh Millenarian with the delightful name of Arise Evans rushed to meet Menasseh. After reading *The Hope of Israel*, Evans had had a revelation in which he discovered that Charles Stuart, the son of the late beheaded Charles I, was the Messiah. When he told this to Menasseh, the rabbi scoffed. He told Evans that Cromwell, or the King of Sweden, might be the Messiah, but not an uncrowned figure like the exiled Charles Stuart. In fact, Menasseh went on, he had it on the best authority from an eminent Frenchman that the King of France was the Messiah. (This is apparently a reference to La Peyrère's theory.) Evans fought back claiming that in fact the King of England *is* the King of France according to all sorts of Medieval treaties.[25]

Late in Menasseh's stay, he published one book in England, *Vindicae Judaeorum*, apparently at the request of Robert Boyle. This work, published in English, answered several of the charges made against the Jews at the time. In this book he offered a peace treaty between Jewish Messianists and Christian Millenarians, a way of harmonizing their views, taken from La Peyrère:

> as a most learned Christian of our time hath written, in a French book, which he calleth The Rappel of the Iewes (in which he makes the King of *France* to be their leader, when they shall return to their country, the Iewes, saith he, shall be saved, for yet we expect a *second* coming of the same Massias; and the Iewes believe that that coming is the *first* and not the second, and by that faith they shall be saved; for the difference consists only in the circumstance of the time.[26]

This trivialization of the difference between Judaism and Christianity was told to Oldenburg on one of his visits to Menasseh and also to the leading Portuguese Jesuit, Antonio de Vieria, when he visited Amsterdam.[27]

Officially, Menasseh's visit to England ended in failure. Despite Cromwell's approval, the English government refused to re-admit Jews legally. Menasseh left dispirited and died en route to Amsterdam. Gradually, however, Jews unofficially began to re-enter England from the Restoration onward. But it was not until the middle of the nineteenth century that they were given full rights of citizenship. And, for reasons that are not clear, the Jewish community in England did not produce the sort of intellectuals that Amsterdam had. After Menasseh departed it was not until the late eighteenth century that an English Jew appeared; he was David Levi, and was taken seriously by the Christian community both as an authority on Jewish beliefs and as a worthy opponent.[28]

As we know the Millennium failed to begin in England in 1656. The believers did not give up hope, however, and when Peter Serrarius wrote to John Dury in April 1657 about his talks with one Rabbi Nathan Shapira of Jerusalem, Dury published a pamphlet with the exciting and excited title, *An Information Concerning The Present State of the Jewish Nation in Europe and Judea. Wherein the footsteps of Providence preparing a way for their Conversion to Christ, and for their Deliverance from Captivity are discovered.*[29] The pamphlet begins with the news that the growing misery of the Ashkenazi Jews in Jerusalem is a sign of their imminent delivery. Their leader, Rabbi Nathan Shapira, a very great Cabbalist, was sent to Europe to get help for his brethren. He got no real support from the Portuguese Jews in Amsterdam. By accident he met some of Dury's Christian friends, who did help and gave money. Dury contended that the increasing misery of the Jews would lead to their deliverance through "enlightening of their mind to acknowledge the true Messiah, Jesus Christ" and by restoring their homeland to them.[30] To show that this was happening, Dury cited the conversations Rabbi Nathan Shapira was having with Millenarians in Amsterdam. Six pages of a letter from Serrarius are then quoted. The rabbi, unlike leaders of the Amsterdam Jewish community such as chief rabbi Saul Levi Morteira and the philosopher Isaac Orobio de Castro, was not polemicizing against Christianity and Christian readings of Scripture.[31]

We are introduced to Rabbi Nathan Shapira when he interrupted a discussion between a Jew and some Millenarians in Amsterdam about whether the Messiah had come. The unidentified Jew said that it was preposterous that a person born in Bethlehem, Palestine, in the year 1, was still alive. Rabbi Shapira reproved his fellow Jew, and said, "Let come whoever will come, our prayers and wishes are to this effect simply, that God would be pleased to reveal him." Having said this, the rabbi was asked how he interpreted the passage in Isaiah 53 which Jewish disputants always insisted was not about Jesus. This is the passage about the suffering servant. There are many, many Jewish polemics about this.[32] Rabbi Shapira simply replied that it was about the Messiah, and gave a Cabbalistic analysis to show this. He then offered his theory that the spirit of the Messiah existed before Adam, and that the Messiah has born men's burdens from the Fall of Adam. The spirit of the Messiah is reborn in each generation, and has appeared at diverse times in various people such as Hezekiah, Hobbakuk, and Jesus. "If men repent and are reconciled with God, they may be admitted once again to God's favor, and if not, the Messiah will finally withdraw his shoulders from under the burden, and leave obstinate and impenitent men to bear their own burden."[33] The Dutch Millenarian, Serrarius, was so impressed that he told the English Millenarian, Dury, "When I heard these things, my bowels were inwardly stirred within me, and it seemed to me that I did not hear a Jew, but a Christian of no mean understanding, who did relish the things of the Spirit, and was admitted to the inward mysteries of our Religion."[34]

Rabbi Shapira startled his new Millenarian friends further by telling them that the Sermon on the Mount contained the fount of all wisdom, and that whoever kept its commandments would be more just than the rabbi of his brethren. He also held the Jews had sinned in putting Jesus to death. When Serrarius and his friends said prayers for the Jews, the rabbi was so touched that he said "If there were but ten men at Jerusalem, who should thus with one heart pray for the coming of the Messiah, that without all doubt he would suddenly come."[35] Serrarius recounted other amazing views of the rabbi, and then asked "What do you think? Is it to be believed that Christ is far distant from a soul this constituted? or that any such thing can be formed without Christ in a man? For my own part, I confess I think, I see Christ in his Spirit."[36] Dury claimed that everything Rabbi Shapira said and did showed that the Lord was preparing for the conversion of the Jews to Christianity. He expected Rabbi Shapira to translate the New Testament into Hebrew and communicate its good news to the Jews of Jerusalem.[37]

Dury and other English Millenarian leaders, such as Henry Jessey and Nathaniel Homes, raised a large sum of money for the relief of the Ashkenazi Jews of Jerusalem and in 1659 sent it to Rabbi Shapira. In their covering letters they told him in no uncertain terms "that our pity argues [not] our Doubt of our Religion" but that it should help bring about the conversion of the Jews and the Millennium. Henry Jessey's biographer tells us that,

> If any be inquisitive to know the success of these large alms, whether they were brought to judge more charitably of Christians; we assure them that it much melted many of them, and indeed it hath been an incontroulable Experiment, that this kind of dealing did ever more winne over the hearts of this kind of people in the primitive Church, than all the force and argument (since used) by moderne Christians could effect to persuade them.[38]

However, there is no evidence at all that either Rabbi Shapira or any of his flock became Christians. Instead the rabbi was seriously condemned by his brethren for taking money from Gentiles.[39]

Rabbi Shapira apparently established a link between the Dutch and English Millenarian leaders and the great Jewish Messianic movement that was beginning. Rabbi Shapira was the teacher of Nathan of Gaza who proclaimed himself Elijah, announcing that Sabbatai Zevi of Turkey was the Jewish Messiah. At the end of 1665 Sabbatai Zevi announced the beginning of the Messianic Age. He appointed his friends and relatives as kings of the earth. He made changes in Jewish Law and Jewish prayers (which only the Messiah is allowed to do). He changed the fast day of the ninth of the month of Ab, the day commemorating the fall of the first and the second Temple and the Expulsion of the Jews from Spain in 1492, into a feast day. Besides the day on which these three catastrophes are supposed to have occurred, it was also Sabbatai Zevi's birthday. His proclamations and actions caused all sorts of wild activities amongst the Jewish communities of the Ottoman Empire in Poland, Italy, Germany, and especially in Amsterdam. Almost the entire Spanish and Portuguese Jewish community accepted him as the Messiah, and became ardent followers.[40]

Rabbi Nathan Shapira is reputed to be the first person to bring the news about Sabbatai Zevi to Europe, reporting it to Jewish leaders in Italy.[41] It began appearing in the non-Jewish world in a series of pamphlets and letters written by Serrarius, published in Holland and England, or sent to friends like Henry Oldenburg of the Royal Society, and Robert Boyle. Serrarius' writings seem to have created a great stir for a while, leading to the formulation of a small, curious group of Christian followers of Sabbatai Zevi headed by Serrarius, and including Jan Comenius, Jean Labadie, Oldenburg and some of the English Millenarians. They speak of the divinity of Sabbatai Zevi, and they develop the Jewish theory of two Messiahs, one from the line of Joseph and one from the house of David, that they learned about from Menasseh ben Israel: Jesus was one Messiah and Sabbatai Zevi the other.[42]

Sabbatai, after being captured by the Sultan and threatened with death, turned Moslem. This created tremendous disillusionment among the Jewish communities in the world, which are still suffering some of the aftershocks. A small part of Sabbatai Zevi's following tried to find a way of explaining his apostacy. He himself lived for ten years after it preaching in mosques and synagogues, and gave secret explanations to various disciples. Some of them offered a theological explanation that only his body had turned Moslem, but his spirit was still Jewish. He had committed the greatest crime, apostasy, in order to take on the sins of all mankind. He will return in his glory as the True Messiah, the Redeemer of all mankind. Even today there is a small group, mainly in Turkey, Greece and Israel, that are outwardly Moslem, but secretly follow a Judaism based around Sabbatai Zevi.[42]

His chief Christian follower, Peter Serrarius of Amsterdam, also remained loyal, and in a letter to Henry Oldenburg said he had been assured that, despite the apostacy, Sabbatai Zevi was still a Jew. The whole episode simply showed that God acts in mysterious ways. Serrarius kept up his dissemination of information about Sabbatai Zevi. The leading Jewish opponent in Amsterdam, Rabbi Saportas, answered some of Serrarius' claims, but the last letter of Serrarius before his death, written to Comenius, says that he will be leaving shortly for Jerusalem to meet the Messiah. Unfortunately he passed away before he could make the trip. With his death the Christian group of Sabbatians seems to have ended. At the same time, John Evelyn published *The History of the Three Impostors*, Padre Ottomano, Mahomed Bei, "and the last, the Suppos'd MESSIAH of the Jews in the Year of the True Messiah, 1666" exposing Jewish credulity.[44]

Judaism in Amsterdam and elsewhere took a long time to recover. It turned inward, and sought to dampen its Messianic expectations in terms of near term fulfillment. From this time on Amsterdam Judaism seems to have lost a good part of its vitality. Its last important theological leader, Orobio de Castro (who had momentarily accepted Sabbatai Zevi), spent his time refuting Christian theology and Spinoza's pantheism. He debated Philip van Linborch about the truth of the Christian religion. (John Locke was present and wrote a long review of the debate.) Orobio died in 1687, the last important theologian of the community.[45] The eighteenth-century leader of Amsterdam Judaism,

Isaac de Pinto, was a public deist, and apparently a private atheist. He was a friend of David Hume and Denis Diderot and gambled, drank and dined with them.[46]

Thus we can see that during the seventeenth century the interaction of Jewish Messianic ideas and Christian Millenarian ones was extremely influential in the religious fervor and expectations of the time. Then Judaism became less imminently Messianic. Millenarianism in England and Holland became less revolutionary, and was absorbed in part by Sir Isaac Newton and his followers, as a neat climax of the order of the world that God had established. Millenarianism as the outcome of natural history was espoused by Newton, the Royal Society, David Hartley and Joseph Priestly.[47] The fertile mix of Jewish and Christian ideas seems to have pretty well ceased in Holland and England, but started up again in the New World, especially in the most tolerant of the colonies, Rhode Island. This colony had been developed by religious refugees from the stern Calvinism of the Massachusetts Bay Colony. The city of Newport in the colony became an important international port, and soon attracted Jewish merchants. In 1652 the Rhode Island colony proclaimed that "all men of whatever nation soever they may be, that shall be received inhabitants of any of the towns, shall have the same privileges as Englishmen any law to the contrary notwithstanding." When the first Jews who had fled from Brazil when the Dutch lost that colony to the Portuguese arrived in New Amsterdam in 1655, they were met by a good deal of hostility. While waiting for the Dutch authorities in The Netherlands to decide whether the Jews should be admitted to the Dutch colony, some of the Jews moved on to Newport and started the Jewish settlement there.

Over the next century many Jews, mostly of Spanish and Portuguese ancestry, made their way to Newport from Holland, the Caribbean and England, as well as from Spain and Portugal. A community was organized. The first synagogue in what is now the United States was opened in Newport in 1763.[48] From the beginning of Jewish settlement in Newport there was general mixing with the rest of the population. Newport attracted a wide variety of religious believers. One very notable visitor was the great English philosopher, George Berkeley, who was to become an Anglican bishop. He stayed in Newport for almost three years, and there wrote his attack on free-thinking, *Alciphron*. In the second half of the eighteenth century Ezra Stiles was the chief congregationalist minister there, until he left to become President of Yale. Stiles,

1727–1795, had studied at Yale, and then became a tutor there, working on science and theology. He was influenced by the fairly prevalent millennialism of the New England Protestants and the wilder form of Newtonian millenarianism that he discovered in some of the works of William Whiston, Newton's hand-picked successor at Cambridge University, works that George Berkeley had donated to Yale before returning to the British Isles. Two of Stiles' reasons for going to Newport in 1755 were that it had one of the best private libraries in North America at the time, and that it was so cosmopolitan, with so many different religious groups.[49] In the mass of diaries, letters, manuscripts, etc., that Stiles left to Yale at his death,[50] most of which are still unpublished, one finds that he actively sought out almost every Jewish scholar who passed through Newport. He conferred with them on various matters, usually including whether they had any information about when the Messiah was going to arrive. Stiles formed very close friendships with some of the Jews, and gently sought to convert them with no success, but developed his own millenarian theory out of his many discussions with them.[51]

When one Abraham Levi arrived in Newport in March 1770 Stiles met him and asked him when the Messiah would come. Levi had no date. Stiles then told him that a German rabbi had said it would be in 1783. Levi simply smiled and said, "they looked for him every day".[52] Stiles was not put off, and kept up his search for information. Altogether Stiles had dealt with six rabbis who came to Newport, plus many other Jews who came through the city. One of Stiles' Millenarian friends, Benjamin Gale, kept hoping that the minister would get a clue about the impending events, then Gale intended to rush to Newport for the news and possible details.[53]

In Newport Stiles started studying Hebrew. He consulted with the rabbis and frequently attended synagogue. One gathers from his papers that two of his very closest friends were the Jewish community leader, Aaron Lopez, and the strange rabbi from Palestine, Rabbi Carigal, who was in Newport in 1773. When Lopez died, Stiles wrote in his diary that he was the greatest merchant in America, that he had no enemies and was beloved by all. His beneficience to everyone was without parallel.

> He was my intimate Friend & Acquaintance! Oh! how often have I wished that sincere pious & candid mind could have perceived the Evidences of Xty, perceived the Truth as it is in Jesus Christ, known that JESUS was the MESSIAH predicted by Moses & the Prophets.

Stiles then went on,

> The amiable & excellent characters of a *Lopez*, of a *Menasseh ben Israel*, of a *Socrates*, & a *Gangenelli*, would almost persuade us to hope that their Excellency was infused by Heaven, and that the virtuous & good of all Nations & religions, notwithstand their Delusions, may be bro't together in Paradise on the Xtian System, finding Grace with the all benevolent & adorable Emmanuel who with his expiring breath & and his deepest agonies, prayed for those who knew not what they did.[54]

Rabbi Haym Isaac Carigal came to Newport on March 3, 1773. Stiles described seeing him and hearing him in the synagogue week after week. Stiles summarized a whole sermon of his, given in Spanish on Jewish Messianic expectations.[55] On June 27 the rabbi came to hear Stiles preach. Carigal was seated in Stiles' pew as the minister set forth his picture of the role of the Jews in Millennial history. Then they went together to Stiles' house. The rabbi said that he had never heard a Christian preach before, though he had been at services at St Peters in Rome and St Paul's in London.[56] The next day they got together again and the rabbi told Stiles his life story. He was born in 1733 in Hebron, studied in Palestine, became a rabbi and Hockam (Jewish wise man) at the age of seventeen. He had travelled widely over much of the Middle East, Italy, Vienna, Prague, Germany, Holland, France, and England, and then to Curacao, where he stayed for two years. He then went back to the Holy Land and in 1768 started out again, going to Paris and London on lengthy stays, then to Jamaica, Philadelphia, New York, and finally Newport (a truly wandering Jew).[57] As their friendship ripened, Stiles finally popped the question on July 7, 1773. "I asked him when he expected the Messiah." Carigal said "daily, probably within about forty years."[58] Stiles asked if he knew anything about the Lost Tribes having been sighted in India, but the rabbi had no information about this.[59] Eight days later Stiles asked Carigal if he had any particular reasons for expecting the Messiah immediately. Carigal said no, "but he thought it was high time for him to come". He added, that "if all Nations were in War and universal Tumult and Confusion, then he should expect him immediately, but this is not being more the Case now than in every current Age".[60] Carigal and Stiles started examining the great Cabbalistic work, the Zohar, together. Stiles sent letters in Hebrew to his rabbinical friend. They discussed minute points about what would happen in the Millennium. For Stiles it was a marvelous way to gain the fruits of Jewish learning and wisdom. Carigal seems to have been impressed by Stiles, for he started addressing him as 'Hocham', a term strictly reserved for Jewish wise men.[61] (One can only become a Hocham by being called a Hocham by someone who is a Hocham. I have preserved a letter I received some years ago from a genuine Hocham in Brooklyn, N.Y., addressed to Hocham Richard H. Popkin. Although I cannot read Hebrew, a rabbi who is hoping to become a Hocham told me that my 'ordination' counts.)

In July 1773 Carigal obtained a much better rabbinical post in Barbados. Just before he left, Stiles started composing a farewell statement for him about the Millennium. This is Stiles' most extended essay on his Judeo-Christian theology. He wrote it in Hebrew and English. It was so important to Stiles that he kept a record of the stages of its composition from July 18, 1773, until it was sent to Carigal on December 8, 1773, and until the Hebrew copy was finished and sent on January 14, 1774.[62] It is addressed "to the venerable Hocham, the learned Rabbi Haym Isaac Karigal".[63] The essay is extremely interesting. Stiles sought first to ground the millennial expectation strictly on Jewish sources. Starting off, "The least of the Disciples of JESUS, Ezra Stiles, wisheth satiety of Delights unto the illustrious and the venerable Hocham the great Rabbi Haym Isaac Carigal, who daily and without ceasing lifteth up his eyes unto things on high, and is enlightened with the Dew of Lights from Jehovah, and has been initiated into the Secrets of Understanding & sublime knowledge",[64] etc. To such a Jew Stiles sought to show that the expectation of the coming of the Messiah and of the Millennium appears in Jewish sources, Biblical, Talmudic, Cabbalistic, the commentaries of the great Jewish authors, Philo, Raschi, Maimonides, Abarbanel, etc. The concept of the Messiah as the eternal Son of God and as Redeemer appears in these sources. Having built up his case out of Judaism, Stiles then said to the rabbi, "Expecting thou expecteth the Messiah, and I also expect him. However you know that all of us Christians believe that he came in the Time of the second Temple." Daniel had prophesied that the Messiah would come in 490 years. Who came in this period? Jesus of Nazareth![65]

Stiles then tried to show that the Jews missed the great events because they expected two Messiahs, first a suffering one, and then a triumphant one. Stiles briefly sought to show some evidence that Jews should have taken note of that would convince them that Jesus was the Messiah. Then, knowing that the rabbi did not believe this, Stiles went on to show that they could both believe in the same future even if they did not believe in the same past. "I believe with you the glorious Appearance of the Prince of Peace on Earth in the latter the future days."[66] He then offered a description of what was to come that both could accept. Stiles, after all of his consultations with Jews, was only willing to place the date when the Millennium would begin at 2300 A.D. Among the things that would happen then is that everybody would speak Hebrew. The Messiah would appear in the clouds of Heaven with the Patriarchs and the Prophets, and he would be recognized by the people of Israel.[67]

Stiles' picture of what will occur is based almost entirely on the Jewish expectations. To leave room for Gentiles, he then pointed out that "on the appearance of the Messiah all the Service of Moses shall be taken away"[68] and people will neither have to be circumcized or baptized. There will be no more marriages or births in the days of the Messiah among those who will be resurrected. (Stiles worried about what would happen to people who had had two or more wives or husbands during their lifetimes.) People then alive (not through resurrection) will continue their family relationships until the end of the Millennium. Here Stiles hoped that the rabbi would agree that during this period the Jews would give up their separateness and allow inter-marriages "so all their children will be holy Seed".[69] Stiles sought to justify his non-denominational, uncircumcized picture of the Millennium solely from Jewish authorities. What will happen thereafter is the conflagration of the World, and a New Heaven and New Earth, described much like the Newtonian picture offered by Whiston.[70] Having pointed this all out of Jewish sources, Stiles near the end brings in a quotation from "the Scriptures of the Christians", *The Book of Revelation*, to indicate the same expectation.[71] He then broadened this to show that Hochamin or Wise Men of all nations said much the same, among Chaldean, Egyptian, Greek and Arabian sources. It also appears in the Jewish Cabbalistic literature (of which he named several books he had read), and in Hesiod, Plato, Zeno, Ovid, Virgil, Seneca, Cicero, Ammonius, the Sibbyls, the Oriental writers Menczius, Confucius, Zoraster, the Druids, the traditions of India, and the views of the Lama.[72] So there is an almost universal expectation of a kingdom of righteousness and a conflagration and renovation of the world. (Stiles worried about the truth value of non-Scriptural sources, and up to his last years was trying to show they were of lesser antiquity than the Bible.[73])

This essay was signed "These are my Words, written at Newport, Rhode Island in America, 28th day of the Month Tammuz in the 5533 year of the Creation. I am the least of the Disciples of Jesus of Nazareth.

Ezra Stiles, July 19, 1773."[74] He made quite a few Hebrew and English copies, and waited anxiously for word from his rabbinical friend to see if they had found common ground. In spite of all of his efforts at basing his millennialism on Jewish sources, presenting his case in an array of Jewish learning, minimizing the importance of the central Christian point that Jesus of Nazareth was the expected Messiah, and putting all of this in Hebrew, Stiles got no reaction from Rabbi Carigal. The rabbi lived until 1777. He kept complaining that he was very busy and did not yet have time to read Stiles' essay.[75] By 1775 Stiles was sending the rabbi detailed reports of the American Revolution (with a map of Boston to explain the battles), and suggesting a Millenarian interpretation of the clash between the colonists and the British.[76] The most Rabbi Carigal would say was that he was glad to see that Stiles' Hebrew was improving.[77] Stiles hoped, as he said in his letter to the rabbi of December 8, 1773, "So far as it respects a suffering Messiah you will differ from me in sentiments – but as to the future glory of the Messianic Kingdom I presume we shall be happy to agree."[78] There is no evidence that with all these concessions, based on his Jewish studies, Stiles was able to arrive at a common position with any of his Jewish friends. He did, however, bring his vast Jewish learning to Yale, when he became its president in 1779, where he said in his oration in 1781 that he was bringing to Yale the wisdom of five of the greatest rabbis, Carigal and others he met in Newport.[79] Both in terms of his teachings, and in terms of the books he accumulated for the Yale, Harvard and Brown libraries, he provided a most important base for Jewish studies in America. Yale today has raised $7,000,000 to start a program of Jewish Studies. But all they really have to do is reinstitute the courses Stiles taught on Jewish literature, history and theology.

Stiles also contributed, mainly by correspondence and personal contact, to the Millenarian interpretation of American history. Although the picture of the emergence of the United States as a product of Enlightenment attitudes still prevails, it is all too slowly being recognized that the United States was being seen as The Promised Land and the Redeemer Nation by many of the revolutionary colonists. Their theological interpretation of events definitely contributed to the formation of the new nation. One of Stiles' disciples and colleagues at Yale (who taught Hebrew there), David Austin, declared that the Son of Man was born in Philadelphia on July 4, 1776. Professor Austin was preparing to gather up the Jews of the world and resettle them in the

Holy Land. He used up his funds building ships to convey the Jews to the restored Israel.[80] Others, then and up to the present day, have seen America as playing a special role in the providential drama leading to the climax of Judeo-Christian history. Elias Boudinot, the President of the Continental Congress and the first Director of the United States Mint, saw the French Revolution as the antecedent step to the Second Advent. His *Star in the West* (1814) portrayed the role America would play.[81] And many evangelical preachers today are proclaming America's vital role in the forthcoming Apocalypse.

Jewish–Christian interactions have had interesting and important results and repercussions. In the Netherlands in the seventeenth century, close contact between Christians and Jews greatly increased Christian knowledge of Jewish source materials and Jewish ideas. As we have seen, this intensified Millenarian theorizing and expectations from during the middle of the century. The free atmosphere of Amsterdam allowed Christians to explore the Jewish world. The Jews did not feel threatened, and could then publicly express their Messianic views.

In England the bizarre visit of one Jew, Menasseh ben Israel, had an enormous effect on the Millenarian thinkers involved in the Puritan Revolution. They found crucial evidence in Menasseh's contacts with them and in his writings of the imminent commencement of the Millennium. Some were willing to accept Menasseh's peace treaty that would allow both Jews and Christians to forget their differences about the past and look forward together to the wonderful conclusion of Providential History.

The very free atmosphere of Newport, Rhode Island, again proved fertile soil for Jewish–Christian interaction. From the open situation, Ezra Stiles was able to imbibe enormous Hebrew learning and became the first great American Hebrew scholar. He also developed his own peculiar formulation of Millenarianism, with a peace treaty much like Menasseh's for joining Jewish and Christian expectations, and he applied his Millenarianism to the events of the time.

The influence of these situations on the Jews was also impressive. In Amsterdam, Jewish thinkers were able to explore their own heritage and examine it in respect to many alternatives. In the ferment of seventeenth-century Amsterdam Spinoza, using elements of the Jewish Christian and pagan worlds, put together the underlying metaphysics of modern secularism. Others, who stayed with Judaism, strained for a free form of Jewish life, and spawned what is now reformed Judaism.

England had a different history, since its Jewish community only gained legal rights in the mid-nineteenth century. It was therefore more defensive, except for the amazing David Levi who, late in the eighteenth century, took on Joseph Priestley, Tom Paine and many others, and argued in English that Judaism was the true religion, and Christianity an unfortunate mistake. England spawned lots of activity toward converting Jews and toward re-establishing them in Palestine. But even to this day England has not developed the same kind of interaction that occurred in either Holland or America.

In America Newport was obviously in the forefront of American toleration and attracted some rather amazing off-beat Jewish characters. As toleration became institutionalized in the U.S. Constitution, free Jewish communities began to develop in New York, Philadelphia, Charlestown and Savannah. By 1818, the leading Jewish writer in America, Mordecai Noah, said that there were only 3,000 Jews in the whole country, and almost all of them had to devote their energies to commerce rather than intellectual issues. As large-scale immigration of European Jewry developed a decade or two later, the U.S. quickly became a center of reformed Jewry, as people sought to modernize their views to fit the tolerant secular milieu of developing America. Thus freedom in America led to large-scale expression of a flexible and less and less Messianic Judaism.

Now, after the devastating events of the Holocaust and the emergence of the State of Irael, Jewish Messianic and Christian Millenarian thinkers are once again intermingling in Israel, and to a lesser extent in America. Fundamentalists, mainly of American background, have found in Israel the signs they were looking for presaging the Second Coming. They have said that Israel is God's timepiece. The Jews have returned to their homeland; they have rebuilt it. They are surrounded by enemies, and the battle of Armageddon is about to begin. In Jerusalem there is an organization of Christian Zionists headed by a Dutch pastor, and containing many American fundamentalists. The Rev. Jerry Falwell and his Moral Majority have become the staunchest supporters of the recent Israeli governments. The Millenarian Christians and the nationalist Israelis are joining forces, for better or worse. A committee has even been formed to raise money for the rebuilding of the Temple. And we are just beginning to see, once again, attempts to formulate a joint version of Jewish–Christian expectation emphasizing what is to come, and diminishing the importance of differences concerning what is supposed to have taken place in the past. It is too early to judge what will happen from this new interaction, but, perhaps, by recalling the experiences of seventeenth-century Amsterdam and London, and eighteenth-century Newport, we shall be more alert to the possibilities both positive and negative. The interaction of Christian Millenarianism and Jewish Messianism has been extremely influential in the past, both in creating hopes and ideals and tolerance, and in creating bizarre and potentially dangerous movements. What may result in the future we do not know, and can only hope that the best results of the past mixture will occur.

Washington University

NOTES

[1] See Christopher Hill, 'Till the Conversion of the Jews', William Andrews Clark Lecture (Oct. 1981), forthcoming; David Katz, *Philosemitism and the Readmission of the Jews to England 1603–1655* (Oxford, 1982), Chap. 3; Mayir Verete, 'The Restoration of the Jews in English Protestant Thought', *Middle Eastern Studies*, **VIII** (1972), 3–50; and R. H. Popkin, 'Spinoza and the Conversion of the Jews', in *Spinoza's Political and Theological Thought, International Symposium Commemorating the 350th Anniversary of Spinoza's Birth, Amsterdam, 24–27 Nov. 1982*, ed. C. De Deugd (Amsterdam, 1984), pp. 171–181.

[2] On Montalto, see Cecil Roth, *A History of the Jews in Venice* (Philadelphia, 1930), pp. 242–244.

[3] Cf. R. H. Popkin, 'The Historical Significance of Sephardic Judaism in 17th Century Amsterdam', *American Sephardi* **V** (1971), 14–27; and Fernand Hayem, *Le Marechal d'Ancre et Leonora Galigai* (Paris, 1910).

[4] On Morteira's career, see his article in the *Encyclopedia Judaica* **XII**, p. 357; and Herman P. Salomon, 'Haham Saul Levi Morteira en de Portuguese Nieuw-Christenen', *Studia Rosenthaliana* **X** (1976), 127–141.

[5] On Menasseh ben Israel's life and career, see the recent article by H. P. Salomon, "The Portuguese Background of Menasseh ben Israel's Father as revealed through the Inquisitorial Archives at Lisbon", *Studia Rosenthaliana* **XVII** (1983), 105–146 (which establishes that Menasseh was born and baptized in La Rochelle, France, in 1604. See also Cecil Roth, *A Life of Menasseh ben Israel* (Philadelphia, 1945); and Henri Mechoulan and Gerard Nahon, 'Menasseh ben Israel: Un regard nouveau sur sa vie et son oeuvre' in Menasseh ben Israel, *Esperance d' Israel* (Paris, 1979), pp. 34–69.

[6] Menasseh ben Israel, *Vindicae Judeaorum* (London 1656), pp. 5ff, where Menasseh discussed the non-Jews who attended the Amsterdam Synagogue.

[7] See the various illustrations of the Synagogue reproduced in Mozes Heiman Gans, *Memorboek* (Baarn, 1977).

[8] See David S. Katz, *Philo-Semitism and the Readmission of the Jews*, Chap. 4; Mechoulan and Nahon, *Esperance d'Israel*, pp. 71–99; and Lucien Wolf, *Menasseh ben Israel's Mission to Oliver Cromwell* (London, 1902).

[9] See Katz, *op. cit.*, Chaps 4–6; and Mechoulan and Nahon, *op. cit.*, esp. pp. 71–117. Mechoulan and Nahon show that Menasseh was denying the claim of the English Millenarians that all of the Indians were Jews, and was just holding that one part of the Lost Tribes had been located in South America.

[10] Menasseh ben Israel, *To his Highnesse, the Lord Protector of the Commonwealth of England, Scotland and Ireland. The Humble Addresses of Menasseh ben Israel, a Divine and Doctor of Physick in Behalfe of the Jewish nation*, n.d, n.p.

[11] See Katz, *op. cit.*, Chap. 6.

[12] See David S. Katz, 'Menasseh ben Israel's Mission to Queen Christina of Sweden 1651–1655', *Jewish Social Studies* **XLV** (1983–84), 57–72; and Roth, *Life of Menasseh*, pp. 169–175.

[13] On Serrarius, see Jan van den Berg, 'Quaker and Chiliast: the contrary thoughts of William Ames and Petrus Serrarius' in R. Buick Knox (ed.), *Reformation, Conformity and Dissent, Essay in Honor of Geoffrey Nuthall* (London, 1977), pp. 186–193; and Ernestine G. E. van der Wall, De Hemelse Tekenen en het Rijk van Christus op Aarde. Chiliasme en Astrologie bij Petrus Serrarius, (1600–1669) in *Kerkhistorische Studien* (Leiden, 1982), pp. 45–64; and note by R. H. Popkin on Serrarius, in K. O. Meinsma, *Spinoza et son cercle* (Paris, 1983), pp. 277–279.

[14] See Paul Felgenhauer, *Bonum Nunciam Israeli* (Amsterdam, 1655).

[15] *Ibid.*, pp. 87–91. See also R. H. Popkin, 'Menasseh ben Israel and Isaac La Peyrère. II', *Studia Rosenthaliana* **XVIII** (1984), 12–14.

[16] See Isaac La Peyrère, *Du Rappel des Juifs* (Paris, 1643).

[17] On Postel, see Marion Leathers Kuntz, 'Guillaume Postel and the World State: Restitution and the Universal Monarchy', *History of European Ideas* **IV** (1983), 299–323 and 445–465; and R. H. Popkin, 'Postel and La Peyrère', *Proceedings of the Postel Congress, Avranches*, 1981 (forthcoming).

[18] See R. H. Popkin, 'Menasseh ben Israel and La Peyrère. II', pp. 13–14.

[19] See Katz, *Philosemitism*, Chaps 5 and 6; and Roth, *Life of Menasseh*, Chap. 9.

[20] See R. H. Popkin, 'Spinoza's Relations with the Quakers in Amsterdam', *Quaker History* **LXXIII** (1984), 14–29; and William I. Hull, *The Rise of Quakerism in Amsterdam*, (Philadelphia, 1938).

[21] See Popkin, 'Spinoza and the Quakers'.

[22] See Katz, *Philosemitism*, Chap. 6 and 'Conclusion'; and Roth, *Life of Menasseh*, Chaps X and XI; and Henry Jessey, *A Narrative of the Late Proceeds at White-Hall concerning the Jews* (London, 1656).

[23] Margaret Fell, *For Menasseh ben Israel. The Calling of the Jewes out of Babylon which is Good Tidings to the Meek, Liberty to the Captives and for the opening of the Prison Doores* (London, 1656). Spinoza apparently translated this from Dutch into Hebrew, but the translation was not published. See R. H. Popkin, 'Spinoza's Relations with the Quakers' and 'Spinoza, the Quakers and the Millenarians, 1656–1658', *Manuscrito* **VII** (1982), 113–133. The Hebrew text presumably by Spinoza will soon appear in *Studia Rosenthaliana*.

[24] See Katz, *Philosemitism*, p. 207; and Roth, *Life of Menasseh*, p. 239.

[25] See Popkin, 'Menasseh ben Israel and Isaac La Peyrère. II', pp. 14–17.

[26] See Menasseh ben Israel, *Vindicae Judaeorum*, p. 18.

[27] Cf. Henry Oldenburg's letter to Menasseh ben Israel, June 25, 1657, in which he recounted what Menasseh had told him. *Correspondence of Henry Oldenburg*, ed. by A.

Rupert Hall and Marie Boas Hall (Madison and Milwaukee, 1966), Vol. I, pp. 123–126; A. J. Saraiva, 'Antonio Vieira, Menasseh ben Israel et le Cinquieme Empire', *Studia Rosenthaliana* **VI** (1972), 43–50.

[28] Cf. R. H. Popkin, 'David Levi, Anglo-Jewish Polemicist', *Harvard Studies in 18th Century Judaism* (forthcoming).

[29] This pamphlet was published in London in 1658. Katz attributes it to Henry Jessey. The Wing Short Title Catalogue lists John Dury as the author. The content indicates it is Dury.

[30] Dury, *An Information*, pp. 1–10.

[31] Serrarius's letter to J. D., April 1657, is pp. 11–16. See also Popkin, 'Rabbi Shapira's Visit to Amsterdam in 1657' in Joseph Michman and Tirtsah Levie, *Dutch Jewish History* (Jerusalem, 1984), pp. 185–193.

[32] Dury, *An Information*, pp. 11–12.

[33] *Ibid.*, pp. 12–13.

[34] *Ibid.*, p. 13.

[35] *Ibid.*, pp. 13–14. The quotation is on p. 14.

[36] *Ibid.*, p. 16.

[37] *Ibid.*, pp. 17–18.

[38] See Popkin, 'Rabbi Nathan Shapira', pp. 197–199; and Anon., *The Life and Death of Mr. Henry Jessey, Late Preacher of the Gospel of Christ in London* (n.p. 1671), pp. 75–76.

[39] Cf. Popkin, 'Rabbi Nathan Shapira', p. 199.

[40] All of this is recounted in Gershom Scholem's Sabbatai Zevi, *The Mystical Messiah* (Princeton, 1973).

[41] *Ibid.*, pp. 478–479.

[42] *Ibid.*, pp. 333–354; Michael McKeon, 'Sabbatai Sevi in England', *Association of Jewish Studies Review* **I** (1977), 131–169; and Popkin, 'Rabbi Nathan Shapira', pp. 204–205; and 'Jewish Messianism and Christian Millenarianism' in Perez Zagorin (ed.), *Culture and Politics from Puritanism to the Enlightenment* (Berkeley, 1980), pp. 79–83. The movement of Christian Sabbatians needs much more investigation.

[43] Scholem, *op. cit.*, Chaps. 7 and 8; and G. Scholem, 'Shabbatai Zevi', *Encyclopedia Judaica* **XIV**, 1219–1254.

[44] Serrarius' explanation of Sabbatai Zevi's conversion is in his letter to Oldenburg, July 5, 1667, *Correspondence of Oldenburg*, III, p. 447. John Evelyn's work, *The History of the Three Impostors*, appeared in London in 1669.

[45] The major work on Orobio de Castro by Yosef Kaplan appeared in Hebrew in Jerusalem in 1982. It is hoped that an English translation will appear soon. It is entitled *From Christianity to Judaism, The Life and Work of Isaac Orobio de Castro*.

[46] On Isaac de Pinto, see Popkin, 'Hume and Isaac de Pinto', *Texas Studies in Literature* **XII** (1970), 417–430; and 'Hume and Isaac de Pinto, Five New Letters' in *Hume and the Enlightenment* (Edinburgh, 1974), pp. 99–127; and Leon Schwartz, *Diderot and the Jews* (East Brunswick, New Jersey, 1981), pp. 130–133.

[47] See Popkin, 'Divine Causality: Newton, the Newtonians and Hume' in *Greene Centennial Studies*, ed. by Paul J. Korshin and Robert R. Allen (Charlottesville, Va, 1984), pp. 40–56

[48] On the background of the Jews in Newport, see M. A. Gutsein, *The Story of the Jews of Newport* (New York, 1936) and the article 'Newport' in Encyclopedia Judaica.

[49] On Stiles, see Edmund S. Morgan, *The Gentle Puritan, A Life of Ezra Stiles, 1727–1795* (New Haven and London, 1962).

[50] There is an inventory to the material in Harold E. Selesky, *A Guide to the Microfilm Edition of the Ezra Stiles Papers* (New Haven, 1978). The microfilms are available in many libraries.
[51] On this see the very interesting article by Arthur A. Chiel, 'The Rabbi and Ezra Stiles', *American Jewish Historical Quarterly* **LXI** (1972), 294–312.
[52] Ezra Stiles, *Literary Diary*, Vol. I, entry for March 16, 1770, p. 62. Yale Microfilm, Reel 8.
[53] On Benjamin Gale, see John F. Fulton's article on him in the *Dictionary of American Biography* and the letter of Gale to Ezra Stiles, June 23, 1773, in the Yale mss. Gale had heard about Rabbi Carigal, who had just arrived in Newport – "am told he is a Man of Learning among the Israelites. Since you doubtless see a lot of him, What are his expectations about when the Messiah will come?"
[54] Ezra Stiles, *The Literary Diary of Ezra Stiles*, ed. by Franklin Bowditch Dexter (New York, 1901), Vol. III, pp. 24–25. The published text of the *Diary* is not complete.
[55] *Ibid.*, pp. 354, 367, 376–377. An English manuscript text of Carigal's sermon (which was given in Spanish) is in the Stiles Ms. Papers 611, 1–13, Reel 16.
[56] *Ibid.*, pp. 391–393.
[57] *Ibid.*, pp. 395–396.
[58] *Ibid.*, p. 394.
[59] *Ibid.*, loc. cit. Rumors about sighting the Lost Tribes in Afghanistan, India and other parts of Asia were common. When the British took over most of India, new reports came back, even from eminent scholars like Sir William Jones.
[60] *Ibid.*, p. 398.
[61] See, for example, Rabbi Carigal's letter "to the Rev. Hocham, the Master of Knowledge, and Understanding & Wisdom, Ezra Stiles", 8th day of Sivan, 5533. Stiles Ms. *Literary Diary*, Reel 8, v.4. The letter is not included in the published text.
[62] This appears in the manuscripts at the end of Stiles' letter to Carigal, July 19, 1773, labelled by typewriter, "Hebrew Copy one". It was obviously written in 1774, when the essay was finished. The text is on Reel 3.
[63] This is labelled by typewriter "19 Jul 1773 ES to Raphael Haim Isaac Carigal (English Copy)". It is on Reel 3 of the microfilmed mss.
[64] *Ibid.*, fol. 3.
[65] *Ibid.*, fols. 3–13. The citation is on fol. 12.
[66] *Ibid.*, fol. 16.
[67] *Ibid.*, fols. 16–20.
[68] *Ibid.*, fol. 21.
[69] *Ibid.*, fol. 23.
[70] *Ibid.*, fol. 25.
[71] *Ibid.*, fol. 31.
[72] *Ibid.*, fol. 33.
[73] Stiles, *Literary Diary*, Vol. III, p. 517, entry for 25 Feb. 1794, where Stiles convinced himself that the Vedas "cant much excede the Times of Ab^{m}(Abraham) tho' they like the Alcoran & the Pentateuch contain Excerpta from Writers of higher Antiquity."
[74] Stiles ms. of his letter to Carigal, fol. 34.
[75] The correspondence for 1773–1775 is all on Reel 3 of the microfilm of the Stiles mss.
[76] Stiles to Carigal, July 7, 1775, fol. 225–232, on Reel 3.

[77] Rabbi Carigal to Ezra Stiles, May 27, 1774, "I am very glad to find you are advancing so well in the Hebrew." Carigal added, "as soon as I am at leisure, I shall examine your text". Unfortunately, the rabbi reported he had a great deal of business "for our nation", that he was tired from his voyage, etc. Reel 3.

[78] Stiles to Carigal, Dec. 8, 1773, Reel 3.

[79] Cf. Chiel, 'The Rabbis and Ezra Stiles', esp. pp. 294–295, and 'Ezra Stiles and Rabbi Karigal', *Yale Alumni Magazine* (March 1974), pp. 16–22.

[80] On David Austin, see Ernest Lee Tuveson, *Redeemer Nation* (Chicago, 1968), pp. 116–119; and David Austin, 'The Downfall of Mystical Babylon', in *The Millennium*, edited by David Austin (Elizabethtown, 1794).

[81] Elias Boudinot, *A Star in the West, or, A Humble Attempt to Discover the Long Lost Ten Tribes of Israel, Preparatory to their Return* (Trenton, 1816). I am working on a study of Boudinot and his role in the beginnings of American fundamentalism.

SIDNEY RATNER

JOHN DEWEY'S PHILOSOPHY OF WAR AND PEACE

Among the books on war and peace written by philosophers and social scientists, especially noteworthy are Kant, *Perpetual Peace*, Clausewitz, *On War*, Marx and Engels, *The Communist Manifesto*, Tolstoy, *War and Peace*,[1] Quincy Wright, *The Study of War*,[2] and Raymond Aron (1966), *Peace and War*. John Dewey never wrote a book on the subject of war and peace comparable to his *Experience and Nature* or his *Logic: The Theory of Inquiry*. Yet his views on this subject, as expressed in essays from the First to the Second World War attracted much attention and aroused both high praise and stringent criticism from different social groups at different times and in varying circumstances. These essays involve complex questions of historical data and judgment, and sharply contrasting theories of the relationship of means to ends. These questions deserve to be analyzed in a more systematic and searching way than most writers on this aspect of Dewey's work have done, such as John Patrick Diggins.[3]

Every college student of American intellectual history knows, or should know, that Dewey vigorously supported American intervention in the First World War. Great American historians like Frederick Jackson Turner and Charles Beard, noted philosophers like George Santayana and Josiah Royce, and iconoclastic economists like Thorstein Veblen and Wesley C. Mitchell agreed with Dewey that Germany and Austria-Hungary were the primary aggressors in the First World War and the major threat to the democracy represented by Great Britain and France. Dewey and the other editors of *The New Republic*, such as Herbert Croly, Walter Lippmann, Walter Weyl, and Alvin Johnson, thought that American intervention into the war meant military and economic aid to the Allies, but on condition that the United States would play a crucial role in shaping the objectives of the war and the peace. Dewey and his *New Republic* colleagues were ardent for war against German imperialism, but *not* for a war to promote Allied imperialism. This point cannot be stressed too much because some of Dewey's critics have misrepresented him and his associates as spokesmen or puppets of capitalist imperialism.[4]

S. Hook, W.L. O'Neill and R. O'Toole (eds.), Philosophy, History and Social Action, 373–390.

In taking this position Dewey broke with his own earlier anti-war and neutrality positions and with pacifist friends like Jane Addams. In 1908 he had written with James H. Tufts in their celebrated *Ethics* textbook: "The argument that war is necessary to prevent moral degeneration of individuals may, under present conditions, where every day brings its fresh challenge to civic initiative, courage, and vigor, be dismissed as unmitigated nonsense" (Dewey, 1978). In 1915 Dewey reaffirmed his opposition to war while making a sharp attack on German militarism (and idealistic philosophy) in his *German Philosophy and Politics*, (Dewey, 1979). This volume, and George Santayana's *Egotism in German Philosophy* (Santayana, 1916), have been severely criticized by critics like C. I. Lewis for being unfair to Kant, Hegel, and other German idealists. But although Dewey's and Santayana's sympathies were with the British as against the Germans in the war, Dewey at this time was not in favor of intervention by the United States.[5]

Dewey's commitment to the use of intelligence in the solution of problems over several decades had convinced many of his friends and former students, e.g. Jane Addams and Randolph Bourne, that he would be opposed to the First World War. In fact, Dewey as late as April 1917 justified the American peoples' hesitation to intervene on the side of France and Britain "until the Allies are fighting on our terms for our democracy and civilization".[6] Nevertheless, once Congress declared war on Germany on April 6, 1917, Dewey supported the war effort along with Charles Beard and many other leading liberals and radicals (Bell, 1952).

Dewey gave the most detailed philosophical justification of his position on war at this time in two essays: 'Force, Violence, and Law'[7] and 'Force and Coercion'.[8] In these essays Dewey distinguished three conceptions of force: force as energy, as coercion or constraint, or as violence. The term 'energy' is used in either a neutral or a eulogistic sense as an effective means of operation, ability, or capacity to execute, to realize ends.

> But it is force nonetheless – brute force if you please, and rationalized only by its results. Exactly the same force running wild is called violence. The objection to violence is not that it involves the use of force, but that it is a waste of force; that it uses force idly or destructively. And what is called law may always, I suggest, be looked at as describing a method for employing force economically, efficiently, so as to get results with the least waste.[9]

Coercive force occupies a middle place between power as energy and power as violence. . . . To use energy to make a man observe the rule of the road is a case of coercive force. . . . Constraint or coercion, in other words, is an incident of a situation under certain conditions – namely, where the means for the realization of an end are not naturally at hand, so that energy has to be spent in order to make some power into a means for the end in hand.[10]

According to Dewey's line of reasoning law or the legitimate use of force is the comparatively efficient and economical use of force for socially approved ends, defining ends as results rather than mere aspirations.[11]

A great Chinese thinker and scholar, Hu Shih, Chinese Ambassador to the United States in the first years of the Second World War, has praised these essays as presenting "an original and truly instrumentalistic theory of force", "the most daring and most original attempt to develop a political philosophy on the foundations of logical instrumentalism" (Shih, 1940, pp. 212, 217). On the other hand, Morton White has criticized Dewey's philosophical analysis of force because Dewey included verbal attempts at persuasion as one form of force or energy. White is correct in asserting that ordinary linguistic usage does not classify 'persuasion' as 'force'. But Dewey, in passages White ignored, clearly explains his deviation from ordinary usage. In discussing the objections of pacifists to the use of gross or brutal physical force, he pointed out that with the advance of knowledge, some individuals or groups used refined, subtle, and indirect forms of force instead of coarse, obvious, and direct forms of force. Hence as he saw it, "the so-called problem of 'moralizing force' is in reality a problem of *intellectualizing* its use"[12] – in other words, using persuasion, economic and social incentives, education and indoctrination as substitutes for physical force or coercion.

On the crucial subject of non-resistance that Ghandi was in the process of making into a weapon for the national independence of India, Dewey argued that except in cases of extreme quiescence, the non-resistance doctrine can mean only that

. . . given certain conditions, passive resistance is a more *effective* means of resistance than overt resistance would be. Sarcasm may be more effective than a blow in subduing an adversary; a look more effective than sarcasm. Only upon such a principle of expediency can the doctrine of non-resistance be urged, without committing ourselves to the notion that all exercise of energy is inherently wrong – a sort of Oriental absolutism which makes the world intrinsically evil.

He went on to say that if pacifists in war and in penal matters

> . . . would change their tune from the intrinsic immorality of the use of coercive force to the comparative inefficiency and stupidity of existing methods of using force, their good intentions would be more fruitful.[13]

These remarks of Dewey on non-resistance reveal an insight on his part into methods of effecting social change that was greater in almost all cases than that of any his critics on this subject. He concluded his masterly analysis of 'force and coercion' by stating explicitly that there is always a possibility that what passes as a legitimate use of force

> . . . may be so wasteful as to be really a use of violence; and *per contra* that measures condemned as recourse to mere violence may, under given circumstances, represent an intelligent utilization of energy. In no case, can antecedent or *a priori* principles be appealed to as more than presumptive; the point at issue is concrete utilization of means for ends.[14]

The importance of these statements by Dewey on the empirical testing of judgments on the legitimate use of force has not been widely understood. After the defeat of Germany, late in 1918, the Bolshevik disclosure of Allied secret treaties on the spoils of war, and the widespread popular reaction against various provisions of the Versailles Peace Treaty, Dewey and other leaders of public opinion like Charles Beard, Herbert Croly, and Walter Lippmann came to regard their support of the First World War with regret and to oppose American ratification of the Versailles Peace Treaty and American entrance into the League of Nations. Some writers have credited Randolph Bourne, a brilliant radical journalist and former student of Dewey's, with having changed Dewey's attitudes on the war through his devastatingly critical essays, 'The War and the Intellectuals' and 'Twilight of Idols'.[15] But this claim by champions of a mythically heroic Bourne is unjustified. Dewey himself told James Farrell two decades later: "Bourne was extremely clever and gifted, but he did not have depth" (Lamont, 1959). Nor did Bourne have freedom from pro-German, anti-British biases or the ability and willingness to give an undistorted account of Dewey's efforts during the First World War on behalf of economic and social reforms on both the domestic and international fronts.[16]

One of the ironies of history is that a strong admirer of Bourne and critic of Dewey, John P. Diggins, has presented a sympathetic picture of Thorstein Veblen's support for American intervention in the First World War in a book on Veblen as America's most penetrating social critic, while castigating Dewey in a magazine article for taking the same position on war as Veblen.[17] Another irony: the belief of liberal and radical scholars in the Allies' responsibility for the war was destroyed in the sixties by Fritz Fischer's *Germany's Aims in the First World War* (published in German in 1961, in English in 1967).[18] This epoch-making work documented the culpability of Germany and Austria-Hungary.

Still another revision, concerning prevailing interpretations of Woodrow Wilson's diplomacy in the First World War and Dewey's support for it, is in order. Many historians up until the last few years have believed that Wilson attempted to steer a middle and futile course between "atavistic imperialism" and a sympathetic position toward revolution in backward areas. Some historians have argued that Wilson was actually trying to aid American economic interests through "Open Door Imperialism". But a group of authorities on the First World War, headed by Arthur S. Link, have presented a new picture of Wilson as responsive to the aspirations of oppressed peoples around the globe and very sympathetic to revolutions against governments that had become "ossified and oppressive". Wilson failed to achieve much that he had hoped for, but he had achieved more than most of his critics in the past have given him credit for. This gives a stronger justification for Dewey's support of Wilson during the First World War than Bourne and other opponents of Dewey have been willing to grant.[19]

Dewey's disillusionment with the Versailles Peace Treaty led him to support a program called the 'Outlawry of War', which had been initiated by an old friend, a lawyer from Chicago, Salmon O. Levinson, as early as March 1918. Dewey threw great energy into championing this program during the 1920s in a series of influential essays in *The New Republic* and other periodicals. In 1928 he supported the Kellogg-Briand Pact that condemned war as a means of solving international controversies and renounced it as an instrument of national policy. The contracting parties agreed "that the settlement or solution of all disputes or conflicts of whatever nature or of whatever origin they may be which may arise among them, shall never be sought, except by pacific means." The pact also asserted in its preamble that a state that violated its prescriptions would not be protected from defensive or policing action by other states; thus the aggressor and the defender would be legally equal. Dewey was not happy about this concession on wars in self-defense but he hoped that the actual facts of a case involving 'self-defense' would be presented to the International Court for a decision as to whether a plea of self-defense was justified. In any case, despite these limitations, Dewey felt that the pact was a fruitful source of international cooperation.

In 1933 he asserted that considering the centuries' old role of war as a recognized means of settling disputes between nations, the change "that has been effected in the legal status of war in a short period of ten years is striking".[20] Although between 1918 and 1921, Dewey's friend, Levinson, and his supporters allowed the use of military force as a sanction against a nation which violated the principles of international peace, after 1921 Dewey and his colleagues on the Committee for Outlawry of War took the position that the sole reliance for the preservation of international peace was the organized moral sentiment of the world and that even if nations should agree to use force against a recalcitrant nation, there would be no assurance save good faith that that compact in turn would be honored. After 1921, even during the 1930s as Japan, Italy, and Germany took military actions that threatened world peace, Dewey and his friends in the outlawry movement opposed any appeal to 'sanctions'. Otherwise Dewey feared that the system of war would be maintained under the guise of sanctions to punish violators of the League of Nations Covenant or the Pact of Paris.[21]

As late as March 1939 in a symposium on war in *Commonsense* Dewey agreed with Charles Beard and Bertrand Russell on neutrality for the United States. He went so far as to say that if Americans believed that war is not inevitable and that if we agree that "no matter what happens we stay out, we shall save the country from the greatest social catastrophe that could overtake us, the destruction of all the foundations upon which to erect a socialized democracy" (Dewey, 1939).

This was a statement for which Dewey has been roundly criticized. But it should be pointed out that two other great philosophers, Bertrand Russell and Alfred North Whitehead, both of them non-pragmatists, took a similar position at this time. Whitehead told a friend of his in December 1939:

> I am firmly convinced that America should keep out. You need about fifty years to settle and judge certain domestic problems which you now seem well on your way to do. If you came in and were deeply involved, it might result in a permanent injury to the future of the world.[22]

Although after the fall of France in June 1940, Dewey still hoped that the United States could stay out of the Second World War, by February

1941 he no longer adhered to his 1939 plea for America to stay out of the conflict no matter what might happen in Europe (Dewey, 1941). During the spring of 1941 John Dewey, along with Thomas Mann, Will Durant, and Van Wyck Brooks, joined an organization favorable to America's intervention in the Second World War, The Friends of Democracy. Dewey refused an invitation to join the America First Committee although two old-time colleagues, Charles Beard and Harry Elmer Barnes, had joined in. In May 1941 Dewey wrote in a notable letter to Selden Rodman, editor of *Commonsense*, that he was not an absolute pacifist. Although he hoped that America might enter the war at the "latest possible time", he asserted that "it is now clear that unless the Nazis are beaten, this country will be permanently militarized".[23] Here is a dramatic statement of Dewey's change in his attitude toward world war as the Nazi danger to world peace and the United States became increasingly clear.

After the Nazi invasion of Russia in June 1941 many of Dewey's closest friends and supporters became full supporters of Franklin D. Roosevelt's policies of opposition to the expansion of Germany, Italy, and Japan. On the night of December 7, 1941, Dewey was scheduled to deliver a speech at Cooper Union entitled 'Lessons from the War [the First World War]: in Philosophy'. When news of the Japanese attack on Pearl Harbor came to him, Dewey told his audience that the current war represented a major turning point in history on the issue of whether free government could survive. Although he had hoped that the experiences of the First World War would cause humanity to avoid wars in the future, he concluded that the dangers of totalitarianism had to be met with force.[24] By December 11, 1941, Germany and Italy declared war on the United States, and America's involvement on all fronts, east and west, was confirmed when Congress declared war on Germany and Italy that same day.

After Pearl Harbor, Dewey not only supported the American war effort but also was critical of excessive praise of Soviet Russia by fellow travellers who used the American wartime alliance with Russia as a means for trying to portray Russia as a democracy. In 1944 Dewey supported Roosevelt for the presidency and from 1945 on backed Truman and the Truman Doctrine. In addition to defending the Marshall Plan, he approved of America's intervention in the Korean War and the United Nations support for the United States in that venture.[25] On June 1, 1952, Dewey died. Although he hated war and for some twenty years was an outstanding champion on the 'Outlawry of War' program, in the First and Second World Wars, and in the Korean War, he believed that circumstances justified the use of military force.

Having charted Dewey's varying positions from 1917 to the end of his life, I should like now to explore the principles on which he based his decisions. Writers on Dewey's philosophy of war and peace have almost all ignored a fundamental principle that Dewey stated in 1919. He mentioned that in dealing with the specific problems of individuals in a specific place, time, and social arrangement,

> . . . the traditionally accepted logic substitutes discussion of the meaning of concepts and their dialectical relations to one another, in place of a logic of inquiry based on generating and testing hypotheses that may be the solution to specific historic phenomena and concrete social difficulties.[26]

Logically, as Hu Shih pointed out over four decades ago, Dewey's view flatly denies any use for a social and political philosophy which, "as a general theory, must discuss, among other things, the state, the individual, major institutions as such, society in general" (Shih, 1940, pp.205–223, esp. pp.205–206), and war in general.

With this guiding principle in mind, we can re-examine key passages in Dewey's seminal essay, 'Force and Coercion'.[27] After distinguishing between energy or force, violence, and war, Dewey made his main point:

> . . . no ends are accomplished without the use of force [or energy]. . . . but antecedent and abstract principles cannot be assigned to justify the use of force. The criterion of value lies in the relative efficiency and economy of the expenditure of force as a means to an end. (p. 787).

In Dewey's eyes passive resistance was justified if it was a more *effective* means of resistance than overt resistance would have been. The implication is obvious. Under certain conditions passive resistance may not be more effective than overt resistance, and in such circumstances war would be justified (p. 788). In all cases of social conflict, including those resulting in war, Dewey felt no antecedent or *a priori* principle can be appealed to as more than presumptive in reaching a decision. The crucial point at issue is the wisdom we use in judging the concrete utilization of specific means for specific ends, and how such actions affect other ends or values that we prize (p. 789).

Using Dewey's guiding principle of on the need for having specific hypotheses for explaining and testing statements on specific problems, we find his varying positions on war in the First World War, the inter-war period, the Second World War, and the Korean War intelligible in terms of the specific problems he encountered. The information that Dewey had on America's role in the First World War was the best then available to him and to such friends and distinguished historians and political scientists as Charles Beard, Carlton J. Hayes, and James T. Shotwell, as well as such scholars as Carl Becker and Frederick Jackson Turner. Dewey was persuaded by the reports from Europe and the analyses of Beard, Croly, J. M. Keynes, Walter Lippmann, and others to reject the Versailles Treaty and to accept as the best alternative to Wilson's League of Nations the Outlawry of War program advocated by Salmon O. Levinson. Here again, Dewey acted on the basis of what seemed to be proof of the overwhelming faults of Wilson's peace negotiations. The arguments of Wilson's critics at that time have been re-evaluated in recent years and found to be open to severe criticism and in certain cases to outright repudiation. Wilson's defenders – e.g. Etienne and Paul Mantoux and James T. Shotwell – have, I believe, been shown to be more in the right on the relative merits of the peace treaties and the League of Nations than Dewey, Beard, Keynes, and Levinson.

Nevertheless, Dewey's Outlawry of War program was worth experimenting with, given the strong international revulsion against war in the 1920s and eary 1930s – witness the great appeal of such literary works as Henri Barbusse's *Le Feu*, Ernest Hemingway's *A Farewell to Arms*, and Erich Remarque's *All Quiet on the Western Front*.[28]

In addition to the historical explanation, Dewey's advocacy of the Outlawry of War should be seen in the light of his social psychology, as he developed it in the 1920s and 1930s. In 1922 he published what some critics consider his best book, *Human Nature and Conduct*.[29] In that book and in a 1932 essay, 'Human Nature',[30] Dewey surveyed the varied meanings given to the term, human nature, and, in the course of human history. He explained how each interpretation related to the institutional and intellectual characteristics of a society and period in which it prevailed. In the emotionally charged debate between those who affirmed the basic fixity of human nature and those who argued for a great range of modifiability, Dewey came out powerfully for the more radical alternative. He had confidence in the possibility of controlled social change that might alter the future of war and of competitive capitalism. These beliefs influenced strongly his position on the economic measures of Herbert Hoover and Franklin D. Roosevelt during the Great Depression and on the outlawry of war.

These views on human nature and war had emerged from the mutually reinforcing work of Dewey and anthropologists like Franz Boas and Ruth Benedict. Ruth Benedict once said one of the sources of inspiration for her very influential *Patterns of Culture* was Dewey's *Human Nature and Conduct*.[31] Dewey, Boas, and Benedict together helped to make prevail in American *avant garde* thought the view that

> . . . each age in society has woven its social patterns, its institutions, out of the primary human drives or impulses under their most diverse conditions, with the firmest conviction that its ideals and standards were categorically determined by certain fundamental human qualities, fixed and unalterable.[32]

The high hopes of Dewey, Boas, Benedict, and social reformers for restraints on war became dissipated as the Great Depression stimulated economic nationalism and imperialist expansion by Japan, Italy, and Germany in the 1930s and early 1940s. Their defiance of the Kellogg-Briand Pact and sanctions by the League of Nations, and their success in carrying out one aggressive act after another finally led Dewey, his Columbia colleagues, and the leader of the Outlawry of War movement, S. O. Levinson, to conclude that American participation in the Second World War was justified. Here again, Dewey's guiding principle of testing his hypothesis about the possibility of the outlawry of war by empirical evidence came into play and finally led him early in 1941 to assert the eventual need of American military action. He may be justly criticized for not changing his position in 1931 as Japan invaded Manchuria, or in 1935 as Italy invaded Ethiopia, or in 1939 when Germany invaded Poland. But similar criticism might be levelled at Bertrand Russell and Alfred North Whitehead, two equally eminent philosophers. The important point is that in every decision on international and domestic affairs there is a combination of general philosophical principles and empirical statements on matters of fact. The best solution, given the apparent feasible alternative courses of action and consequences, is more difficult to establish than some historians, for example, John P. Diggins, with perfect hindsight have been willing to admit. Decision-making as a specific discipline has developed only since 1944, with John von Neumann and Oskar Morgenstern's epoch-making *Theory of Games and Economic Behavior*.[33] The seeming inconsistencies in Dewey's writings on war and other political or social topics reflect the fact that he was concerned with specific problems and events and hence centered on "the needs and tensions of a particular time and place". In a 1939 letter to Hu Shih about the draft of an essay by the latter on Dewey's political philosophy, Dewey wrote:

> You are quite right in suggesting that my political philosophy has changed in emphasis at least at various times, according to what was uppermost at the time. The economic collapse, the reaction from the capitalistic orgies of the twenties made me more socialistic than I had been – as in *Individualism Old and New* [1930] as well as in *Liberalism* [*and Social Action* 1935]. Probably the best balanced of my writings is *The Public and Its Problems*, and also the most 'instrumental', at least by implication, for it brings out the principle of relativity. . . . In my last change of emphasis [on the decisive role of individuals in society in *I Believe*, ed. Clifton Fadiman (New York: Simon and Schuster, 1939), pp. 347–54], to which you refer, I have again been actuated by the principle of relativity.[34]

The remarks that Dewey made on the "principle of relativity" that he employed on economic issues applies also to his writings on war, especially his 1916 essays, 'Force, Violence, and Law' and 'Force and Coercion'. At the time of their publication, these essays won a warm reception among authorities on international relations. In the same year, Hu Shih, a student of Dewey's, already a rising young scholar and later an outstanding Chinese philosopher, developed Dewey's theory in 'Is There a Substitute for Force in International Relations?' This was published by the Carnegie Endowment for International Peace as a prize-winning essay in *International Concilation*, June 1916.

In 1940, Hu Shih wrote to Dewey: "Do you object to my considering these essays as more important than 'The Public and Its Problems'? And more instrumentalist" Hu Shih continued: "I have never understood why you have dropped the line of thought developed in those two essays of 1916. Have you any objection to my reviving them now?"[35] Dewey replied: "You certainly have my permission to publish passages from my letter." He went on to say that he was inclined to think "that after I have given sufficient time and thought to the matter I could make a statement not inconsistent with what I have previously written . . .". He ended by expressing his interest in Hu Shih's question as to why he hadn't followed up the leads in his 1916 papers on force. "Doubtless there is again an element of 'relativity'! For of course the year (1916) brought the question of force to the foreground. But that I am sure is not in this case the whole of the matter."[36]

In my judgment, Dewey's 1916 essays, 'Force, Violence, and War' and 'Force and Coercion', deserve to be regarded as notable contributions to the theory of ends and means as applied to the problems of war and peace. Yet Dewey's writings on this subject cannot be compared to such comprehensive treatises of the past as Kant's *Perpetual Peace* and Clausewitz's *On War*, or to such recent impressive works as Quincy Wright's *A Study of War* (Wright, 1965) and Raymond Aron's *Peace and War* (Aron, 1966). These two later works represent great advances in both theoretical and empirical analysis over all prior work in the area of international relations. Aron's book has been highly praised and deserves to be studied in its own right.[37] But since it does not discuss Dewey's philosophy in any way,[38] I shall not analyze its main theses or contributions, as these have been discussed with great skill by others.

On the other hand, Quincy Wright pays attention to Dewey and uses some of his ideas for building up his own elaborate theory and analysis. He agrees with Dewey's thesis in *Human Nature and Conduct* on the modifiability of human nature and social institutions (p. 13 n. 14, p. 519). Wright also agrees with Dewey's assertion in a 1939 essay that peace will exist according to the extent to which cultural conditions are established that will "support the kinds of behavior in which emotions and ideas, desires and appraisals, are integrated".[39]

Wright's own theory of war implements Dewey's ideas by developing a four-factor model of the cultural conditions that affect the origins of war. The four factors are: (1) technology, especially as it applies to military matters; (2) law, as it relates to war and its initiation; (3) social organization, as it bears on such general-purpose political units as tribes, nations, empires, and international organizations; and (4) the distribution of opinions and attitudes on basic values. These four factors correspond to the technological, legal, socio-political, and biological-psychological-cultural levels of human life, respectively. Whenever a major change occurs at any level, the established adjustments and control mechanisms may become strained and break down. Hence any major change in the world increases the risk of war, unless that change is offset by compensating political, legal, cultural, and psychological adjustments. Peace therefore, if it is to be preserved, or restored after a partial or world-wide breakdown, requires constant new efforts, new arrangements, and often new institutions, such as Dewey and other innovators proposed.[40]

On the 'Outlawry of War' issue, Wright has been far more favorable to the program than either Aron or most writers on that subject. In a magisterial essay, 'The Study of War',[41] Wright discusses the outlawry of war in different international conventions from the League of Nations

Covenant of 1920 and the Kellogg-Briand Pact of 1928 to the United Nations Charter. Wright points out that restraints on hostilities between relatively minor powers have been effective, but similar attempts to hold back major powers from aggression have been unsuccessful.

In his judgment, "history has made it clear that outlawry of war has not eliminated the possibility, or even the probability, of hostilities or of war". He shares with Dewey the convictions that the cold war between the communists and Western states has created justified anxieties on both sides of the 'iron curtain' that a third world war is possible.[42]

The conclusion to Wright's reflections on war might have been written by John Dewey, to whose philosophy he has been sympathetic. The problem of war cannot be solved by developing the art of war, but only by developing the art of peace. War arises because each politically organized system has guided its behavior by paying attention to its own internal interests and motivations without adequate understanding or concern for the probable reactions and interests of others. On the other hand, peace is achieved when there is a general desire to create and preserve it, based on a correct image of the world as an interdependent whole, and on the guidance of political decisions and actions by sound psychological, sociological, political, economic, and technological knowledge of the probable reactions of the different political systems or units that are capable of precipitating hostilities.[43]

John Dewey and Quincy Wright would have shared with Albert Camus the hope that men throughout the world will finally aspire to be 'Neither Victims Nor Executioners'. The success of the non-violence movement led by Gandhi in India and by Martin Luther King in the United States demonstrated that the resort to force could be effectively renounced on domestic issues in countries where the ruling groups could be influenced by appeal to human rights and by democratic governmental processes. In totalitarian countries, however, such movements have so far not succeeded. Efforts to expand such non-violent techniques into international disputes so far have not been effective where major powers are involved. The plasticity of human nature, in which Dewey had such faith, has manifested itself since the Second World War in the avoidance of conventional war among the Western powers and of nuclear warfare by the super-powers. But numerous international conflicts elsewhere have erupted. And more such wars loom on the horizon – witness at present the Iran–Iraq War, the tensions between Israel and its neighboring Arab states, and the conflict between Soviet Russia and the United States over the future of Central America and other danger spots in Asia and Africa.

Since the Second World War the defense of democracy against the expansion of communist dictatorships has been complicated by the dangers of nuclear war. Dewey himself had prized the preservation of freedom for the individual above the costs of the Second World War and the Korean War. Since the Soviet acquisition of nuclear weapons, champions of freedom have had to face difficult choices on how to thwart the spread of communist dictatorships without bringing about a nuclear war. Soviet Russia and its allies have resorted to propaganda, espionage, subversion, and guerrilla warfare to expand their dominion and yet avoid nuclear disaster for themselves. Dewey did not live long enough to grapple with this problem and to formulate a program for dealing with it. Quincy Wright, one of his leading disciples, in 1955 suggested that international peace might be promoted if certain general ethical standards be adopted throughout the world involving belief in humanity, in liberty, in science, and in tolerance. In his eyes, this culture of humanism, liberalism, pragmatism, and relativism could become the basis for freedom of communication and other human rights in all the nation-states comprising the United Nations (Wright, 1955). Unfortunately religious fundamentalists and communists blocked the global adoption of this very Deweyan program in the 1940s and 1950s. In 1961 Wright again made a bold proposal for promoting international peace through super-power acceptance of genuine peaceful coexistence and stringent reduction of weapons, *inter alia* (Wright, 1961). But none of these and other good-willed proposals has had much success, if any, given Soviet expansionist tendencies, past and present. In short, Dewey's admirers have to apply his means–ends philosophy, as expounded in his brilliant 1916 essays on force, to each crisis as it arises, and, if possible, before it arises. Each specific case has to be met and judged on its own merits. There is no substitute for eternal vigilance and the constant use of our best and most intelligent judgment.[44]

Rutgers University

NOTES

[1] Cf. W. B. Gallie, *Philosophers of War* (Cambridge: Cambridge University Press, 1978).
[2] See Wright (1965); the first edition was published in 1942.

[3] John Dewey in 'Peace and War', *The American Scholar*, **50** (Spring 1981), 213–230. Two more scholarly discussions of this subject are Gary Bullert, *The Politics of John Dewey* (Buffalo: Prometheus Books, 1983) and Charles F. Howlett, *Troubled Philosopher: John Dewey and the Struggle For World Peace* (Port Washington, N.Y.: Kennicat Press, 1977). My approach differs from theirs in both the scope and focus of my inquiry.

[4] Cf. Bullert, *op. cit.*, pp. 33–68; Howlett, *op. cit.*, pp. 20–42; Carol S. Gruber, *Mars and Minerva* (Baton Rouge: Louisiana State University Press, 1975), pp. 42–45, 90–95, 110–115; Alvin Johnson, *Pioneer's Progress* (New York: Viking Press, 1952), pp. 248–249, 261–268.

[5] Cf. *Collected Papers of Clarence Irving Lewis*, ed. John D. Goheen and John L. Mothershead, Jr (Stanford, Cal.: Stanford University Press, 1970), pp. 55–65. A more recent stringent criticism of Dewey's *German Philosophy and Politics* is by Sidney Hook in his 'Introduction' to John Dewey (1979), *The Middle Works*, Vol. 8, pp. ix–xxxvi.

[6] 'In a Time of National Hesitation', *The Seven Arts*, May 1917; reprinted in John Dewey, *Characters and Events*, 2 vols, ed. Joseph Ratner (New York: Henry Holt, 1929), II, 446. Note that although this essay was printed in May 1917, it was written in April 1917.

[7] *New Republic* V (Jan. 22, 1916), 295–297, reprinted in *Characters and Events*, II, 636–641.

[8] *International Journal of Ethics* **26** (April 1916) 359–67; reprinted in *Characters and Events*, II, 782–789.

[9] *Characters and Events*, II, 637.

[10] *Ibid.*, II, 784.

[11] *Ibid.*, II, 789, 639.

[12] *Characters and Events*, II, 787.

[13] *Ibid.*, II, 788.

[14] *Ibid.*, II, 789.

[15] Randolph Bourne, *The History of a Literary Radical* (New York: S. A. Russell, 1956), pp. 205–222, 241–259. Cf. George Dykhuizen, *The Life and Mind of John Dewey* (Carbondale, Ill.: Southern Illinois University Press, 1973), pp. 165–167; Howlett, *op.cit.*, pp. 36–37; John Adam Moreau, *Randolph Bourne: Legend and Reality* (Washington, D.C.: Public Affairs Press, 1966), pp. 146–148, 154–162, an all-out defense of Bourne. A more balanced and critical judgment appears in Charles Madison, *Critics and Crusaders*, 2nd edn. (New York: Frederick Ungar Publishing Co., 1959), pp. 435–441.

[16] Cf. Allen Cywar, 'John Dewey: Toward Domestic Reconstruction 1915–1920', *Journal of the History of Ideas* **30** (1969), 385–340, and 'John Dewey in World War I: Patriotism and International Progressivism', *American Quarterly* **31** (1969), 578–594.

[17] John P. Diggins, *op. cit.*; and *The Bard of Savagery: Thorstein Veblen and Modern Social Theory* (New York: Seabury Press, 1978), pp. 186–191.

[18] Cf. V. R. Berghahn, *Germany and the Approach of War in 1914* (London: Macmillan, 1973); Fritz Fischer, *Germany's Aims in the First World War* (New York: Norton, 1967); Laurence Lafore, *The Long Fuse*, 2nd edn. (New York: Lippincott, 1971); Dwight E. Lee (ed.), *The Outbreak of the First World War*, 4th edn., (Boston: Heath, 1975); Jack J. Roth (ed.), *World War I: A Turning Point in Modern History* (New York: Knopf, 1967).

[19] Arthur S. Link (ed.), *Woodrow Wilson and a Revolutionary World, 1913–1921* (Chapel Hill, N.C.: University of North Carolina Press, 1982). On J. M. Keynes' indictment of Wilson at Versailles, see Étienne Mantoux, *The Carthaginian Peace or The Economic Consequences of Mr. Keynes* (New York: Scribner, 1952); Paul Mantoux, *Les délibéra-*

tions du Conseil des Quartre (24 mars-28 juin 1919), 2 vols. (Paris: Centre National de la Recherche Scientifique, 1955); for a partial defense of Keynes, see R. F. Harrod, *The Life of John Maynard Keynes* (London: Macmillan 1951), pp. 254–284. A much more severe criticism of Keynes' *Economic Consequences of the Peace* is to be found in Robert Skidelsky, *John Maynard Keynes*, Vol. 1 (London: Macmillan, 1983).

[20] See *Encyclopaedia of the Social Sciences*, s.v. 'Outlawry of War', 11, 510.

[21] John Dewey, *Intelligence in the Modern World*, ed. Joseph Ratner (New York: Modern Library, 1939), pp. 466–604. Cf. Bullert, *op. cit.*, pp.113–126, 147–170; John E. Stoner, *S. O. Levinson and the Pact of Paris* (Chicago: University of Chicago Press, 1942), *passim*. Most diplomatic historians in recent years criticize the Pact of Paris for not being effective in preventing war, e.g., Selig Adler, *The Isolationist Impulse* (London and New York: Aberlard-Schuman, 1957), pp. 228–239. But Quincy Wright, a great authority on war, defended the Pact of Paris for its beneficial influence in international affairs, 'Introduction', Stoner, *op. cit.*, IX–XI.

[22] *Dialogues of Alfred North Whitehead*, ed. Lucian Price (New York: New American Library, 1956), p. 108. On Bertrand Russell's position in the late 1930s and the Second World War, see Ronald Clark, *The Life of Bertrand Russell* (New York: Knopf, 1976), pp. 460–487.

[23] John Dewey to Selden Rodman, May 24, 1941, Selden Rodman MSS, University of Wyoming Library, University of Wyoming; cited by Bullert, *op. cit.*, pp. 164, 168.

[24] John Dewey, 'Lessons from the War in Philosophy', John Dewey Papers, Morris Library, Southern Illinois University, Carbondale; Bullert, *op. cit.*, pp. 165, 169. Cf. Howlett, *op. cit.* and Arnold A. Offner, *The Origins of the Second World War: American Foreign Policy and World Politics, 1917–1941* (New York: Praeger, 1975); on the complexity of the debate among historians, see Warren F. Kimball, *Franklin D. Roosevelt and the World Crisis, 1937–1945* (Lexington, Mass.: Heath, 1973).

[25] *New York Times* (Nov. 19, 1950), p. E14; cited by Bullert, *op.cit.*, pp. 199–203.

[26] See *Reconstruction in Philosophy* (New York: Holt, 1920), reprinted in John Dewey, *The Middle Works, 1899–1924*, Vol. 12, ed. Jo Ann Boydston (Carbondale: Southern Illinois University Press, 1982), p. 188.

[27] *Characters and Events*, II, 782–789.

[28] Cf. Charles De Benedetti, *The Peace Reform in American History* (Bloomington: Indiana University Press, 1980); Merle Curti, *Peace or War: The American Struggle 1636–1936* (Boston: Canner, 1959); C. Roland Marchand, *The American Peace Movement and Social Reform, 1898–1918* (Princeton: Princeton University Press, 1972). On Wilson, see Whittle Johnston's essay in Link (ed.), *Woodrow Wilson and a Revolutionary World*, pp. 190–231 and his forthcoming book, *Wilson, His Critics and World Politics*.

[29] See John Dewey, *The Middle Works*, 1899–1924, Vol. 14, ed. Jo Ann Boydston (Carbondale: Southern Illinois University Press, 1983).

[30] See *Encyclopaedia of the Social Sciences* **8**, 531–536.

[31] (Boston: Houghton Mifflin, 1934), p. 271; oral statement by Benedict to Sidney Ratner, *circa*. 1935.

[32] Sidney Ratner, 'Patterns of Culture in History', *Philosophy of Science* **6** (Jan. 1939), 88–97, at 89. On the nature versus nurture debate in the United States during the 1920s and 1930s, see Merle Curti, *Human Nature in American Thought* (Madison: University of Wisconsin Press, 1980), pp. 294–312.

[33] Cf. Karl Deutsch, *The Nerves of Communication*, 2nd. edn. (New York: Free Press, 1966); *Encyclopaedia of the Social Sciences*, s.v. 'Decision Making'; Robert Jervis, *Perception and Misperception in International Politics* (Princeton: Princeton University Press, 1976); Herbert A. Simon, *Reason in Human Affairs* (Stanford, Calif.: Stanford University Press, 1983); John Spanier and Eric M. Uslaner, *How American Foreign Policy Is Made* (New York: Praeger, 1974).

[34] John Dewey to Hu Shih, October 27, 1939 Dewey Papers, Morris Library, Southern Illinois University, Carbondale, Ill. Dewey's punctuation and capitalization corrected by the author.

[35] Hu Shih to John Dewey, March 2, 1940, Dewey Papers, Morris Library.

[36] John Dewey to Hu Shih, March 6, 1940, Dewey Papers, Morris Library.

[37] See Stanley Hoffmann, *The State of War*, (New York: Praeger, 1965), pp. 22–53; and François Bondy, 'Aron's Ideas', *Encounter* **62** (Feb. 1984), 21–24.

[38] On the Kellogg-Briand Pact, see Aron (1966), pp. 710–725, esp. pp. 710–713.

[39] *Theory of Valuation* ('International Encyclopaedia of Unified Science', Vol. 2, no. 4 [Chicago, 1939], 65; cited by Wright (1965), pp. 11–16.

[40] Karl W. Deutsch, 'A Preface to the Second Edition', in Wright (1965), pp. xiii–xiv.

[41] See *Encyclopaedia of the Social Sciences*, s.v. 'War: The Study of War'.

[42] Wright, 'War: The Study of War', pp.453–454.

[43] *Ibid.*, p. 466.

[44] Albert F. Eldridge, *Images of Conflict* (New York: St Martin's Press, 1979); Lewis S. Feuer, 'Horace M. Kallen on War and Peace', *Modern Judaism* **4** (May 1984), 201–214; Robert E. Osgood and Robert W. Tucker, *Force, Order, and Justice* (Baltimore: Johns Hopkins Press, 1967); Kenneth N. Waltz, *Man, the State and War* (New York: Columbia University Press, 1959). On the differences between totalitarian and democratic definitions of 'freedom' and the difference between 'negative peace' and 'positive peace', see *Encyclopedia of the Social Sciences*, s.v., 'Freedom' and 'Peace'.

REFERENCES

Aron, Raymond. 1966. *Peace and War*, Transl. Richard Howard and Annette Baker Fox. Garden City, N.Y.: Doubleday.

Bell, Daniel. 1952. 'The Background and Development of Marxian Socialism in the United States', *Socialism and American Life*, 2 vols, Donald Egbert and Stow Persons (eds.). Princeton, N.J.: Princeton University Press, I, 316–20.

Dewey, John. 1939. 'No Matter What Happens – Stay Out', *Commonsense* (March), 11.

Dewey, John. 1941. 'Here at Home', *Progressive* (Feb. 15), 20.

Dewey, John. 1978. *The Middle Works*, Vol. 5, Jo Ann Boydston (ed.). Carbondale, Ill.: Southern Illinois University Press, p. 431.

Dewey, John. 1979. *The Middle Works*, Vol. 8, Jo Ann Boydston (ed.). Carbondale, Ill.: Southern Illinois University Press, pp. 135–204, 421–446.

Lamont, Corliss (ed.). 1959. *Dialogue on John Dewey*. New York: Horizon Press, pp. 25–30.

Santayana, George. 1916. *Egotism in German Philosophy*. New York: Scribner.

Shih, Hu. 1940. 'The Political Philosophy of Instrumentalism', *The Philosopher of the Common Man*, Sidney Ratner (ed.). New York: Putnam.

Wright, Quincy. 1955. *The Study of International Relations*. New York: Appleton-Century-Crofts, p. 451.

Wright, Quincy. 1961. *The Role of International Law in the Elimination of War*. Manchester: University Press, pp. 41–43.

Wright, Quincy. 1965. *A Study of War*, 2nd edn. Chicago: University of Chicago Press.

VERA COOPER RUBIN

TO L. F. FROM V. C. R., 1984

Dear Lewis,

It seems appropriate for me to re-establish our friendship after more than thirty-five years by writing a letter to you on the occasion of your seventieth birthday celebration. In 1948, I was a senior at Vassar College; you were my friend and the serious professor under whom I studied the Philosophy of Science. After all these years, details of two classes stand out. In one, I picture myself at a blackboard as I attempt in 45 minutes to explain the mathematics of the Lorentz transformations to the class. In another, you return a term paper (subject long since forgotten) with the note: "A–, this should have been better." This latter comment is a true assessment of how you taught me. In your own work, you demanded a high level of scholarship, and you attempted to instill the same high aspirations in your students.

And so the occasion of your celebration becomes a time for me to recall how very much has changed in the field of astronomy, and to tell you about these changes. I entered Vassar College with one aim: to become an astronomer. But it was many years, several degrees, and much hard work later that I finally was satisfied to pronounce myself 'an astronomer'. The world of doing astronomy has altered dramatically over these many years. I used to stand at the telescope during a cold, dark, long 12-hour night, solitary, intense, often bored, always philosophical. I did believe, as did most observers, that each incoming photon needed me there, or the exposure could not continue. Frequently, as I stood guiding a 6-hour exposure on the nucleus of the Andromeda galaxy, I would wonder if an astronomer on a planet in Andromeda was observing us.

Now I sit in a warm, over-bright room, surrounded by electronics, computers, technicians keeping watch over complex television monitors. Often, I still guide the telescope by keeping a star fixed on a cross wire, but now I view the star on a TV screen. Some important things are the same, however. The view of the sky at a dark site on a moonless night, so filled with stars that even familiar constellations are not recognizable, is to me the most remarkable scene from the earth. And I

S. Hook, W.L. O'Neill and R. O'Toole (eds.), Philosophy, History and Social Action, 391–393.

still stand and wonder how anyone could choose to do other than study the stars.

The Universe which astronomers knew in the 1950s was a majestic, slowly evolving stellar Universe. Knowledge of stars and galaxies came from ground-based optical observations. But galloping technology plus instruments above the earth's atmosphere have permitted us to expand the accessible spectral region to the radio, infrared, x-ray, γ-ray, and ultraviolet. Astronomers now know that we live surrounded by a veritable zoo of objects as diverse as quasars, coronae about cool stars, matter accreting onto black holes, pulsating stellar remnants from supernova explosions, dense molecular clouds surrounding, cocoon-like, the sites of new-star formation. Although life elsewhere in the Universe has not yet been detected, the vast regions between the stars in our Galaxy contain complex chemical compounds, some of which are the fundamental constituents of living things on earth. An exotic menu is routinely available for study, and the Universe is known to be immeasurably richer, more varied, and more violent than would have been dreamed even twenty years ago.

Stars are born from the interstellar gas, spend their lifetimes, alchemist-like, producing heavier elements which they spew back into the interstellar gas at the ends of their lives. All of the elements (except hydrogen and helium) which make up our Universe, and which make up our bodies, were synthesized in the interior of some star. And galaxies too evolve; they use up their gas, they tidally distort and even cannibalize outer stars belonging to their neighbors, they accrete gas, and in dense cores of clusters of galaxies they lose their identities after multiple mergers.

In the 1940s and 1950s, we studied stars; in the 1960s and 1970s, galaxies. Now we attempt to study cosmology and the Universe. We know that galaxies are receding from each other: the Universe is expanding. Is there enough mass in the Universe to ultimately halt the expansion, causing the galaxies to approach each other and collapse together in a 'big crunch'? This is the question we cannot yet answer.

We cannot answer because Nature has played a big joke on astronomers. From detailed studies of motions of stars in galaxies, many of them carried out by my colleagues and myself, we have learned that most of the mass in the Universe is non-luminous, and is clumped, halo-like, around individual galaxies. Observations in a variety of spectral ranges are unable to detect this dark matter. So those of us who set out to study the Universe by its light now understand with some amusement that we study only the 5 or 10 percent of the Universe which is luminous.

The presence of the dark matter is revealed by its gravitational attraction on stars in galaxies, and by the mutual gravitational attraction of binary galaxies and galaxies in clusters. We know that the dark matter is not composed of normal or low mass stars, or of gas. Guesses as to its composition include Jupiter-like planets, massive black holes, mini-black holes, neutrinos, monopoles, gravitinos, axions, and other exotic particles. Only when we learn the spatial distribution and the composition of the dark matter can we predict whether the Universe is of sufficiently high density that the expansion will ultimately be halted.

For the non-believers in heavy halos and invisible mass, an alternative way of explaining the observations is to modify Newton's law of gravitation for objects which are enormously far apart. Now there's a good topic for a Philosophy of Science class!

And thus, Lewis, the astronomy of today differs profoundly from the astronomy of 1948. Undoubtedly, the concepts in sociology and philosophy which you study are similarly altered from those of 1948. But independent of the subject matter, you have imbued your students with a lasting love of intellectual pursuits. We who were fortunate enough to study with you will attest to the great value of this accomplishment. We hope that the rewards for you were equally great.

Bob, whom you knew as a weekend visitor to Vassar and who has been my husband for over thirty-five years, joins me as we send our best wishes and congratulations to you on this happy occasion.

Vera

Carnegie Institution of Washington

HAROLD Z. SCHIFFRIN

CHINA TODAY: RETREAT FROM MAO AND RETURN TO MARX?

Mao Tse-tung: "I want to surpass the U.S. before I go to see Marx!"[1]
Karl Marx: "And even when a society has got upon the right track for the discovery of the natural laws of its movement . . . it can neither clear by bold leaps nor remove by legal enactments the obstacles offered by the successive phases of its normal development. But it can shorten and lessen the birth pangs."[2]

As Lewis Feuer and others have pointed out, the attraction exerted by Marxism in Asia and its diffusion there "cannot be explained within the framework of the Marxist system".[3] Marx and Engels riveted their attention on Europe and gave only passing attention to Asia and other underdeveloped areas. To be sure, when they did mention Asia, they did not postulate an exact, stage-by-stage recapitulation of the pattern of historical development that they discerned in Europe. Their thinking was too sophisticated for such a crude mechanistic formula. "Only vague hypotheses", wrote Engels, could be formulated for the sort of stages that the 'semi-civilized' countries would have to pass through on the way to socialism.[4] But while they did not discount the possibility of different and shorter routes to the Promised Land in, for example, Russia, these conjectures were always predicated upon the *prior* victory of socialism in the West.

A reversal of roles was unimaginable. At least on one occasion Marx expressed the fear that the backward areas, once opened and sucked into the world market, would spawn new strongholds of capitalism and threaten socialist success in the West. "On the continent," he wrote to Engels in 1858, "the revolution is imminent and will immediately assume a socialist character. Will it not of necessity be crushed in this little corner, since over a much vaster terrain the movement of bourgeois society is still in the ascendancy?"[5]

The unimaginable happened and roles were reversed. Lenin and Stalin showed that socialist revolutions could triumph more easily, not in places where the proletariat was strong, but where the bourgeoisie was weak. However, the bracketing of socialism with underdevelopment meant the underdevelopment, if not the complete distortion, of socialism itself. It required monolithic party rule, totalitarianism and a

S. Hook, W.L. O'Neill and R. O'Toole (eds.), Philosophy, History and Social Action, 395–404.

forced pace of industrialization that exacted an unprecedented cost in human suffering.

It is in this perverted form, best called Leninism–Stalinism,[6] that Marxism attracted so many intellectuals in Asia. As Feuer has suggested, the vanguard party concept fit in well with the intellectuals' elitist predilections. While fascinated by the Marxist appeal to science and rationality, they filled Marxist terms with a new content: foreign imperialists replaced the non-existent 'capitalist'; the oppressed people as a whole became the 'proletariat'; and national liberation took the place of 'class struggle'. Socialism, in effect, was reduced to anti-imperialism or nationalism.[7] (Without this anti-imperialist content, socialist movements made little headway in Asia. This was the case in Japan, which realized her nationalist goals quite early and escaped the long period of humiliation endured by Asian victims of colonialism and imperialism.)

But while they have been successful vehicles for nationalism, Asian socialist movements have had even more difficulty than their Soviet predecessors in meeting the needs of development.

This difficulty is especially pronounced in China, an impoverished, backward country with over a billion people – almost one-quarter of the world's total population. There is no precedent for modernization on so monumental a scale. Despite more than thirty years of intensive industrialization under the command economy of the People's Republic, 800 million Chinese are still classified as peasants. Yet, as Feuer predicted, Chinese leaders continue to speak in the Marxist idiom. To what purpose, is the subject of this paper.

First of all, the commitment to Marxism in post-Mao China is still required to legitimize the monolithic rule of the Communist Party. Having all but scuttled the legacy of Mao Tse-tung, and having departed from the Stalinist model for economic development, the party can only fall back on Marx and Lenin. Marxism holds the promise of a classless and bountiful society, and Leninism, the justification for deviating from the original model under the guiding hand of a prescient, elitist party. "Marxism", proclaims a party ideologue, "is the crystallization of human wisdom and a universally applicable truth."[8] And Karl Marx is the new Confucius: "the greatest thinker in history."[9] By identifying with the grand vision of this universal sage who predicted the doom of capitalism and the triumph of socialism, the party attempts to show that it has steered China into the mainstream of world history armed with the knowledge of how history works and in what direction it is heading. This ideological underpinning acts like a gyroscope. Missteps, however unavoidable, can only be temporary.

Quoting Engels, post-Mao leaders, like Lenin and Mao before them, emphasize that Marxism "is not a doctrine but a method. It does not provide ready-made dogmas, but criteria for further research and the method *for* this research."[10] Waving this license, Teng Hsiao-p'ing can summarize the task of 'socialist modernization' – a task that makes no sense in the context of classical Marxism – for the benefit of the Twelfth Party Congress (1982): "to integrate the universal truth of Marxism with the concrete realities of China, blaze a path on our own and build socialism with Chinese characteristics . . ."[11]

The blend of universalism and particularism re-echoes a familiar theme in modern Chinese nationalist thinking.[12] From the very beginning, it was a major source of the Marxist appeal, and presumably still attracts many Chinese intellectuals. Moreover, there is no reason to believe that Teng and other Chinese leaders have lost faith in the Marxist world-view, just as there is no reason to assume that Lenin ever discarded it, however much he twisted the original concept.

Yet, the most serious political problem in post-Mao China is the credibility of the Communist Party. The major resource of any government, and particularly one that claims to monopolize historical truth and wisdom, is its credibility. In that sense, the Chinese Communist Party has depleted its capital to a dangerously low level. No other Communist Party has ever steered such an erratic course or has admitted to having perpetrated such disastrous errors. Natural conditions of geography and demography were bad enough to begin with. But the party itself, by its own admission, bears directly responsibility for the worst catastrophes that have afflicted the People's Republic.

In 1958 the party launched the Great Leap Forward that flattened the economy for several years and brought back the horrors of famine that had haunted old China. In 1966 the party, now dominated by radical Maoists, prescribed chaos: the Great Proletarian Cultural Revolution convulsed the entire country for at least three years and did not end completely until the death of Mao in 1976. By then it had permanently scarred the lives of millions of Chinese families. The modernization effort was set back for a decade. Thus, during its stewardship the party has spent much of the time repairing damages of its own making. What went wrong?

It has been noted that, whereas communists attribute the ills of capitalist societies to the capitalist *system*, they never blame the *theory* but its misuse for the failures of communist countries. Bad leaders, bad imperialists and bad weather are generally the main scapegoats. To their credit, it must be said that, since 1976, the Chinese communist leaders have been unusually frank in acknowledging the party's culpability.

This is not difficult because Teng Hsiao-p'ing and other senior leaders of the post-Mao era had opposed the most disastrous of the courses plotted by the Great Helmsman. They paid for their opposition to Mao. With his own son paralyzed for life after being pushed out of a window – only because he was his son – Teng needs no other reminder of the ravages of the Cultural Revolution.

While blaming the party – and specifically, its radical, 'Leftist' faction – for the excesses of the Maoist period, Chinese leaders and ideologues are making strenuous efforts to salvage the authority of Marxism. They absolve Marxism from any responsibility for party errors by emphasizing the obvious: Marx never spelled out the detailed operations of a socialist economy under any conditions, not to speak of the conditions of poverty and backwardness that still prevail in China. This is how it is put by a leading Chinese economist:

> Socialism is a new system . . . Marx and Lenin showed us the laws governing the transition from capitalism to communism through socialism. Their scientific predictions remain the guide to our study of questions of the socialist economy. However, the classics they authored are insufficient for a study of the socialist economy because socialism never actually existed in their lifetime. History proves that the Marxist theory of socialism and communism can only develop through practice . . . China is already a socialist country but one with a backward economy and culture. We have had our successes and failures. A look at the history of the past thirty years shows that it is by no means easy to build socialism and achieve modernization in a country like ours.[13]

Successes are attributed to the correct adaptation of Marx's 'universally applicable truth', and failures, to the misunderstanding of it. Since the party must claim that its overall record is positive, it must also bestow a positive evaluation to the overall historic role of Mao, who charted its path to victory and its initial steps toward socialism. Therefore, 'Mao Tse-tung Thought' – now defined as the 'collective wisdom'[14] of the party – is linked with Marxism–Leninism in formulating China's unique path to socialist modernization. By definition, then, Mao's 'gross mistakes'[15] in the Cultural Revolution "were obviously inconsistent with the system of Mao Tse-tung Thought".[16]

According to party ideologues, a long list of personal failings accounts for Mao's mistakes: 'arrogance', 'pride', 'smugness', 'subjectivism', 'individualism', 'arbitrary work style', and 'placing himself above the Party Central Committee'.[17] These tendencies, the indictment continues, led to what the party must consider his most serious error, namely, "to depart somewhat from the Marxist path in theory and practice and to put forward a number of 'Left' concepts".[18] This has to be the cardinal sin. Only by exposing Mao's errors from a Marxist point of view, can the party legitimize its reversal of his policies and still lay claim to the mandate – one is tempted to say, 'Mandate of Heaven' – derived from its identification with Marxism. In short, the party needs Marxism to exorcise the ghost of Mao.

Consider first the dramatic changes that have been taking place since the death of Mao in 1976 and especially since the famous Third Plenum of the Eleventh Party Congress of December 1978 that institutionalized Teng's leadership: partial decollectivization of agriculture, work incentives and wage differentials, a quality-oriented, even elitist, educational system, the dispatch of thousands of students abroad, catering to intellectuals and technical experts, partial liberalization of cultural activities, openness to foreign advisers and investors, devolution of economic decision-making, encouragement of small entrepreneurs, the establishment of 'special economic zones' to entice overseas Chinese capital and, above all, a new emphasis upon production and efficiency for the immediate and simple purpose of giving the Chinese people a higher standard of living.

The substance and mood of leadership directives has changed entirely. No more mass drives for quickly overtaking the industrialized nations through superhuman acts of will, but rational planning in pursuit of moderate economic goals to which even the pundits of the Harvard School of Business Administration and Milton Friedman can conceivably contribute.[19]

Given this transformation, it has become fashionable for outsiders to speak of Teng's 'pragmatism' in comparing him with Mao. If they are referring to the dictionary meaning of 'pragmatic' in the sense of 'practical as opposed to idealistic', I suppose the term is appropriate. It is misleading, however, if it implies that Teng is less of a Marxist–Leninist than Mao.

Of course Teng is improvising. So did Lenin, Stalin, Mao, and everyone else who has tried to apply theory to conditions for which it was never intended. Comparing Mao and Teng on a Marxist scale, therefore, is virtually meaningless. Nevertheless, if one makes the comparison, it is the post-Maoists who come out less guilty of violating orthodoxy than their predecessors. Mao, in fact, boasted of his 'pragmatism',[20] whereas Teng and his theoreticians resort to the deterministic or dogmatic aspects of Marxism to repudiate him. If quoting the classics is the name of the game, the post-Maoists obviously have the upper hand.

The theoretical issue is quite straightforward. Mao insisted that backwardness is an *unqualified* advantage and that the transformation of the relations of production and the super-structure can or even should precede the development of the productive forces. For Mao, "The important question [was] the remolding of the people."[21] To post-Mao theoreticians, this kind of thinking – believing "that anything could be accomplished at the will of those authority"[22] – smacks of metaphysics, and in refuting it they go straight back to Marx as, for example, in *A Contribution to the Critique of Political Economy*:

> A social order never perishes before all the productive forces for which it is broadly sufficient have been developed, and new superior relations of production never replace older ones before the material conditions for their existence have matured within the womb of the old society. Mankind thus inevitably sets itself only such tasks as it can solve, since closer examination will always show that the task itself arises only when the material conditions for its solution are already present or at least in the process of formation.[23]

This is Marx in his most deterministic vein, and it is the Marx that post-Maoists have been clutching as they hammer relentlessly at the vestiges of Maoist ideology. But is quoting the classics really the name of the game? For most of the Chinese people, presumably not. They abhor Maoism because it brought disaster, not because it violated Marxist fundamentalism. Herein lies the danger for the post-Mao leadership.

Having repudiated Maoist metaphysics and having pledged to prove the superiority of the socialist system by improving the *material* life of the people and only in *that* way, awaken their socialist consciousness, the party must produce tangible results or lose completely what is left of its credibility.

At the same time, the path presently chosen to pursue the 'Four Modernizations' – in industry, agriculture, science and technology, and national defense – may bring in its wake other dangers for the party. First, there is the danger of residual Maoism.

Granted, almost any system that restored order would have been preferable to the chaos of the Cultural Revolution. Granted, too, the policies of the past seven years have improved living standards in general and, at least in the short-run, have narrowed the disparity between urban and rural populations by raising the prices of agricultural products and through partial decollectivization.

However, the party's policy of spurring productivity with material rewards may, in the long-run, widen the gap between the 'haves' and the 'have-nots'. Under the prevailing socialist system of "from each according to his ability and to each according to his work", there are bound to be losers. Teng has told the peasants to get rich. But those who stay poor and all the others who fail to meet the new standards of quality and efficiency may long for the 'iron rice-bowl' of Mao, who was more concerned with inequality than poverty. This concern is not completely senseless in a country where the struggle for bare subsistence is not yet over. There will be those who will remember and endorse Mao's strictures against bureaucratism, against the sleek living style of the elite, against the pampering of intellectuals and experts, and his ridicule, not only of examinations, but of lengthy, formal education in general. The potential constituency of residual Maoism also includes the 20 million party members – about half the total membership – who have been removed because of their failure to meet post-Maoist criteria. To them must be added thousands of purgees from the ranks of the army and bureaucracy. If there is sufficient cause for a feeling of relative deprivation, and especially if the general level of subsistence does not continue to rise, the siren-song of Maoist utopianism may yet be revived.

A further and more fundamental problem is the contradiction between economic liberalization – now deemed essential for achieving developmental, modernizing goals – and totalitarian politics. Demands for the 'Fifth Modernization' of democracy and human rights have already been sounded and, though stifled, will surely mount.[24] The familiar contention that, because the concept of individualism had no roots in traditional China, it will be of little significance in modern China, cannot be substantiated. Ever since China entered the twentieth century, there have always been intellectuals and students who considered democracy an intrinsic human value and a necessary prerequisite to modernization. This value was sacrificed on the altar of nationalism, which gave higher priority to building state-power. But the post-Cultural Revolution relaxation revealed that there is indeed a dissident movement echoing the earlier call for democracy and human rights. And "The Chinese Communists must . . . deal with the dissidents' key argument – that the Four Modernizations cannot succeed without the Fifth Modernization . . ."[25] Armed with more authentic information about the outside world, dissidents can argue that the examples of Japan, the United States, and Yugoslavia prove that freedom accelerates the economic and technological growth that the party is striving for.

It has sometimes been argued – in a way reminiscent of the Marxist interpretation of history – that, because the economic basis of society inevitably impinges upon and transforms the political super-structure, rigid totalitarianism will wilt under the pressures of economic modernization.[26] This assumption has yet to be proved. Relief from Stalinist terror and Cultural Revolution mania do not necessarily indicate secular tendencies toward de-totalitarianization in Russia and China. Taken as a whole, the post-Stalin period cannot be characterized by an irreversible trend toward pluralism and the institutionalization of limits to one-party dictatorship. And in the case of China, Teng Hsiao-p'ing has made it quite clear that he did not assume power in order to preside over dissolution of the party dictatorship.

Teng's treatment of dissidents, condemned by Amnesty International and other libertarians, reveals the ruthlessness that characterized his earlier political style, especially when he helped Mao suppress intellectuals in the 1950s. He demonstrated a similar efficiency in suppressing the democracy movement in 1978–1979, when popular condemnation of the 'Gang of Four' spilled over into criticism of the system that had spawned them.[27]

Recurrent warnings against manifestations of 'bourgeois liberalism' and 'spiritual pollution' serve as reminders that the 'pragmatism' of the post-Mao leadership does not extend to politics and human rights. The party still insists upon the following criteria for defining the limits of acceptable behavior. One must adhere "to the socialist road, to the people's democratic dictatorship (i.e. the dictatorship of the proletariat), to the leadership of the Communist Party and to Marxism-Leninism and Mao Tse-tung Thought."[28]

Perhaps more than anything else, the involuntary birth-control campaign reveals the potential for ruthlessness in Chinese totalitarianism. Though the size of China's population is, undeniably, a most serious problem, the one-child-to-a-family-goal is unrealistic. And, despite the attempt to invoke the authority of Engels,[29] the methods chosen to achieve this goal and their side-effects cannot be justified by any standards of human decency.[30]

Finally, one can only share Feuer's hope that, "as freedom is reborn in Eastern Europe and Asia, it will speak in the Marxist idiom and try to disenthrall the universal humanist bearing of Marx's ideas from their Stalinist perversion."[31] This is exactly what some Chinese dissidents have been trying to do[32] and what Ch'en Tu-hsiu, an original founder and the first head of the Chinese Communist Party, tried to do in the last years of his life. In 1940, two years before his death, and when he had long parted company with Mao, Ch'en came to realize "that even a socialist society should accept some Western democratic political devices and experience, such as parliamentary government, legal procedures in guarantee of civil rights, such as freedom of thought, speech, the press, and workers' right of strike, protection of opposition parties, and free elections."[33] In current party literature Ch'en still carries the brand of 'Right capitulationism',[34] the euphemism for carrying out the Comintern directives that brought disaster to the party in 1927. He has not been posthumously rehabilitated.

Truman Institute, The Hebrew University
Jerusalem, Israel

NOTES

[1] From a speech in December 1958; in *Miscellany of Mao Tse-tung Thought* (Joint Publications Research Service, Arlington, Va., 1974), Part I, p. 148.

[2] From *Capital: A Critique of Political Economy*, quoted in Lewis S. Feuer (ed.), *Basic Writings on Politics and Philosophy: Karl Marx and Friedrich Engels* (New York, 1959), p. 136.

[3] See Feuer, *op. cit.*, p. xix.

[4] Quoted in George Lichtheim, *Marxism: An Historical and Critical Study*, 2nd rev. edn (London, 1964), p. 362.

[5] *Ibid.*, p. 139.

[6] The term 'Leninism-Stalinism' is Lichtheim's. See *Ibid.*, p. 371.

[7] See Feuer, *op. cit.*, pp. xix–xx; and George Lichtheim, *Imperialism* (Harmondsworth, 1974), p. 139.

[8] Quoted in Ma Hong, 'Marxism and China's Socialist Economic Construction – Written to Commemorate the Centenary of Marx's Death', *Social Sciences in China* IV, 3 (1983), 95.

[9] *Ibid.*, p. 95.

[10] Quoted in *ibid.*, p. 96.

[11] See *Twelfth National Congress of the CPC* (Beijing, 1982), p. 3.

[12] For an excellent analysis of the pre-Marxist nationalist mood, see Michael Gasster, *Chinese Intellectuals and the Revolution of 1911: The Birth of Modern Chinese Radicalism* (Seattle and London, 1969). See also Robert A. Scalapino and Harold Z. Schiffrin, 'Early

Socialist Currents in the Chinese Revolutionary Movement', *Journal of Asian Studies* **18** (1959), 321–342; and Harold Z. Schiffrin, 'The "Great Leap" Image in Early Chinese Nationalism', *African and Asian Studies* **3** (1967), 101–119.

[13] See Xue Muqiao, *China's Socialist Economy* (Beijing 1981), pp. iv–v.

[14] See Hu Yaobang, 'Speech at the Meeting in Celebration of the 60th Anniversary of the Founding of the Communist Party of China' in *Resolution on CPC History (1949–81)* (Beijing, 1981), p. 94.

[15] See 'Resolution on Certain Questions in the History of Our Party Since the Founding of the People's Republic of China' in *Resolution on CPC History (1949–81)*, (Beijing, 1981), p. 56.

[16] *Ibid.*, p. 33.

[17] Official assessments of Mao, including these personal failings, can be found in *ibid.*, pp. 1–86. See also Ma Hong, *op. cit.*, p. 99. For a more objective evaluation, see Stuart Schram's introduction in Stuart Schram (ed.), *Mao Tse-tung Unrehearsed: Talks and Letters: 1956–71* (Harmondsworth, England, 1974), pp. 7–47.

[18] See Ma Hong, *op. cit.*, p. 99.

[19] For a concise account of these changes, see Steven M. Goldstein, Kathrin Sears and Richard C. Bush, *The People's Republic of China*, 4th edn (New York, 1984).

[20] See *Miscellany of Mao Tse-tung Thought*, Part I, p. 96.

[21] From 'Reading Notes on the Soviet Union's "Political Economic" ', in *ibid.*, p. 258.

[22] See Xue Muqiao, *op. cit.*, p. 313.

[23] Quoted in Xue Muqiao, *op. cit.*, p. 298. For a slightly different translation, see Feuer, *Basic Writings . . .* , p. 44.

[24] See James P. Harrison, 'The Problem of the Fifth Modernization – A Review Article', *The Journal of Asian Studies* **XLII** (1983), 869–878.

[25] *Ibid.*, p. 869.

[26] This is essentially the thesis of Richard Lowenthal in his paper, 'Totalitarianism and After in Communist Party Regimes', presented at the International Colloquium in Memory of Jacob Talmon, Jerusalem, June 1982.

[27] See Jonathan Mirsky's profile of Teng, 'A Survivor for all Seasons' in *The London Times*, (April 12, 1984), p. 10.

[28] See Hu Yaobang in *The Twelfth National Congress of the CPC*, p. 14.

[29] Engels' preface to *The Origin of the Family, Private Property and the State* in which he includes "the production of human beings themselves" as an aspect of production in general, is cited in Liu Zheng, Song Jian and others, *China's Population: Problems and Prospects* (Beijing, 1981), p. v.

[30] See Nick Eberstadt, 'Peking's Family Policy', *The New York Times* (weekly review), p. 7.

[31] See Feuer, *Basic Writings . . .* , p. xx.

[32] See Harrison, *op. cit.*, p. 376.

[33] Chow Tse-tung, *The May Fourth Movement: Intellectual Revolution in Modern China* (Cambridge, Mass., 1960), p. 232. For an analysis of Ch'en's conversion to Marxism and his subsequent opposition to Mao and Stalinism, see Benjamin Schwartz, 'Che'en Tu-hsiu and the Acceptance of the Modern West', *Journal of the History of Ideas* **XII** (1951), 61–72.

[34] See *Resolution on CPC History (1949–81)*, p. 4.

HAROLD ISSADORE SHARLIN

LIFE AND WORK: A BIOGRAPHY OF LORD KELVIN RECONSIDERED

My reason for undertaking the task of writing the biography of William Thomson, later Lord Kelvin, that paragon of nineteenth-century science, was not dictated by some 'great man of history' idea but, on the contrary, I wanted to demonstrate that great achievements in science were accomplished by ordinary mortals. I had an opportunity to expound my theory of the biography of scientists in a review article (Sharlin, 1963). That article was a manifesto and soon afterward my theories were to be tested in the process of writing the biography of Lord Kelvin.

Now I have an occasion for a retrospective testing of my theories about how to write a biography of a scientist. My convictions have not changed. The biography of a scientist that analyzes in great detail the achievements of the scientist but relegates the life story to a few quaint anecdotes is not a biography in any sense of the word and is bound to leave the reader exasperated and with no better insights into the scientific enterprise than before. Biographies of scientists are no different from biographies of poets, novelists, politicians, military leaders, or artists. All biographies, whether of scientists or non-scientists, have the same objective: to place the achievements of an extraordinary person within the context of a life and the times. The difficulties are the same: there is never enough information from which to draw a complete portrait.

As examples of the unsatisfactory state of writing the biographies of scientists, my 1963 review article selected some recently published books as well as some older works newly republished. John Tyndall's *Faraday as a Discoverer* (New York: Crowell, 1961) was first published in 1868, the year Michael Faraday died. Tyndall, who was Faraday's assistant and his successor at the Royal Institution, wrote that he did not intend to write a life of Faraday, as he said, "in the ordinary acceptation of the term". Rather Tyndall intended to give only such personal traits "as may be necessary to the completion of the picture of the philosopher". He admitted that the picture that he was going to paint would by no means "be adequate to give you a picture of the *man*". In very

S. Hook, W.L. O'Neill and R. O'Toole (eds.), Philosophy, History and Social Action, 405–417.

explicit terms, Tyndall was making a dichotomy between the 'philosopher', or scientist, and the life. That model has been followed over and over again by biographers of scientists.

Even when Tyndall did have information about Faraday that gave insight into the scientist *and* the man, the biographer gave only tantalizingly brief glimpses. Faraday had written a letter in 1858 in which he said, "Do not suppose that I was a deep thinker or was marked as a precocious person. I was a very likely imaginative person and could believe in the *Arabian Nights* as easily as the *encyclopedia*. But facts were important to me, and saved me." Tyndall tells the reader very little about what that self-evaluation might mean for Faraday as a scientist. He interweaves quaint anecdotes about Faraday with serious discussion about his scientific achievements. The biographer makes no attempt to relate the life with the work.

Tyndall argued that the way Faraday did his work told much about the man and, although this is undoubtedly true, the object of a biography is to obtain insights into the work through the story of the scientist's life. The starting point, after all, of a biography is the wish to discover the life that is behind unusual achievements. The importance of the subject's work is what first attracts the biographer who tries to find out what it was about the life that will help the reader to understand the achievement.

Lewis Campbell and William Garnett's, *The Life of James Clerk Maxwell* (London: Macmillan, 1882) was another book in my 1963 survey of scientists in biography. The authors' objective was to portray the "influences which formed the cradle of his [Maxwell's] young imagination, the channels through which ideas reached him from the past, the objects which most challenged his observation and provoked his invention, his first acquaintance with what permanently interested him in contemporary speculation and discovery, and the chief moments of his own intellectual progress in earlier years." The goal was an ambitious one, but the authors hastened to add that they were only able to bring Maxwell's life in 'shadowy outline'. Perhaps knowing the subject through personal acquaintance inhibits the imagination of the biographer from filling out a full characterization of a personality from the bits of available evidence. An acquaintance with the subject, such as Tyndall as well as Campbell and Garnett had, might limit the imagination that would be provided through historical perspective.

The historical imagination is not an automatic product of time, but rather requires some professional training as well as a certain amount of creativity. One biographer whose work I reviewed recognized that writing a biography required both an ability to synthesize a great deal of material and a creative imagination to fill in the gaps. In a new series, British Men of Science, Angus Armitage wrote a volume on William Herschel which he called, *William Herschel, The Pioneer of a New Era in Astronomy* (Garden City: Doubleday, 1963). Armitage restricted the biographical portion of the book to a brief outline in order to be able to devote most of the space to explaining Herschel's discoveries and ideas. But even that brief outline gave little in the way of insight into Herschel's work. Armitage wrote:

> After tracing Herschel's story, reading through his published correspondence, and studying his scientific papers . . . one should be able to form some conception of what manner of man he was. But this is not easy, for his age, though not yet remote in time, is separated from ours by the gulf of an intellectual and social revolution which makes it impossible for us to enter into his mental experience or to share in imagination his outlook on life.

In that admission, Armitage explained well what the chore of the biographer was and, at the same time, he admitted that he was unable to rise to the challenge.

Another book that I reviewed, by Cyril Bibby, was called *T. H. Huxley, Scientist, Humanist, and Educator* (New York: Horizon Press, 1960). This biography, I found, was like the others in that the work was explained and analyzed in detail against a background made up of a few events from the subject's life.

I finished that review essay, written over twenty years ago, with these words:

> The biographer-scientists have put together factual material with an explicit or implicit admission that they cannot truly grasp the image of the subject. The biographer must rather seek to understand the personality, to integrate personality with achievement, and to interpret the result for its meaning in general history.

One might gather that the reviewer, while expressing discontent with the state of biography of scientists, was making plans to write a biography of his own. I was.

CHOOSING A SUBJECT

Having issued my manifesto on writing biographies of scientists, I was asked, and I asked myself, when was I going to write a biography following those admonitions? My experience in writing the history of science and technology in the nineteenth century led me to want to integrate the science of the nineteenth century and to establish links between science and technology. I believed that the right choice of a scientist would enable me to write a biography that was both integrative and explanatory. I believed, to use James Clerk Maxwell's metaphor, that the scientist's life was a scaffolding from which scientific theory was constructed.

The scientific work was the basis for my choice, and I assumed that, whichever scientist I chose, the biography would inevitably reveal the links between the scientific theories, and that there would be strong indications of the connection between science and technology.

My first choice was James Clerk Maxwell. I had read a number of his papers while doing research for my first two books, and I was impressed with the lucidity of his writing as well as being intrigued by the theories on which he worked. How fascinating it would have been to explore the biographical scaffolding from which Maxwell's electromagnetic theory of light had been constructed. But, unfortunately, I learned that there were very few unpublished Maxwell papers in existence.

Then someone suggested that William Thomson, who had been both a scientist and a successful engineer, would be a good subject, and I agreed; Kelvin had given me the title and theme of my book, *Convergent Century* (Sharlin, 1966). Letters, diaries, and drafts of scientific papers – in fact, the major part of a large collection of uncatalogued and unpublished Kelvin papers – were on deposit at the Cambridge University Library. The choice of Thomson was natural and obvious.

THE LIFE'S WORK

William Thomson, later Lord Kelvin, was the nonpareil scientist of the nineteenth century. He was the leading figure in the most important scientific changes of the time. Thomson's most original and path-breaking paper, written when he was seventeen, gave the mathematical analogy between heat and electricity. The analogy was one of the starting points for modern physics in that it was the basis for concentrating on energy as the fundamental concern. Thomson's idea greatly influenced James Clerk Maxwell, whose work culminated in the electromagnetic theory of light, the theory that ushered in the modern period of electrical theory and electrical technology.

Thomson made several other fundamental contributions to the abstract mathematical science of the nineteenth century, primarily in the fields of electricity and thermodynamics. His was one of the more precise statements of the second law of thermodynamics. In the realm of geophysics, Thomson was interested in everything from the figure of the earth to the cooling of the earth, atmospheric electricity, and ocean tides. For the study of atmospheric electricity and ocean tides, Thomson invented and manufactured his own instruments of observation. He played a leading role in the conclusion of classical physics in its quest to replace the mystical, imponderables, such as caloric (heat) and the electrical fluid, with the concept of energy produced by particles in mathematically describable motion.

Thomson became an engineer and entrepreneur by interpreting his own scientific theories. He became a partner in a manufacturing company that produced laboratory instruments. The company manufactured Thomson's mariner's compass that compensated for the effect of a metal ship on the compass heading. He also invented and produced a series of instruments for measuring and predicting the tide level in any harbor in the world.

Thomson's mathematical theory for the flow of electricity in a long wire led him to invent a telegraph receiving instrument used on the original Atlantic Cable. This invention brought him a fortune as well as fame. He was knighted for his work on the Cable and in 1892 rose to the peerage as Baron Kelvin of Largs. As an outgrowth of his work on the Atlantic Cable, he became part of a consulting firm that worked on the world-wide submarine cable network that the British built at the end of the century.

Kelvin, as he is now referred to, produced a tremendous amount of work in his lifetime, including twenty-five books, over 600 scientific papers, and seventy patents. Anyone writing a biography of Kelvin would have to be selective. Since my objective was to write a biography that combined a life with the work, my system of selection had two criteria: (1) works that had made Kelvin an important scientific-engineering figure in the nineteenth century, and (2) works that best illustrated his intellectual development.

My selection was consciously historical in that I was most interested in the works that established Kelvin's reputation in his own time. In my conversations with physicists, I learned that their knowledge of Kelvin was anachronistic. They were well aware of some of Kelvin's work that presaged modern physics, but they were prejudiced about what they termed Kelvin's obstructionism. His work is of interest to modern-day physicists as a minor curiosity and these historically minor works were passed over in my biography. He challenged the emerging modern era of atomic physics, quantum mechanics, and relativity. That challenge deserves serious attention as an example of the role that the older generation of scientists play in the questioning of the new emerging theories. I thought it necessary to reply to some of the harsh ridicule and criticism levelled at Kelvin in his later years. Unfortunately for his reputation, many of the not-so-comical anecdotes survived well into the twentieth century, and physicists, who are notoriously poor historians, completely neglected his earlier contributions to the development of nineteenth-century physics.

Since biography requires that the subject be placed in his time, for Kelvin that meant dealing with his role in the development of nineteenth-century physics as well as nineteenth-century engineering. For example, few scientists today know, or care, that Kelvin was James Clerk Maxwell's mentor. Kelvin originated the ideas that Maxwell brought to fruition in his electromagnetic theory of light.

Kelvin had many major scientific achievements to his credit, such as the absolute temperature scale, the analogy between heat and electricity, his contribution to the second law of thermodynamics, his landmark paper on the mathematical theory of magnetism, and a long list of his contributions would include the dynamical theory of heat. But none of these contributions survived to be a mainstay of twentieth-century physics. Kelvin's work was important because he influenced other scientists, and the way that influence was exerted was a major theme in my biography. The biography was, therefore, part of the history of science and not a humanizing of modern science.

The life of a major figure in any creative activity is like a connecting tissue that gives continuity to the world of original work. A creative person is influenced by the past and the times in which he or she lives, and that person in turn, influences the future of creative thought. The purpose of my biography, then, was to trace those lines of influence.

Are some creative works so monumental that they stand alone and that the works themselves, rather than the lives of the creative individual, become of single historical interest? Is the life of Shakespeare, Elizabeth Barret Browning, Van Gogh, Newton, or Einstein of any interest except to the curious reader? The plays, poems, paintings, and scientific theories by themselves are a force in history, so why biographies of these great figures? Regardless of how monumental the achievement, the influence on creative personalities is of importance as a means of explaining the work in that every great achievement is, by its very nature, a synthesis.

Why can't the influence on a great achiever be a disembodied one? An original mind can be influenced by the works of its predecessors without ever coming into contact through any kind of personal exchange. In Kelvin's case, two of the most important and long-lasting influences on his scientific works were books: one by Joseph Fourier and the other by George Green. Kelvin never met the men, yet it was through human contact that their books influenced him. At fourteen, his physics teacher introduced him to Fourier, and at Cambridge his mathematical tutor gave Kelvin a difficult-to-obtain copy of Green's treatise.

I don't think that the relationship between Kelvin's life and work was unique, but the relationship was unusual in that much of the influence on his thinking, as well as the influence that he had on others, was through personal contact. For example, Kelvin was present at a British Association for the Advancement of Science meeting when James Prescott Joule gave one of his landmark papers on the mechanical equivalent of heat. Although Kelvin was skeptical of Joule's idea, he engaged him in conversation after the paper (Kelvin was the only one to take notice of Joule at the meeting) and asked him for copies. Thus began a long correspondence with Joule on his scientific research. Michael Faraday, another important influence on Kelvin's work, discussed and explained some of his experimental results with Kelvin when he visited Faraday in London.

All the major figures in nineteenth-century physics – British, Continental, and American – came within Kelvin's purview. But his relationship with G. G. Stokes was probably the most significant in Kelvin's life, as well as being important to the development of Stokes's ideas. The two men, both Cambridge University graduates, had a life-long correspondence about scientific ideas. Stokes prodded Kelvin to become involved in the controversy over transmission of signals through the not-yet-built Atlantic Cable. In turn, it was Kelvin who was a constant stimulus to Stokes's work on hydrodynamics.

Kelvin was a world traveler and a prodigious letter writer. As an active man, the physical aspect of his life placed him in intellectual contact with the mainstream of nineteenth-century science. Everything he did, including his leisure, was used to develop new ideas or new technology. Even his 126-ton schooner yacht, *Lalla Rookh*, purchased ostensibly for his holidays, was employed in Kelvin's submarine cable undertakings and led to original work in wave theory, prompting him to develop a mariner's compass, and interested him in tides. As a result, he invented tide-measuring and tide-predicting equipment. The physical, that is the active, part of Kelvin's life, tied his many scientific activities together. In order to understand how he accomplished so much in scientific and technological undertakings, the biography traced the intricate network of paths that he followed.

But the comings and goings – the activities in a scientist's life – are the most accessible and are most easily related to his work. The scientist was here in January 18—, there March 18—, and at a scientific meeting in July of the same year. He met scientist X, wrote to scientist Y, and that summer began work on his paper dealing with the theory of blank. Does the reader want to know more about a scientist than the physical aspects of his life? Tyndall attempted to tell the reader something about Faraday as a 'philosopher', that is, how his character affected his work.

I was told by more than one publisher's reader that any attempt to integrate the life with the work would fall between two stools. Either write a scientific biography that analyzed Kelvin's work, which would be pleasing to scientists, or write a popular biography that would deal primarily with Kelvin's life and personality. But any attempt to link the two would dissatisfy both groups of readers: a popular biography would strike scientists as superficial, and a technical biography would befuddle the average reader. Besides, as one reader argued, what was there about a scientist's life, that is, his personality, that in any way helped the reader to understand his work? One understood that a poet who had been in love could write love poems, or a poet wrenched by the death of someone close could write a poem about death. But how are a scientist-engineer's emotions and personality relevant to his work on abstract scientific fields or impersonal technology?

Historians of science often try to 'humanize' science by telling amusing anecdotes about famous scientists. One of the best known of these is the one about Newton being hit on the head with an apple. A more recent humanistic science story is the touching one about Einstein doing the mathematics homework for the little girl next door. Although these tales are amusing, they have little to do with scientific work. The man in the white coat, detached and unfeeling, is closer to the popular image and is just as misleading as the quaint stories.

Have scientists as a group achieved the state that T. S. Eliot thought an artist should aspire to? He said, "The more perfect the artist, the more completely separate in him will be the man who suffers and the mind that creates; the more perfectly will the mind digest and transmute the passions which are its material" (Donoghue, 1984). Rainer Maria Rilke observed that great artists sometimes allow their ordinary lives to wither away, "like some organ they no longer require". Rilke himself was said to be lacking in vividness as a person even though he was reputed to have a powerful 'fluorescent' presence.[1]

While struggling with the problem of how to bring the life – that is, the personality – into a relationship with Kelvin's work, I found little help in the books and articles on writing biography. No key insights come from biographies themselves. In one guide to the writing of biography, the task is seen as "an almost impossible act of synthesis". Directions for collecting biographical material, organizing it, and presenting the material abound, but "what he [the biographer] must do is easy to describe; how he is to do it in a particular case defies explanation" (Garraty, 1957).

Perhaps that demonstrates that writing biography is more an art than a science that can be taught. Either the biographer, in describing the way he or she works, declares the impossibility of combining life and work or else speaks of some special technique mastered but unable to describe. Leon Edel talked about the "figure under the carpet", and in this quest for "the life-myth" the biographer treads "on dangerous, speculative and inferential ground, ground that requires all our attention, all of our accumulated resources" (Edel, 1979).

Justin Kaplan, in trying to come to grips with the life of Walt Whitman that served as the scaffolding for the writing of *Leaves of Grass*, wrote that a biographer ought to be resigned to merely describing "with as much precision, credibility, and passion as possible, the moments and years preceding and following the miracle". He warned that one does not try to explain a miracle like *Leaves of Grass*, only describe it (Kaplan, 1979).

But the idea of finding a relationship between the life and work is a prevalent one in the thinking and writing of biography. The urge to connect life with work is irresistible, given the fact that the biographer was originally attracted to the subject because of the subject's work. The biographer of Edith Wharton spoke about the relationship between life and work as "a central one for the literary biographer". He added that, like most other writers, Mrs Wharton drew upon her life; *Ethan Frome*, for example, is one of the most autobiographical stories ever written (Lewis, 1975).

Scientists do use their life experience in the selection, approach, and methodology that they employ in their research. The relationship to their lives is more subtle than that of a novelist. The subtlety has eluded biographers of science, and I believe Lewis S. Feuer has made an important breakthrough in this area. His *Einstein and the Generations of Science* imaginatively and convincingly links Einstein's time of life in Zurich and Berne with his revolutionary scientific theories. Einstein, as Professor Feuer points out, found the counter-culture of his day very compatible and a stimulus to challenging the accepted theories in science. At least one of Einstein's friends listened to his new theory of relativity, and Einstein said, "I could not have found a better sounding-board in the whole of Europe." What Professor Feuer did so well in this book (Feuer, 1974) was to show the causal connection between the culture that Einstein was immersed in and his daring scientific imagination. Nowhere else have I seen a biographer make the life-work connection so brilliantly and yet in so straightforward a manner.

Knowing about the life of a creative person, whether an artist, musician, or scientist, helps one understand the work, but the relationship between the work and the life of the person who creates is not a literal one. True, novelists find material for books in their own lives and some novels are virtually autobiographical. But that kind of relationship is not as profound or revealing as the one I was thinking about when I wrote my biography of Kelvin. The simple and obvious fact is that all creative people produce work based upon their own experience. One of the major purposes of a biography is to show that the great work was created by another human being and not by a disembodied genius.

One of the questions that haunted me from the outset was: Where did Kelvin's creativity come from? If the man who wrote twenty-five books, over 600 scientific articles, and was granted seventy patents was not super-human, what were the very human traits that accounted for this remarkable output? Fortunately, there was a large cache of letters that Kelvin wrote home when he was a youth, as well as youthful diaries and other unpublished papers. Strangely, the source material for drawing a picture of Kelvin's youth was much more complete than the material about his maturity and old age. I suspect that the dearth of material dealing with his later life was due to unwitting destruction by his family and heirs.

But the disproportionate amount of material dealing with Kelvin's youth fitted in with my plan to devote a large part of the biography to Kelvin's early life and the sources of his originality. I found that, like other child prodigies such as John Stuart Mill, Kelvin was strongly influenced by his father. He learned his mathematics at his father's knee, and it was his father who convinced him to keep a mathematical notebook which he maintained all his life. With a father whom he adored, and who was a college professor of mathematics as well as a prolific writer of mathematical textbooks, Kelvin had a running start as a scientist. One of the first papers that he published, at the age of seventeen, was a landmark paper in the history of physics, and it was one of the pillars of Kelvin's life-long investigation.

Two mathematical works, one by Joseph Fourier and the other by George Green, had a strong influence on the youthful Kelvin's thinking, so much so that these early acquired concepts shaped his approach to all scientific questions. Fourier's work was on the mathematical analysis of the motion of heat and Green's book was a mathematical analysis of electrical attraction. Those concepts for the analysis of forces – the conceptualization of forces being the result of matter in motion – led Kelvin to develop his mathematical analogy between heat energy and electrical attraction. He used this approach in his research on all other forces, including magnetism.

Kelvin's first paper, the argumentative tone of which was softened by his father, brought him into controversy with one of the leading British mathematicians of the time. Kelvin's competitive spirit manifested itself early in his youth. He won several first prizes and a gold medal in student competitions at the University of Glasgow. In winning those awards he bested his older brother James. At Cambridge University, Kelvin trained like an Olympic athlete for the Tripos examinations and barely missed being senior wrangler, although he was first Smith's Prizeman. At Cambridge, Kelvin also competed successfully for the silver sculls in an all University competition.

The story of Kelvin's youth provides an explanation for his early remarkable creativity. It describes his intellectual powers for scientific work, identifies a competitive cast of his personality that assured him early success in scientific research, and explains his involvement in controversies all through his life.

Were I to revise the Kelvin biography, I think I would make more of the competitive and controversial aspect of his personality; that was the strongest life-long characteristic of the man. From that first essay – in which the young Kelvin attacked a leading British mathematician over Fourier's ideas – to Kelvin's controversy over the operation of the Atlantic Cable, his argument with the Atlantic Telegraph Company over the amount of compensation for his work, and the controversy with T. H. Huxley over Darwin's theory, Kelvin's life was peppered with controversy. My first conclusion was that he disliked controversy and avoided it, but in this retrospective review I now realize that competition and controversy were an integral part of Kelvin's personality and these traits help to explain much about his approach to both scientific and technological work.

CONCLUSION

What was I looking for when I tried to integrate Kelvin's life and work? Why did I make the attempt? Did I succeed? I was searching for the human being behind the scientific and engineering work. Before I began the biography, my impression of Kelvin was like a child's view of God: an unapproachable figure with a long white beard. I was looking for the life that could replace the disembodied scientist-engineer, and I think I found him.

I must confess that the biography is not a completely filled-out portrait of Kelvin.[2] But I was only looking for those parts of his life that help explain why he did the work that he did, and why he did it. In that effort I think I was partially successful.

The biography did serve as an integrative tool that helped tie together such disparate scientific ideas as those in the fields of hydrodynamics, thermodynamics, electricity, magnetism, the dynamical theory of heat, and the fundamental nature of matter in the universe. All these subjects and more were dealt with by Kelvin, and one understands through the biography how one person could be interested in, and successfully contribute to, such widely diverse fields. I also obtained some valuable insights into abstract science as practised by Kelvin, and I was able to follow the way this abstract science was turned by him into useful inventions.

Writing a biography, any biography, whether it be of a musician, poet, artist, or scientist, takes imagination, daring, and a certain amount of cunning. The article that I wrote over twenty years ago for the

Bulletin of the Atomic Scientist was an audacious one, and I confess that the work on the Kelvin biography was humbling yet exhilarating.

Washington, D.C.

NOTES

[1] See *New York Times* (20 June, 1984), review by Anatole Broyard of *Rilke: A Life* by Wolfgang Leppman.

[2] Let the reader decide. See Harold Issadore Sharlin, in collaboration with Tiby Sharlin, *Lord Kelvin, The Dynamic Victorian* (University Park and London: Penn State Press, 1979), 272 pages.

REFERENCES

Donoghue, Denis. 1984. 'The Man Who Suffers, the Mind that Creates' *New York Times Book Review* (March 11), p. 1.

Edel, Leon. 1979. 'The Figure in the Carpet', *Telling Lives, The Biographer's Art*, Marc Pachter (ed.), Washington, D.C.: New Republic Books/National Portrait Gallery, p. 25.

Feuer, Lewis S. 1974. *Einstein and the Generations of Science*. New York: Basic Books, see esp. pp. 49–55.

Garraty, John A. 1957. *The Nature of Biography* (New York: Alfred A. Knopf), p. 258.

Kaplan, Justin. 1979. 'The Naked Self and Other Problems', *Telling Lives, The Biographer's Art*, Marc Pachter (ed.). Washington, D.C.: New Republic Books/National Portrait Gallery. p. 55.

Lewis, R. W. B. 1975. *Edith Wharton, A Biography*, New York, Evanston, San Francisco, London: Harper and Row, p. xiv.

Sharlin, Harold I. 1963. 'The Scientist in Biography'. *Bulletin of the Atomic Scientist*, **19**. (Nov.), pp. 27–28.

Sharlin, Harold I. 1966. *The Convergent Century: The Unification of Science in the Nineteenth Century* (London, New York, Toronto: Abelard-Chuman), 299 pages.

JOSEF ŠKVORECKÝ

THE CASE OF LEWIS S. FEUER, CRIME WRITER

Frankly, Lewis, *The Case of the Revolutionist's Daughter*, viewed as a detective novel, is not, as Holmes's son Nero would say, satisfactory. Why, you haven't got a corpse, Lewis! Surely you know that there simply must be a corpse. No, don't tell me you *have* got one. Eleanor is a case of suicide, even though Aveling may have been a sort of murderer by sweet-talking the confused lady into a double exit, and letting her go solo. "To end an odyssey of sleuthing with such an anti-climax (as a suicide) is to hoodwink the trusting and kind-hearted reader." I don't have to tell you who wrote this. And what about Holmes's work? He is hired by Engels to find the missing girl which he does in a rather chancy way, but then, instead of deducing what lies in store for Eleanor at the hands of her monster husband, and instead of doing everything he can to save her from that mortal danger, he gets mixed up with his old sweetie Irene who never gave him anything but an inferiority complex.

No, the case of poor Eleanor cannot be regarded as one of the illustrious successes of the sleuth of Baker Street. Dr Watson told Laski that the reason he relegated this particular adventure among the 'other cases' he never bothered to publish was a reluctance to "involve Holmes's name in political acrimony". Unconvincing, my dear Lewis! Dr Watson obviously sensed that his friend had made a mess of the affair and hesitated; like Pilate he washed his hands over the manuscript and left the decision to Laski. You should have – in a dialectical way – imitated Max Brod and suppressed this story, in spite of Dr Watson's and Laski's expressed wish to have it published in the 1980s. As a tale of ratiocination it stands deep below your own *The Case of the 'Darwin-Marx' Letter* where you so brilliantly demonstrated what worthy disciples of Holmes serve today among the Baker Street Irregulars. With what merciless logic, Lewis, you unmasked, among others, that late nincompoop Arnošt Kolmàn whom it took over sixty years, throughout which he lived in Moscow, in the very heart of things Marxistic, before, at the age of eighty, he finally saw through it all, defected to Sweden and returned his party card to Brezhnev. And even then, the old fool, when, with my tongue in my cheek, I signed a letter to him "Her British

S. Hook, W.L. O'Neill and R. O'Toole (eds.), Philosophy, History and Social Action, 419–423.

Majesty's Loyal Subject J.Š.", he responded angrily with a long diatribe against the institution of monarchy. In this case, Lewis, you proved yourself a pastmaster of logical deduction.

All right, let's assume that the role of a reversed Brod didn't suit you and you felt reluctant to destroy Dr Watson's half-baked manuscript. If that was the case, why didn't you re-write it yourself? Why didn't you put in a nice corpse, perhaps that of Fred Demuth, and trace the crime to – well, you would find someone. You know everything about that crowd that used to hang around Marx's household, and surely you would have no trouble locating two or three characters with solid motives. Never mind, Lewis, that Demuth was never murdered. The King of Bohemia, at the time when that scheming bitch Adler made a fool of Holmes, was not an Ohrenstein – Dr Watson apparently had little German and misspelled the name as Ormstein – but Franz Josef II. He was fifty-eight in 1888 and had other things on his mind than a compromising photograph; his son, the crown prince Rudolf, had just committed a rather compromising suicide. Well, I guess one cannot expect a man who was unfamiliar with the Copernican theory to know much about a people in Central Europe "of whom we know nothing". After all, he was a genius.

What a pity you did not tamper with the manuscript, Lewis!

And yet, strangely enough, I enjoyed the book. I wonder why. As I said, there is no real crime in Dr Watson's narrative, only a missing person who soon ceases to be missing. And as for the cliff-hanger rescue of Adler – well, in my opinion, the shady lady was not worth rescuing.

Perhaps it is the end of the story which, I admit, *is* interesting. I had no idea that Vladimír Ulyanov had consulted Holmes about Stalin. That he suspected the Georgian of going cahoots with the old Professor Moriarty's gang came as a surprise to me. With Moriarty of all crooks! The man who claimed that, by applying Marx's method, he could, from the nature of economic developments, forecast the future development of crime; who believed that the "individually acting criminal belongs to the petty bourgeois stage of production"; and that "the new order of crime will be as centralized as modern monopoly." This is, I grant you, food for thought. Judging by the results of history, Lenin's suspicion seems to have been correct. I only hope that Dr Watson's manuscript on *that* case *will* be unearthed, as you promise. Although, I'm afraid, it will prove to be just another report of failure of Holmes. The Georgian, apparently, was never apprehended and eventually had his way. The criminal, as Professor Moriarty expected, became the ideal type of revolutionist, and the result was some 30 or 60 million corpses . . .

Wait a minute. Lewis! You've *got* corpses in your story after all!

But who is the murderer? The Georgian? No, that cannot be. It would go against the basic commandments of Father Knox. Remember? "The criminal must be someone mentioned in the early part of the story." Someone who is present throughout the action, with whom the reader has had an opportunity to become familiar. And the Georgian is mentioned only on the very *last* page.

Well, logical deduction rests on a belief in the causal nexus. One takes one's tangible clue – the corpses of the Gulag – and logical reasoning from the clue will eventually lead one to the perpetrator. Which is what Holmes is obviously doing in the story you did not tamper with, Lewis. That's why – I see it now – the sleuth does not concern himself much with the crooked Aveling. The rascal clearly belongs to the "petty bourgeois stage of the production of corpses". Holmes concerns himself mainly with – Marx!

Yes, that bearded sage, that icon, that near-mythical atheist-saint is mentioned in the early part of the story, and makes frequent appearances ever since, commenting on persons and theories, and being commented on by persons holding theories different from his own. Marx, not Aveling is the rival, the counterpart of Holmes in the case; *he* is apparently cast in the traditional role of the perpetrator.

But what was his crime? I could think of one. As you know, Lewis, I lived for twenty years under the communist system which, after all, is the brain child of old Karl. According to the communist constitution in my old country, Marx's crime was clearly that of social parasitism. He was never gainfully employed; he had no visible means of support; in fact, he was Engels's kept person. Quite a few contemporary Czech scholars had been forced to lead that sort of existence, and were punished for it. But of course, Victorian England did not view such a way of life as criminal. So that must be dismissed.

There was a much worse speck on Marx's character, as far as the Victorians were concerned. Freddie Demuth. However, under Queen Victoria, that wasn't a crime either. Moral blemish – sure. A kind of skeleton in the closet – yes. Something that had to be hushed up – most certainly. But not a punishable offence. And anyway, logical reasoning from our tangible clue does not lead to either of these misdemeanors.

Holmes, apparently, sensed this only too well. He saw that there was no physical, material guilt, no blood on Marx's own hands. Therefore, as Dr Watson indicates, he looked elsewhere; subtly, I see now, and convincingly. The sleuth from Baker Street searched not for tangible clues that, as we know, point from the crime to the criminal; he looked for intangible matters pointing in the other direction.

Lewis: on re-reading Dr Watson's narrative which you – I admit now – fortunately did not tamper with, I think I found those clues. That is: I can see now that Holmes lighted on them. There it is, on page 158 in Dr Watson's story, and it is Holmes who is speaking:

> Marx . . . was powerfully righteous in condemning the capitalist system, but he never explored closely what would be the consequences of the system he advocated. I don't think he wanted to know. He avoided reading what John Stuart Mill and Herbert Spencer wrote . . . Spencer . . . once wanted to meet Marx and raise his objections. But Marx begged off with some hollow remarks about Spencer piling up facts the way a shopkeeper arranges his merchandise. Actually I think Marx was scared. Lankester tells me he always avoided meeting real scientists; he preferred to be surrounded by a circle of uneducated admirers, a shaman for the rebarbarizers.

Wasn't that his crime, Lewis? Isn't that a grave crime for a scientist, the gravest, really? To become, like a true religious believer, deaf to reason, to refuse to re-examine one's beautifully consistent theories? To care for a slogan-like, well-coined phrase more than for the prosaic truth? To hold a coterie of admirers closer to one's heart than the community of scientists? To abandon the proven facts of history, which Mill and Spencer saw – that a society completely controlled by bureaucrats enthrones the hegemony of the third rate, that it leads to military-bureaucratic despotism; that behind the facade of every revolutionary social scheme lurks the will of a liberticide? To make a glib transition from a knowledge of society's past development to prophecies that cannot be deduced from that development, such as "When history will pose such problems (the danger of socialism deteriorating into a military-bureaucratic tyranny) to the working class, it will at the same time give them the tools for solving them."

That, I see it now, Lewis, was the crime Holmes solved in *The Case of the Revolutionist's Daughter*. Marx borrowed the idea of socialism from the noble utopians and tainted it with his poorly defined and dangerous concept of the dictatorship of the proletariat. All it took was a cynic like Lenin to take the nebulous abstraction, dispel the fog and uncover the ominous dictatorship of the party-police complex. Then came the Georgian and manipulated that complex for a thorough murder job.

But, Lewis, Marx, toward the end of his life, sensed that something was not entirely right with his predictions. "I can only say with Kant", he tells Holmes by way of an old-age confession, "that my motive was the highest – my will was good, yet my goodness may have been compounded with folly."

I apologize, Lewis. The story you did not tamper with meets all the requirements of a good detective novel after all. The criminal is not someone without any redeeming qualities.

Please, forget my criticism which opens this article. It was correct that you did not tamper with Dr Watson's report on the case of Karl Marx.

University of Toronto

STEPHEN J. WHITFIELD

AFTER STRANGE GODS: RADICAL JEWS IN MODERN AMERICA*

After the Yom Kippur War of 1973, the foreign-born Jewish intellectual then directing American foreign policy paid his first visit to Saudi Arabia, where Henry Kissinger was obliged to listen to what was widely known as King "Faisal's standard speech. Its basic proposition was that Jews and Communists were working in parallel, now together, to undermine the civilized world as we knew it." The Secretary of State wondered whether Faisal was "oblivious to my ancestry – or [was] delicately putting me into a special category". But in any event, "Faisal insisted than an end had to be put once and for all to the dual conspiracy of Jews and Communists". A later meeting allowed the king to repeat his warning to Kissinger, who concluded that Faisal's absorption in "the epic conflict between good and evil" made issues like an armistice on Israel's northern front seem trivial by comparison.[1]

What mattered to Faisal should, in this case, also intrigue historians of modern Jewry as well. For the association between political radicalism and the Jewish people continues to smack of certainty, to ring with the conviction of dogma. That the Balfour Declaration and the Bolshevik Revolution both occurred within five days of one another did not strike some American foreign policy experts as coincidental, and during the First World War they warned that a Jewish commonwealth would nurture radicalism in the Middle East.[2] It is also worth remembering that the Führer, whose country Kissinger himself later fled, could not distinguish between Jews and Bolsheviks and believed that the headquarters of world Jewry was in Moscow. If such a belief in the intertwined nature of Jewry and radicalism can penetrate all the way to the desert kingdom of the Saudis, it is a subject worth investigating, even though limitations of space are an excuse to restrict these observations to the United States. It is also a ticklish topic, as William B. Saxbe, an Attorney General under Nixon, discovered in 1974 when he committed the fallacy of composition, expressing relief that Jewish intellectuals were no longer "very enamored of the Communist Party".[3]

Even at this moment, when the passions of American radicalism seem exhausted, when Jerry Rubin is a Wall Street analyst and Eldridge

S. Hook, W.L. O'Neill and R. O'Toole (eds.), Philosophy, History and Social Action, 425–460.

Cleaver and Rennie Davis are pursuing even more sacred endeavors, the social roots of Marxist as well as anarchist beliefs deserve to solicit scholarly attention. Much of it is quite recent. The fullest account to date of the American Jewish proclivity for socialism, Arthur Liebman's *Jews and the Left*, appeared in 1979. Its scope stretches from the Czarist Pale of Settlement through the New Left of the 1960s – and beyond, for the author predicts that Jewish self-interest will find in socialism a remedy for the crisis of capitalism. Since then Percy Cohen, based in England, and W. D. Rubinstein, based in Australia, have published books that express far greater reservations about the Jewish propensity for radicalism. In 1982 Stanley Rothman and Robert Lichter drew upon psychological data to corroborate their view that the *Roots of Radicalism* could be located in the differing dynamics of Jewish and Gentile families in the United States. European Jewish radicalism from the mid-nineteenth century until the Second World War has also been the focus of Robert Wistrich's portraits of *Revolutionary Jews from Marx to Trotsky* (1976) and other works, and of Jonathan Frankel's prize-winning *Prophecy and Politics* (1981). Briefer discussions and analyses have also appeared.[4] But it is the aim of this paper to offer a historical overview of the relationship between American Jews and radicalism, to suggest its cultural and psychological texture, and to evaluate the theories that have been proposed to account for this relationship.

As Lewis S. Feuer has pointed out, the impression of such a copula can be traced virtually to the birth of the modern era itself. In seventeenth-century England, as the Puritan revolution of the saints was liquidating the legitimacy of monarchy, Oliver Cromwell denounced his left flank, the millenarian Fifth Monarchy men, as 'Judaisers'.[5] This term of reproach was hardly noticed in the midst of the simultaneous cataclysms of Jewish history, such as the Chmielnicki massacres in Poland and the Ukraine and the messianic movement of Sabbatai Zevi that swept through communities from Yemen to Hamburg. But John Calvin had also dismissed the image of Utopia as the 'foolish' concoction of the Jews,[6] so that well over a century before emancipation in France and the Germanies – the prototypes of the Jewish encounter with the modern world – the descendants of the Biblical prophets were perceived as harbingers of social change and messianic hope. More often Jews were merely liberals, as was the *Allgemeine Zeitung des Judentums*, which presumed that the 1881 pogroms in Czarist Russia were sparked by Jewish radicalism. This main organ of liberal German Jewry therefore advised its Russian brethren to abstain from 'secret conspiracies and subversive activities' which would discredit Jewry in general.[7]

But in the first two waves of Jewish immigrants to the United States, dissidents like Ernestine Rose and the abolitionist David Einhorn were rare; and the fragmentary Jewish communities established in the United States in the eighteenth and nineteenth centuries harbored few radicals. Only the arrival of two million Jews fleeing persecution and poverty in Eastern Europe fostered the rise of radicalism – a polysemous term which in this paper is intended to encompass the varieties of Marxian socialism, anarchism and the more estranged forms of youth activism in the 1960s. The ideologies that the *fin-de-siècle* immigrants professed, the causes that they espoused, and the institutions that they created established the standards against which subsequent versions of Jewish radicalism have been judged. That first chapter has often been told – by Melech Epstein, by Moses Rischin, and with magisterial intelligence by Irving Howe – and needs only brief mention here. For even after allowances have been made for the conventional, moderate politics and even indifference of perhaps the bulk of the Jews, what was special about the first generation or so of Eastern European immigrants to the Northeast and Midwest was how authentically they consolidated political radicalism with other manifestations of Jewish life. On its masthead the Jewish *Daily Forward* proclaimed slogans like "Workers of the world, unite," and "The freeing of the working class must be the task of the working class itself". At its peak the circulation of the *Daily Forward* was a quarter of a million, making it the most widely read Yiddish newspaper on earth. (With twice as many Jews in the United States today, the neo-conservative magazine *Commentary* has about 40,000 paid subscribers.) In the pages of the *Daily Forward, De Proletarischer Magid* preached red-hot propaganda, and lessons were given in the *Red Sedre*. (Perhaps it was not too precipitous a step after all to Arthur Waskow's *Freedom Seder*, a haggadah that introduces such fresh voices as 'Rabbi' Cleaver.) Daniel Bell concluded that "the *Forward* bound together the Jewish community and made it socialist". The newspaper usually put ethnic claims above ideological loyalties, and honored such claims whem socialism was mute, as when – early in its history – the *Daily Forward* supported the bourgeois Dreyfusards in France and did not oppose the imperialist Spanish–American War.[8]

Socialism may have helped satisfy some of the idealistic yearnings of the Jewish community, but that community did even more for the radicalization of the American labor movement and for making the economic system more humane. The garment trade unions and their associated Workmen's Circle provided the Socialist Party of America with much of its organizational and financial strength. They maintained their allegiance to socialism long after other militant unions had fully reconciled themselves to the capitalist order. The International Ladies Garment Workers Union, one labor historian notes, "used its considerable vote at AFL conventions on behalf of radical resolutions . . . more consistently than almost any other trade-union". In opposing immigration restriction (more for ethnic rather than class reasons), in supporting women's suffrage and Negro rights, and in encouraging the growth of socialism in Europe, unions like the ILGWU exhibited

> a humanitarian, idealistic, and deeply held desire for equality and social justice. . . . But perhaps the most distinctive contribution of the ILGWU was its various provisions for the health, education, and general welfare of its members, which went far beyond the sick, death, and other benefits provided by the most other American trade-unions. It was on the educational program of the union, however, that the greatest emphasis was laid.

Even in the 1920s, despite the pressures of the open-shop movement and the bitter factional struggle with the communists, the ILGWU continued to foster the revival of post-war radicalism and to contribute to the coffers and the morale of the socialist movement. Only when it endorsed Franklin D. Roosevelt for President in 1936 (as did the *Daily Forward*'s editor Abraham Cahan himself) did this relatively long chapter in American radical history come to a close. For other imperative and values within the Jewish community sapped a sustained working-class militancy. The promises of the wider society and the impulses of upward mobility among its own membership weakened the ILGWU's commitment to class consciousness. Italian-Americans transmitted the skills of the needle trade to their children, and therefore became increasingly important in such unions. But Jewish workers generally recommended educational advantages and property acquisition instead.[9] Despite the stirring words of the *Daily Forward*'s masthead, the Jewish working class did not free itself. Instead it taught its own children to free themselves *from* the working class.

Yet if any single figure incarnated that immigrant generation, bursting with idealism and hope and creative energies that not even the ordeal of the sweatshops could drain, it was Emma Goldman. She was not, of course, a socialist. But she was, after Eugene V. Debs, the most impressive figure to emerge from radical politics in the United States in the early twentieth century, and her legacy may be even more enduring (though not in a rigorously programmatic sense). Her mind was far less shallow than Debs'; and her sensitivity to the threats to human freedom, especially the greater consistency with which she opposed the tyranny of Bolshevism, make her even more relevant. She also lived long enough to denounce the menace of Nazism, for which she was expelled from jittery Holland in 1933, even though her own vaguely pacifist sentiments spiked any effective opposition to Nazism. She shared the pre-war radical dream of a humanity lifted above the divisions of class and race and nationality (which is why her magazine, *Mother Earth*, printed at least one article critical of Zionism). But even though she repudiated the normative life of the Jewish masses and the institutional world of the immigrants, it is hard to separate Goldman from that environment if her life is to be historically understood. Even though she was shunned as an *apikorista* (free-thinker) by the Jews of Worcester, Massachusetts, where she had opened an innocent ice cream parlor (a double scoop for five cents was what it cost boys like the future playwright S. N. Behrman), even though her radicalism led a Rochester synagogue to excommunicate her father, even though she broke out of her Yiddish in order to propagate anarchism among the American masses, Goldman was primarily an inhabitant of a Jewish world. The important association of her life was with Alexander Berkman, and a couple of her subsequent lovers were also Jewish. Her feminism was not so implacable that she could not admit that the only men she ever met who treated women as equals were Russian Jews. Exile was her fate, and she knew its burdens perhaps more terribly than most Jews. After her expulsion from the United States, she lived in Russia, in France, in England; stateless, she wandered the earth as much as Trotsky, her neighbor in France in 1935. In 1940, her wish was granted to be buried back in the United States, a country which – even in exile – she loved far more than its officials could bring themselves to reciprocate. Goldman was also Jewish in her responsiveness to the avant garde and to cosmopolitanism, and was a pioneer American interpreter of Shaw and Strindberg and Ibsen.[10] She thus embodied the radical intellectual – cosmopolitan, unpatriotic, polygot, an emissary of the power of the word, a believer in the pressure of the civilized conscience.

Her influence on American society was healthy. Having attended lectures by Freud in Vienna in 1895 and in Worcester in 1910, she championed reproductive freedom in an age of unwanted pregnancies and tragic consequences. She expanded the contours of free speech. Goldman so affected a young social worker named Roger Baldwin that he founded the American Civil Liberties Union, an organization that remains, in an indirect way, her cenotaph. Her life was devoted to the assertion of rights that were frequently denied in practice, which is why a feminist group that once hurled a pie at ERA opponent Phyllis Schlafly called itself the Emma Goldman Brigade. No wonder then that her deportation, sneakily engineered by J. Edgar Hoover of the General Intelligence Division in 1919, prompted an attorney to lament: "With Prohibition coming in and Emma Goldman going out, 'twill be a dull country."[11]

Dullness was something that Jewish radicals were determined to prevent. Indeed they had trouble staying out of the line of fire. A number of free speech cases, which the U.S. Supreme Court almost always decided in ways that *weakened* freedom of expression, made Jewish radicals into victims of injustice. One important test of the Constitution was the right of socialists to be elected and serve in the state assembly of New York. That test was flunked, even though a Republican who had served on the U.S. Supreme Court and had run for the Presidency, Charles Evans Hughes, represented the Albany 5, three of whom were Jewish. Victor Berger, a former congressman, was sentenced to twenty years in prison for violation of the Espionage Act. Though his conviction was overturned by the Supreme Court, Berger was nevertheless denied a congressional seat won in a free election – because of his unpopular opinions. Militant socialists like Jacob Abrams, convicted under the Espionage Act, and Benjamin Gitlow, author of the *Left Wing Manifesto*, led Attorney General A. Mitchell Palmer to warn against their pamphleteering: "It is not good reading late at night when you are at home in your own house. It gives you the creeps . . ." Almost the entire citizenry would have found these tracts so impenetrable, their guiding assumptions and argot so remote, that they could just as effectively been inscribed in cuneiform. Their authors were nevertheless jailed during and after the Great War to save the world for democracy.[12] Yetta Stromberg, a member of the Young Communist League, broke a California anti-syndicalism law while she was a summer camp supervisor. Each morning she raised a red flag, emblazoned with hammer and sickle, and led the salute and pledge of allegiance. She was imprisoned too.[13]

But the last of the major free speech cases to burnish Justice Holmes' reputation as a liberal dissenter involved Rosika Schwimmer, a Hungarian-born immigrant who in 1915 had helped convince Henry Ford to send a 'peace ship' to a Europe devastated by war. Ford complied with her plea for this dramatic gesture even while blaming the war on the Jews. Had he done so oblivious to Schwimmer's ancestry, or was he delicately putting her into a special category? It is imposible to say. In any event this feminist and suffragist, a co-founder of the Women's International League for Peace and Freedom, this former Hungarian ambassador to Switzerland (probably the first female ambassador of modern times), this pacifist eloquent in several languages, this refugee from the Horthy regime filed for citizenship papers in 1926. Her principles compelled Schwimmer to announce that she would not bear arms. Holmes, who *had* borne arms to preserve the Union, was impressed by her integrity, and compared her to the Quakers, whose adherence to the Sermon on the Mount – he observed – did not make the Friends objectionable. She struck Holmes as "a woman of superior character and intelligence, obviously more than ordinarily desirable as a citizen of the U.S.". But six of his colleagues disagreed and denied her petition, which incidentally revealed no explicit indebtedness to Judaism as a source of her "cosmic consciousness of belonging to the human family". When she died in 1948, the year a Jewish state was reborn, Rosika Schwimmer was still an alien.[14]

The denial of Judaic inheritance was also common among the communists whose beliefs and methods have haunted the American radical tradition. Bolshevism contributed to the factionalism which, even without the resilience and repressiveness of the American political system, has postponed the day of revolution. For present purposes, it is unnecessary to calibrate the differences that separated communists, Trotskyists, Lovestoneites, Shachtmanites, Oehlerites, or Fieldites in the 1930s and 1940s – to say nothing of the later factionalism of the New Left. Like mayflies, these *groupuscules* were born, mated, split and died, often quite briefly. Typical was the sect founded by Albert Weisbord, the son of Russian immigrants. He was elected to Phi Beta Kappa at the City College of New York, became a national secretary of the Young People's Socialist League, graduated from Harvard Law School, and was inspired toward greater proletarian militancy by the writings of

Lenin. At the tender age of 25, Weisbord became one of the heroes of the dramatic Passaic textile strike in 1926, which involved as many as 13,000 workers. But Weisbord drifted away from the communists, who expelled him for 'individualist conceptions' in 1930, a year after he participated in the Gastonia textile strike. He split from the Trotskyists, and then in 1931 formed his own group, which Weisbord fiercely designated the Communist League of Struggle. It lasted seven lean years, with all seven of its other members eventually quitting. His two-volume contribution to Marxist theory, grandiloquently entitled *The Conquest of Power* (1937), was his last radical gesture before becoming an AFL organizer in New England and the Midwest.[15] Weisbord's career – punctuated with the stardust of immigrant dreams, with the desperation of the Great Depression, with factionalism and puffed-up theorizing, finally with accommodation to moderate reform of the capitalist system – was virtually a parable of the course of leftism.

Its course was to be a melancholy one, but there is no denying the appeal that communism exerted upon Jews before that particular god failed. Even the Joint Distribution Committee, one of the philanthropic pillars of 'uptown' Jewry, backed development schemes in Russia in the 1920s. The JDC invested millions of dollars in Jewish resettlement in the Crimea and the Ukraine, not only because of its desire to deflect the strategy of political Zionism, but also because of its appreciation of the improvement of Jewish life under communism. (Only a generation later the purge trials anathematized the 'Joint' as a Western espionage agency.) In the 1920s the circulation of the Yiddish *Freiheit* exceeded that of all other communist newspapers in the United States, including the English-language *Daily Worker*. One of the garment trade unions, the Furriers, was largely under communist control; and its fiery leader, Ben Gold, did not relinquish control until the combination of the Taft-Hartley Act and McCarthyism gave the union no alternative.[16]

Of the intellectuals who publicly proclaimed their support for the communist ticket of William Z. Foster and James Ford in 1932, almost a third were Jewish. They were thumbing rides on what was later realized as 'the road to serfdom'. Even Sidney Hook, the first Marxist professor in the United States, explained to readers of *Modern Monthly* "Why I am a Communist" (1934). The Stalinist group that Whittaker Chambers later charged Alger Hiss with having joined, and which was burrowed in the Department of Agriculture, included John Abt, Lee Pressman and Nathan Witt. At the end of 'the red decade', leading fellow travellers signed an ad criticizing those who believed that totalitarianism linked the Soviet Union and Nazi Germany. The ad was unluckily timed to appear shortly before the Nazi–Soviet pact was signed in Moscow; many of the signers were Jews. (So, it must be added, were most of their opponents in such organizations as Americans for Intellectual Freedom, including – by this point – Hook himself.)[17] In the 1940s and 1950s the HUAC investigations of Hollywood and other nests of subversive thought ineluctably uncovered a disproportionate number of Jews, exposure which anti-Semites like Congressman John Rankin (D-Miss) quickly exploited. The historian Daniel Boorstin, by 1953 a conservative who told HUAC that his support of the Hillel Foundation at the University of Chicago indicated his hostility to communism, had been a Party member for a year or so before the Molotov–Ribbentrop Pact undermined his reasons for joining – his opposition to Nazism and anti-Semitism.[18] Such motivations were typical.

The extent of Jewish involvement need not be elaborated for the years when the Comintern supported Arab rioters against Jewish settlers in Palestine, when the Stalinist purges claimed uncounted thousands of innocent Jewish victims, when Foreign Minister Molotov proclaimed in 1939 that fascism was 'a question of taste', when the NKVD handed Jewish prisoners over to the Gestapo, and when Yiddish culture and its custodians were systematically extinguished under Stalin. The anti-Semitism of Stalinism caused the membership and support of the Party to be depleted of thousands of Jews who had been attracted to the promise of a militantly effective universalism and 'progressivism'.

But what is curious is the favorable light in which American communism and its satellites continue to be held. What Ernest Renan said of early Christianity is applicable: "Enthusiasm and love know of no hopeless situations. They play with the impossible, and rather than despair, they violate reality."[19] Even when one would have thought that all the evidence was in, that all the precints had reported, retrospect had lavished upon the Jewish communists varying measures of sentiment, sympathy, even enthusiasm and love. It takes the form of 'oral history', as in Vivian Gornick's *The Romance of American Communism* (1977), whose *dramatis personae* are mostly Jews from New York. Avoiding the pregnable ground of policy and focusing entirely upon human qualities, the author describes mostly warmth, gusto, idealism. Several of the ex-comrades are even physically attractive. Politics had once been, in Benjamin Gitlow's phrase, 'the whole of their lives'; and those lives, Gornick's account implies, were honorable and decent.

The Rosenberg case also continues to summon its sympathizers, none more tenacious than Walter and Miriam Schneir in their *Invitation to an Inquest* (1965), which even in its revised edition conveniently ignores all the FBI evidence that impugns the Rosenbergs' professions of innocence. The Hollywood communists have their defender in a far more nuanced and subtle book – quite excellent in its own right – by Victor Navasky, *Naming Names* (1980). And the late Lillian Hellman was quite successful in imposing her interpretation of McCarthyism and of the left on many American readers. She never really repudiated her support of Stalinism. Hellman could be said also to typify an important strain in such 'progressivism' which takes a little or no account of Jewishness. Her anti-Nazi play, *Watch on the Rhine* (1941), omits the topic of anti-Semitism, for example. Her recollection of a visit to the death camp of Majdanek does not bother to mention that virtually all the victims were Jews. They were, presumably, otherwise unidentifiable victims of fascism. And even in her introduction to the memoir of dissident Lev Kopelev, *To Be Preserved Forever*, Hellman underplays the resurgence of Jewish feeling and opposition to the Soviet state, and neglects to mention the disastrous impact of the communism that she supported upon Jewish life and culture.[20] The bad faith of such a position is reminiscent of those communists who sniffed anti-Semitism in the prosecution and conviction of the Rosenbergs but failed to criticize the concurrent 'purge trials' in Eastern Europe that condemned to death as 'Zionists' communists of Jewish birth insufficiently subservient to Moscow.

Even more than Stalinism, the movement led by Leon Trotsky won the allegiance of a disproportionate number of Jews. The heresiarch himself was, of course, born Lev Davidovich Bronstein. He was a multilingual intellectual nicknamed 'Pero' – The Pen. But Trotsky's interest in literature and his fluency of expression were overshadowed by the flair for action – political and even military – that his remarkable career exhibited. It is true that James Cannon shared with Max Shachtman the leadership of the Trotskyist movement in the United States, until their split in 1940. But the rank-and-file were, apart from the truckers of Minneapolis, probably almost exclusively Jewish. Several of the younger members later rose to prominence as intellectuals, critics and scholars, among them the late Martin Diamond, Leslie Fiedler, Gertrude Himmelfarb, Irving Howe, Irving Kristol, Seymour Martin Lipset, Marvin Meyers, and the late Isaac Rosenfeld. Even

Bernard Cornfeld, the erstwhile 'Midas of mutual funds', was a Trotskyist in his youth. It therefore made sense for Saul Bellow to send Augie March to Mexico intending to serve as Trotsky's bodyguard, and to make Hyman Lustgarten, the black marketeer in post-War Europe in 'Mosby's Memoirs' (1968), a former Trotskyist. American Trotskyism taught Kwame Nkhrumah the operations of an underground organization, before he returned to Ghana and became its Osayegfo ('redeemer'). Near the end of his life, Malcolm X was tentatively interested as well.[21] Nevertheless, for all the efforts primarily of Jews to keep this version of revolutionary Marxism alive, the verdict on it must be the same as that which Trotsky himself pronounced upon Martov and the Mensheviks. "You are miserable isolated individuals" was Trotsky's curse in 1917. "You are bankrupt. You have played out your role. Go where you belong, to the dustheap of history."[22]

It was a place to which more than the Trotskyists were consigned. Even though a disproportionate number of American radicals were Jews, it is also worth asking why so few Americans were deeply dissatisfied with their society and eager to seek alternatives. The Jewish contribution to radicalism is all the more striking because of its unattractiveness to so many other Americans. There are many reasons – perhaps as many as 27, according to one scholarly survey of why there is no socialism in the United States. But one aspect that ought to be highlighted here shows the character of American society as well as any other. That is its resilience, its relative openness to political reform and social change, its capacity to roll with the punch. Consider the editorial published on November 15, 1909, in the socialist daily, the New York *Call*, which listed immediate demands to end the exploitation and misery of the working class. This agenda was 'so revolutionary', the editorial insisted, that only through revolution was it attainable. The demands included labor laws, employer liability, the right of unions to control their own treasuries, legalization of the boycott, the elimination of the labor injunction, government inspection of workplaces, reduced working hours for women and children especially, and serious efforts to bring to an end the terrible loss of life in mines and factories. But the joke on the New York *Call* and other socialists was that not only were these demands conceivable; they were after much struggle attained, in most instances within about one generation.[23]

The role of reform as the nemesis of radicalism can be measured through Jewish biography. In his youth the Dutch–British immigrant

Samuel Gompers had flirted with socialism which, in the United States, he considered unnecessary and even dysfunctional; and his stewardship of the American Federation of Labor had much to do with making the immediate demands of the New York *Call* conceivable. He claimed to share the ideals of socialism, differing – at least initially – only over tactics. A trade unionist here, Gompers asserted that he would have been a nihilist in Russia, a socialist in Germany.[24] In Russia Sidney Hillman and David Dubinsky were radicals. In America they were militant trade union leaders whose leadership of strikes did not land them in jail – or in the equivalent of Siberia. Abraham Cahan escaped from Russia because Czarist police discovered his involvement in a revolutionary organization. By 1936 he too was supporting Franklin D. Roosevelt.[25] Another immigrant, Frank Tannenbaum, came to the United States in 1905, quickly became involved with the Wobblies, and in 1914 was convicted of disturbing the peace when he tried to lead New York's unemployed and hungry in seeking refuge in churches. He was imprisoned on Blackwell's Island. The warden of Sing Sing encouraged Tannenbaum to go to college, so he went to Columbia, graduated Phi Beta Kappa and became one of its most distinguished professors. And when Tannenbaum published a book about the warden of Sing Sing, Franklin D. Roosevelt wrote the introduction.[26]

So common is this pattern that perhaps an unfamiliar illustration can be offered to suggest how the relatively benign effects of American society could temper the radical spirit. The choreographer Jerome Robbins (*né* Rabinowitz) encapsulates a particular shift from radicalism to ethnicity, a realization of the opportunities for subcultural creativity within a pluralistic society. From 1943 until 1947 Robbins was a communist, who was nevertheless offended by a comrade's question about how "dialcctical materialism help[ed] me do *Fancy Free*", Robbins' first major ballet. The Party attracted him because it "was very much for minorities and for advancing their causes. . . . I had had, prior to my joining, several instances of very painful moments because of minority prejudice." Robbins had also believed that communism would provide security and freedom for artists. But he became disillusioned, as he told the House Committee on Un-American Activities in 1953. He realized that communism was by no means opposed to anti-Semitism and he was shocked by the purges and anti-Semitic forces in Eastern Europe. Two summers working with dance groups in Israel also affected him. It was about that time that Robbins first conceived of a stage work that would update the story of Romeo and Juliet, pitting Christians against Jews instead of having Montagues fight Capulets. It was to take place at Easter and Passover. But he realized that in America such a conflict lacked a basis of plausibility, so the groups were shifted to Puerto Ricans and 'older' white ethnics. Blazing with the music of Leonard Bernstein and the lyrics of Stephen Sondheim, Robbins' *West Side Story* (1957) captivated Broadway and beyond.[27] It contains no references to Jews. But because the play is urban in its setting, satiric and firmly liberal in its slant, respectful of 'high culture' and upbeat in its hope that, however fragile, love might transcend the barriers of bigotry, *West Side Story* may be categorized as Jewish in the sense that the composer Aaron Copland fancied his own music Jewish – "because it's dramatic, it's intense, it has a certain passionate lyricism in it. I can't imagine it written by a goy."[28] Robbins also directed and choreographed *Fiddler on the Roof* (1964), thus exposing (or exploiting, according to critics) the roots of his own past. Here, in part, revolutionary politics and anti-Semitism were put outside America, in the Pale of Settlement (where they made historical and theatrical sense). Thus was Robbins partly responsible for triggering an ethnic revival whose force was so powerful that it has not yet been spent. Indeed some Jewish women in the audience were so moved by *Fiddler on the Roof* that, at least according to apocrypha, operations were performed to restore their noses to their original dimensions.

Yet oddly enough the 1960s marked not only the emergence of ethnic assertion among Jews but also of radical rejuvenation. Somehow many young Jews still felt a little like strangers in the land of Egypt, expressing their marginality through civil rights activism, organizing against poverty and against the Vietnam War, and through feminism and support of Third World causes. They helped edit magazines like *Liberation*, the *Village Voice, New Politics, Democracy, Radical America, Socialist America*, and *Studies on the Left*, among many others. A New Left perspective was amplified in the pages of the *New York Review of Books*, edited by Robert Silvers and Barbara Epstein. Its brilliant caricaturist, David Levine, calls himself 'a radical unfactionalized socialist'; and his early cartoons appeared in the *Daily Worker*.[29]

Jewish radicals were no more successful in avoiding trouble in the 1960s and 1970s than were their predecessors during the Great War and the Red Scare that followed. The indicted co-conspirators of Dr Benjamin Spock and the Reverend William Sloane Coffin were Mitchell

Goodman, Marcus Raskin and a Unitarian, Michael Ferber. The Chicago 7 included Abbie Hoffman, Jerry Rubin and Lee Wiener. Though Daniel Ellsberg was never a radical, strictly speaking, his exposure of the Pentagon Papers, his unsavory prosecution (and the plumbers' rifling through his psychiatrist's files) and vindication helped bring to an end both the Vietnam War and the Nixon administration itself.

Even in unexpected places, Jews cropped up. Martin Luther King, for all his militancy and activism, was a Baptist preacher, not a radical. One of his key advisors nevertheless was Stanley Levison, a pivotal money-manager for the Communist Party and a close associate of its top leadership prior to his intimacy with King. Their friendship became the justification for the FBI's squalid bugging campaign against the civil rights tribune.[30] The organizing struggle of Cesar Chavez in the fields of California drew upon the resources of Catholic commitment and Mexican–American ethnic solidarity (*La Causa*). But Saul Alinsky, the son of immigrant Orthodox Jews, helped tutor him in the tactics of peaceful but forceful confrontation. The pacifist and non-violent mood of the folk music movement was much influenced by Joan Baez, who was in turn influenced by Ira Sandperl. And the letters from Folsom Prison which gave shape to Eldridge Cleaver's nascent militancy were addressed to his attorney, Beverly Axelrod, to whom *Soul on Ice* is dedicated. According to one journalistic account, the first wave of civil rights workers to go South in the early 1960s was more likely to quote Harry Golden that Gandhi or Thoreau.[31]

It is also worth noting the tactics of SNCC's Stokely Carmichael, who bravely went into Lowndes County, Alabama, to encourage voter registration. None of the county's 12,000 blacks was listed on the voter registration rolls, whereas the whites were so enthusiastic about 'participatory democracy' that 117 percent of them were eligible to vote. To break down the climate of fear that enveloped the black community, Carmichael taunted the sheriff, mocked his stride, and insulted him with a Yiddish term, an anatomical allusion unsuitable for a scholarly publication. The graduate of Bronx High School of Science was quite sure that Southern whites did not understand Yiddish, which he used in Lowndes County as a disruptive force of delegitimization. In any case Carmichael had felt little respect for his Southern white adversaries after his jailers had confiscated his Camus and DuBois paperbacks but left him in his cell with C. Wright Mills' apologia for Castroism, *Listen Yankee!*, which they mistook as a warning to Northern agitators.[32] Not that the influence of Jewish radicals couldn't be found in the Caribbean too. In the early 1960s in then British Guiana, whose population was about evenly divided between East Indians and blacks, the leader of the People's Progressive Party was a Marxist dentist named Cheddi Jagan. His wife was a former member of the Young Communist League from Chicago, Janet Rosenberg; and Jagan once told President Kennedy how much he admired the *Monthly Review*, particularly the writings of Leo Huberman and Paul Baran.[33]

But the last bloody clash of the civil rights movement took place in Greensboro, North Carolina, where the peaceful sit-ins had begun almost two decades earlier. On November 3, 1979, five members of the Communist Workers Party were murdered by the Ku Klux Klan and its associates. The dead included two Jews; another Jew was seriously wounded.[34] Thus was demonstrated in tragic fashion the continuity of the Jewish commitment to radicalism, even amid prosperity, effective reform and the patent decline of anti-Semitism.

In some ways the most baffling aspect of that commitment involved the student power movement. Here the Jewish propensity for radicalism was most striking, because its militancy could not be explained as easily as the immigrant working-class socialism of the New Leftists' grandfathers, or in terms of the desperation engendered by the Great Depression. The disproportionately Jewish student radicals who rebelled at Berkeley, Columbia, and Harvard, and who were also prominent on other Ivy League and Big Ten campuses, were privileged. They were unmotivated by material self-interest, unhampered by prejudice and discrimination. Jews constituted about a tenth of all college students in the 1960s, yet they were often half or more of the radicals on leading American campuses. Under the auspices of the American Council of Education, a survey of 1966–1967 asserted that the most accurate single predictor of protest was the matriculation of a substantial minority of Jewish students. Even during the Berkeley sit-in of 1964, according to one report, *Hatikvah* was sung; and SDS was packed with Jews, whose Jewish identity was often disguised or downplayed. Few made their origins an incentive to self-consciousness or reflection about their motives or their ideals; and when it came to camouflaging, Hollywood did the rest. The best-known memoir to emerge from the campus rebellions was probably James Simon Kunen's *The Strawberry Statement*. Yet in the film version the Columbia campus of the book was shifted to a mythical middle America, and the swarthy urban rebels who seized buildings at Columbia became the sorts of blondes whom Jesus is generally depicted as resembling. The Judaization that Oliver Cromwell had linked to radicalism was therefore bleached out. Hollywood's grip on the social reality of the 1960s can also be gleaned from the casting of Stanley Kramer's *RPM* (hyperbolically an abbreviation for 'revolutions per minute'), in which the university president is played by Anthony Quinn and a graduate student is played by none other than Ann-Margret.[35] Such casting thwarts sociological understanding.

For the prototype of the 1960s radical did not look like Ann-Margret, but more like Abbie Hoffman. He was a novelty item in the history of American radicalism. Like Norman Mailer, who contributed a very favorable introduction to Hoffman's autobiography, Hoffman grasped the irrelevance of older notions of revolutionary purity and integrity. "The Communists disdain to conceal their objectives", Marx and Engels had written in their *Manifesto*. Hoffman disdained to conceal his pursuit of fame. He was less interested in power than in publicity – even timing his ascent from almost seven years in the underground to coincide with the publication of his autobiography.[36] Hoffman was unattracted to democratic centralism but rather by the celebrity that democracy seemed eager to bestow. He specialized not in the *putsch* but in the *put-on*. For all of their alleged conservatism, the Founding Fathers knew that they were staking their lives, their fortunes and their sacred honor. Hoffman, who relished vexing the respectable, realized that the 'revolutionaries' of the 1960s had no such stakes to play. He was closer to Madison Avenue than to James Madison, who was at least preoccupied with the proper distribution of power. One of Hoffman's favorite professors at Brandeis University, Herbert Marcuse, had propounded the paradox of 'repressive tolerance'. Hoffman acted it out, or thought he did, proving how flaccid American society was and yet how ultimately resistant to serious change it remained. Sometimes he trapped himself in his own anarchic amoralism, as when the author of *Steal This Book* accused a collaborator of plagiarism. (Aficionados of literary lore will doubtless be reminded of Edgar Allan Poe's charge that Lewis Gaylord Clark had stolen one of Poe's own ideas for a hoax – in other words, that Clark had been untruthful in revealing the source of his own lie.)[37] In contrast to SDS, whose contribution to Western political thought included the slogan, "Screw the ass of the ruling class". Hoffman preferred to goose it instead, and thus exhibited the most nimble mind and the most inventive stance among the young radicals of the 1960s.

He was also probably their most 'Jewish' representative as well. Never mind the offensiveness that his actual opinions presented for most Jews, including his nose-thumbing support of the Palestinians, and his distaste for his religious upbringing (an experience that hardly made Hoffman unique). For he shared with many other Jews a sense of estrangement, a spirit of dissidence, an identification with the oppressed that took him on dangerous trips to the black South. His forty-seven arrests gave his parents 'the ultimate in Jewish nightmares'. He also endured 'the ordeal of civility' that certain other Jews have borne in getting used to – and in getting at – Western culture. His autobiography returns to the subject of his Jewishness in an almost compulsive manner, with wry and bittersweet irony, without defensiveness or vindictiveness. He was probably the first American radical to be influenced by a comedian – in this case, Lenny Bruce. No other young radical was so witty, in a recognizably Jewish manner. No other exuded so much charm, which is why – before it's too late – Dustin Hoffman should be encouraged to play him. For 'dear Abby' realized that he came into this world "acutely aware of being Jewish and [I] am sure I'll go out that way".[38] Note the contrast with another political memoir published a year before *Soon To Be a Major Motion Picture*. Even though written by the editor of a magazine sponsored by the American Jewish Committee, even though its author had attended the Jewish Theological Seminary, Norman Podhoretz's *Breaking Ranks* makes little effort to relate the transformation from the radical sympathies to neo-conservatism to the patrimony of Judaism or Yiddishkeit. Little emphasis is placed on Podhoretz's affiliation with the Jewish people and its interests. Podhoretz could not find – or at least chose not to report – how his shift to the right could be buttressed by Jewish values.[39] There is nothing in principle inconsistent with neo-conservatism or conservatism in Judaism. But it is odd that no one else – including other former editors of *Commentary* like Irving Kristol and Nathan Glazer, nor the defense intellectuals like Herman Kahn or Harold Brown or Simon Ramo – *explicitly* associated himself with Judaic tradition, any more than did the 'non-Jewish Jews' whom Isaac Deutscher located on the left.

Something of the texture of the Jewish involvement in the radicalism of the 1960s can be seen in Roger Simon's *The Big Fix*, also a film, a detective story whose protagonist is himself a former political activist from Berkeley. A secondary character is based on Abbie Hoffman, thus already transfigured into legend, imagined as having a disguised identity in which 'Howard Eppis', chairman of the Free Amerika Party and the author of *Rip It Off*, makes it big in the advertising business. In the film, the protagonist was played by Richard Dreyfuss, whose thespian career had begun in a Hanukah play and who was later granted conscientious objector status because of his opposition to the Vietnam War. The protagonist has an aunt who is still a Stalinist while living in a Jewish old-age home. In the movie one of the books shown on Dreyfuss's shelf is a novel by Alan Lelchuk, *American Mischief*, itself a depiction of a radical rebellion among Jewish students, led by Brooklyn-born Lenny Pincus, at a university strikingly like Brandeis University (called Cardozo College in the novel). In the Lelchuk novel, Pincus shoots Norman Mailer to fulfill the logic of literary radicalism.[40] The lines could scarcely be more taut.

But it is much easier to tabulate the impact of the Jews upon radical movements and ideas in the United States than to account for such influence. Four theories have been devised to explain this historical propensity for the left, but none has the compass or ingenuity – the formulaic authority – to compel assent.

One theory stresses Judaic culture itself, with its yearning for *tikkun olam* ('repair of the world'), its commitment to social justice and its insistence on speaking prophetic truth to power. Lawrence Fuchs has identified other ingredients in the history of Judaism that are also pertinent. A love of learning encouraged the capacity to transcend the immediacy of the environment, the possibility of imagining alternatives to an often beleaguered and degraded present. Fuchs also reports a Jewish immunity to asceticism, as an acceptance of the pleasures of this world fostered the hope that redemption need not be postponed into the world to come.[41]

But Fuchs advanced this theory to account for the tenacity of liberalism – not radicalism – among American Jews, though it has been adapted to the latter purpose. Thus the appeal to the Jewish cultural and religious tradition cannot distinguish liberalism from radicalism, between the desire to effect reform and moderate change and a more dramatic revision of the existing order. A theory that is supposed to explain both Louis Brandeis and Emma Goldman may nevertheless explain too much. Arthur Goldberg and Mike Gold can scarcely be said to have had much in common; the progressive barrister and the incendiary communist *littérateur* scarcely inhabited the same political universe.

Other problems weaken this theory as well. The appeal for social justice is one strain in Jewish tradition, but it is certainly not the only one; and the long history of the Diaspora prior to Emancipation fails to disclose many antecedents for modern radicalism or even modern liberalism. The rabbis and sages who wrote the commentaries on the Bible and the commentaries on the commentaries, who codified the laws and explicated the customs, would not have received 100 percent ratings from the Americans for Democratic Action, much less been vulnerable to red-baiting. The terrors of much of pre-modern Jewish life, with its incessant threats of extortion, pariahdom, pogroms and expulsion, promoted a wary conservatism that cannot convincingly be invoked by Jews in quest of sanctions for their radicalism. Radicals in the post-Emancipation era have distanced themselves from both pious and impious homes. But it has become a commonplace that the most observant are rarely radical, and the most radical are rarely observant. The more radical the Jew, the less he or she is likely to know (or care) about normative Jewish life.[42]

Charles Liebman has also discovered another flaw in the theory. Fuchs

> may be correct in assuming that Jewish liberalism finds its explanation in Jewish religious values, but he may be wrong about the values. . . . In order to demonstrate that liberalism is accounted for by traditional religious values, it is not enough to show that one can find values in the religious culture which promote liberalism; one would have to show that values which lead to political conservatism are absent in Judaism. Otherwise [Liebman adds] all that could be said is that Jews are selective about their religious values, choosing or being influenced by only those that lead to liberalism. But in that case there is no theory, because the crucial question still remains: Why do Jews choose or why are Jews more influenced by the religious values which promote liberalism rather than those which promote conservatism?

Albert Einstein, himself a socialist, believed in a connection between socialism and Judaism, for both are based upon the sense of "solidarity of all human beings". But he could not claim a logically necessary connection between a radical ideology and a religious source. It is more likely, as Nathan Glazer argued, that the social ideals of contemporary Jewry owe more to nineteenth-century liberalism and socialism than to the distant Biblical prophets.[43]

If the Fuchs theory cannot entirely explain liberalism, it is even less likely to account for radicalism, which has attracted far fewer Jews – especially after the eclipse of the immigrant generation that had directly encountered Czarist autocracy and obscurantism. Even those who professed to find in holidays like Pesach and Chanukah the inspiration for subsequent struggles against tyranny, as Julius Rosenberg did in his letters to his wife in Sing Sing, failed to acknowledge the nature of the despotism that they themselves served. They identified with the executioners, not the victims of Stalinism, which only corroborates Liebman's challenge that what needs to be explained is how, say, leftist Jews have selectively applied their religious heritage.

But the skepticism that this thesis engenders does not require its complete abandonment. For it allows us to take ideas and values seriously; and if what has distinguished the Biblical Hebrews and their descendants from those around them has been their religion, then perhaps something in the value system of the Jews has made them susceptible to modern radicalism. As an independent variable in the equation of human experience, suffering alone is an insufficient explanation. Misery, for example, propelled neither American blacks nor Russian *muzhiks* to the varieties of Marxism. Among the Jews, as indicated by the privileged partisans of the New Left, suffering is not even a necessary – much less a sufficient – cause of attraction to radicalism. That cause should be sought elsewhere. Denial of the importance of ideas in the formation of the Jewish people would defy common sense. To be sure, in a universe which is reputed to be curved as well as pockmarked with black holes, an appeal to common sense is risky. Nevertheless some tropes, values, and passions must have been transmitted in a such a way that Jews even in the United States nourished more radicals than comparable groups. One does not have to be an intellectual historian to consider an analogy with Yogi Berra's description of baseball: "90 percent of this game is half-mental."

The observation that orthodoxy coexists with radicalism far more rarely than religious ignorance and indifference is also problematic. Much depends upon the implication that *Halacha* (Talmudic law) has frozen Judaism into a rather unchanging mold, that Reform Judaism, for example, places undue weight upon the evolutionary aspects of Judaism and excessive stress upon prophetic ideals of social justice. It may well be the case, as Alexander Orbach has pointed out, that only this feature of Judaism – the stress on *tikkun olam* – got transmitted to emancipated Jewry. In any event scholarship does not require the acceptance of only an orthodox definition of Judaism. Perhaps Gershom Scholem can be quoted out of context: "I do not . . . view the events of Jewish history from a fixed dogmatic standpoint . . . [nor do I] know exactly whether some phenomenon is 'Jewish' or not." Scholem rejected the "assumption that there is a well-defined and unvarying 'essence' of Judaism, especially not where the evaluation of historical events is concerned", and found it impossible to know *a priori* which sorts of beliefs were tolerable within Judaism.[44] The argument that messianic hopes have been injected into politics therefore cannot be summarily rejected.

The hospitality of Judaism to diverse beliefs can be underscored by a comparison with Roman Catholicism. Papal encyclicals against socialism by Pius IX and Leo XIII may well have retarded the growth of socialism in the late nineteenth century. The Church – what Lenny Bruce later called 'the only *the* Church' – was ideologically opposed to systematic political efforts to redistribute or abolish private property. So hostile was it (even in some cases to trade unionism itself), so frequent were the warnings uttered by parish priests all the way up the hierarchy, that ecclesiastical opposition helps account for the failure of socialism to sink deep roots in American soil. The movement was condemned as 'a heresy, a rampant heresy', and to be both a Catholic and a socialist was considered a contradiction. But even though some socialists were anti-clerical and anti-religious, even though Jewish radicals sometimes deliberately flouted traditional rituals and piety, American rabbis did not consistently oppose radicalism.[45] Unlike rabbis, Catholic priests and Protestant preachers were occasionally radicals. But rabbis did not categorically stigmatize socialism either, or make adherence to its creeds in any way incompatible with Judaism or with membership in *klal yisroel* (the people of Israel). Indeed a major strain of Zionism was based upon a socialist ideology, and emphasized a natural affinity between the destiny of the Jewish people and an egalitarianism and collectivism pivotal to socialist doctrine.

A second theory is anti-Semitism. Robert Michels explained in 1915 that "The legal emancipation of the Jews has not . . . been followed by their social and moral emancipation." He added that "everywhere in the Jewish race there continues to prevail an ancient and justified spirit of rebellion against the wrongs from which it suffers; and this sentiment, idealist in origin, animating the members of an impassioned race, becomes . . . transformed into a revolutionary impulse towards a grandly conceived improvement of the world."[46] Faced with the disabilities of anti-Semitism, the Jews sought to alter societies that denied their rights and disparaged their merits. And since the Jews were the pariahs of

Europe, radicalism and revolutions were intelligible responses to the frustration of full equality. In the United States, anti-Semitism has been far milder, especially by the time the New Left emerged in the 1960s. But the experience of discrimination and the fear of bigotry, abroad as well as in the United States, nevertheless motivated Jews to subscribe to and support radical causes, especially communism in the 1930s and 1940s. The advantage of this theory is that some Jews themselves explained their attraction to communism in the light of its militant opposition to Nazism. The HUAC testimony of Daniel Boorstin and Jerome Robbins, already cited, may be taken as representative.

Again the problem with this theory is its fudging of the distinction between liberalism and radicalism – quite apart from other reactions to anti-Semitism, such as Zionism. Nor is there any way of matching the extent of such deprivation to the popularity of radicalism among the Jews. The intensity or scope of persecution does not correlate with the extent of adherence to radicalism, and those who have been the most talented or the most ambitious or the most wealthy do not necessarily turn left when they are turned down or turned away. The Lower East Side, which was the burned-over district of American socialism, was probably far less aware of 'status deprivation' and the incongruities of social and civic position than were the liberal and even conservative 'uptown' Jews of the same era.

Moreover, if status deprivation were so decisive a motivating factor, one might have expected American Jewish radicals to hope that the Jews themselves, so stung by prejudice and discrimination, would become an important constituency for political change. This was rarely, if ever, the case. The 'Jewish' garment workers unions were not consolidated along ethnic but rather economic and class lines, and made a point of their alignment with others in the garment trades, like the Italians. No American equivalent of the Bund emerged, although Jewish labor leaders appreciated its work and accomplishments and were touched by its fate. But a Jewish separatist labor socialist movement did not emerge in the United States; battles were conducted instead from the left flank of the American Federation of Labor. Though there were Jewish sections of the Communist Party in the 1920s, it is not too hyperbolic to state that Jews became experts on the revolutionary potential of every oppressed group in the United States except their own. The Party simply could not speak (even in Yiddish) in a special way to Jewish group interests. Israel Amter's famous salutation, 'Workers and peasants of Brooklyn!', was unlikely to rouse the masses. If such radicals thought that status deprivation made their fellow Jews embittered and indignant, such knowledge was successfully repressed.

A third theory, again originally propounded to account for liberalism (by Werner Cohn), has been extended by W. D. Rubinstein to incorporate radicalism as well. These scholars argue that the portside tilt of modern Jewry is due to the historical circumstances of post-Emancipation Europe and beyond. The right was inhospitable and uncongenial because of its allegiance to tradition, hierarchy, and privilege, and (in Maurice Barrès' terms), to *la terre et les morts*. The right was nostalgic for an era in which Christianity was pre-eminent, which means that it was no accident that the first Western thinker to uncouple religion from politics was a Jew, Spinoza. Rubinstein notes that, at least until recently, a certain set of historical circumstances kept the Jews away from the right: its anti-Semitism, its association with elites that excluded them, its defense of privilege at a time when most Jews were impoverished and disadvantaged. The left was, by contrast, impelled to reduce the particularities of religion, tradition, and even class.[47] Here the connection with Jews was evocatively drawn in the mid-nineteenth century by a German Jewish socialist, J. L. Bernays, whose people had

> rescued men from the narrow idea of an exclusive fatherland, from patriotism. . . . The Jew is not only an atheist, but a cosmopolitan, and he has turned men into atheists and cosmopolitans; he has made man only a free citizen of the world. . . . The Jews took their revenge upon the hostile world in an entirely new manner . . . by liberating men from all religion, from all patriotic sentiment . . . from everything that reminded them of race, place of origin, dogma and faith.

This connection between universalist hope and revolutionary Jews was, for Bernays, a source of pride, reflecting the distance from rightist politics, which were defined in opposition to such atheism and cosmopolitanism.[48]

This set of historical circumstances no longer prevails, at least not as sharply as in the nineteenth century. Until 1945 the lethal enemies of the Jewish people were primarily on the right. Since then the political constellation has shifted considerably. The Holocaust taught many of its surviving remnant of the failure of European emancipation, the violation of the hopes once invested in the promise of civil equality. Israel was the most notable result of the disillusion with atheism and cosmopolitanism, for it reminded Jews of precisely what Bernays boasted that his people had liberated Europe from: race, place of origin, dogma and faith. Now Jewish interests and welfare, without which Jewish moral values cannot be perpetuated, are dependent upon a secure and thriving Israel, whose enemies in the world arena have usually emanated from the left, from the putatively revolutionary regimes of the Third World. Anti-Zionism in Europe is almost entirely a phenomenon of the left and especially of communism. The European center of anti-Semitism has been relocated to where it was, under Czarist auspices, in the nineteenth century: in Moscow. Jews have increasingly realized that a tyranny so vast that it encompasses eleven of the world's twenty-four time zones, so that a city like Vladivostock is closer to Honolulu than to Leningrad, is brazen enough to support the Arab claim that Jewish sovereignty encompasses too large a corner of the Middle East. Moreover, the conservative forces in the United States have been detoxified of much of their anti-Semitism, and barriers have been lowered to others besides Henry Kissinger. The Jewish proletariat has largely disappeared, thus eliminating the class basis of socialist ideology.

And yet, while the historical circumstances that might have produced the Jewish slant toward radicalism have largely evaporated, the Jews remain more on the left than on the right. Wherever radicalism in the United States can still be found (apart from black separatism), Jews are within its ranks. Otherwise they have largely contented themselves with supporting and voting for liberal candidates for public office. The most leftist major party nominee for the Presidency after the Second World War, Senator George McGovern, had a reputation for only lukewarm support of Israel. He still received two out of every three votes of American Jewry. Had the rest of the electorate voted in 1972 as the Jews did, McGovern would have been swept into office with the greatest landslide in American history. Admittedly there has been a noticeable shift in Jewish political behavior to the right. But it is easy to overestimate its velocity and to exaggerate its tempo. If, as Rubinstein and others argue, Jewish interest dictates support of the right, Jews themselves are remarkably slow and reluctant to perceive such an interest. Leftist traditions – and perhaps even the Judaic value system – still block the reorientation of Jewish politics. What caused immigrants to be socialists may not resemble what drew some of their grandchildren toward the New Left, but social and economic conditions have neither changed so drastically nor overcome inertia in leading most Jews away from their allegiance to liberalism and their tolerance for radicalism. If historical circumstances alone produced the Jewish propensity for the left, the disproportionate number of Jews coagulating in the New Left, could scarcely be explained.

There is no reason to assume that historical change since the eighteenth and nineteenth centuries somehow skipped the Jewish family, whose dynamics have given rise to the fourth theory for the origins and perpetuation of Jewish radicalism. As religious authority yielded to the lure of secular Western civilization, as some forms of traditional piety became more rigid and opportunities in business and the liberal professions widened, as ghetto walls tumbled down and the winds of modernity rustled through the *stetl*, the family was obliged to bear an increasing burden in reconciling the tensions and antinomies of Jewish life. The roles of fathers and mothers and children altered, sometimes subtly, sometimes searingly; and even when the tensions ran silent, they ran deep.

As the Jewish family struggled to gain and restore its cohesiveness, it sanctioned political heterodoxy. "To other people, I'm a professional radical," Saul Alinsky once observed. But to his mother, "the important thing is, I'm a professional".[49] Literary evidence can also be cited. Sholom Aleichem's Tevye realizes the destruction of tradition and faces – both figuratively and literally – a new world. Nevertheless he cannot abide the romantic involvement of one daughter, Chava, with a Gentile, and mourns her inclination to accept the consequences of putting private happiness ahead of peoplehood. But another daughter's marriage to a revolutionary is not a disgrace to Tevye, however baffled he may be; for Hodel and Pertschik remain part of the Jewish people. Similar indulgence is granted in America as well. In Jules Feiffer's most recent play, *Grownups*, a New York couple bubble with pride that their son is a journalist for *the New York Times*. That is their emblem of making it, the certification of success. A jarring note is injected when the son informs them that he is writing a book blasting the political system as corrupt and immoral, but the parents don't skip a beat in exuding *naches* (joy). It does not seem to matter whether the system is rotten, so long as their son wins recognition for exposing it.

Nathan Glazer has argued that Jews were disproportionately attracted to the New Left simply because they grew up in families which already nurtured earlier dissidents: "Jewish political traditions, traditions of liberalism and socialism, open young Jews to the *possibility* of influence by liberal and socialist views. Young Jews are more likely to be aware of a variety of left-wing political creeds." Glazer suggests that the presence in so many Jewish families of older radicals has given the young a sense of political alternatives. In such families upward mobility was authorized and encouraged in various ways, but Marxists were still hanging around – undoubtedly more adhesively than in Gentile families. No character is more greedy and ambitious than Mordecai Richler's Duddy Kravitz, but the Uncle Benjy who runs a dress factory is shown – in both the novel and the film – as wealthy but still a Stalinist.[50] Compare, on this point, two novels of upward mobility, in which young men who start impoverished become wealthy, only to taste the ashes of emptiness and disenchantment. *The Rise of David Levinsky* (1917) allows its protagonist to realize that he was not necessarily destined to become a successful clothing manufacturer. Had he chanced to hear a speech by Karl Marx, Levinsky speculates, he might well have become a socialist (as was his creator, Abraham Cahan). Yet the eponymous protagonist of *The Great Gatsby* (1925) never imagines a similar alternative. Or take the two most famous Beatnik writers of the 1950s. Jack Kerouac's heritage was French Canadian, which may have helped make him surprisingly old-fashioned, from his vices (alcohol) to his politics. Kerouac was, if anything, a Republican who like Ike, preferred Nixon in 1960, and reacted to the televised McCarthy hearings by announcing that the Wisconsin senator had "the real dope on the Jews and fairies". Allen Ginsberg's father was a poet who had published in the bohemian socialist *Masses*. His mother was a communist whom Ginsberg memorialized in perhaps his greatest poem, *Kaddish*, attributing to her the haunting advice: "The key is in the window, the key is in the sunlight at the window. I have the key. Get married Allen don't take drugs." (The advice was ignored, but Ginsberg nevertheless became so rich from poems denouncing Moloch that he has incorporated himself for tax purposes.)[51]

The phenomenon of the 'red diaper baby' appears to be almost exclusively Jewish; and, while the case is undoubtedly exceptional, the sons of Julius and Ethel Rosenberg, Michael and Robert Meeropol, were themselves involved in SDS. For every Mark Rudd and Abbie Hoffman, whose fathers were fairly conservative if not apolitical, there was a Bettina Aptheker, a Berkeley activist whose father is a communist historian and publicist; or a Kathy Boudin, whose father has been prominently engaged in the legal defense of communists and fellow-travellers. And even Rudd's parents made a point of supporting him during the Columbia insurrection, bringing him food and referring to him as their son, the revolutionary.[52] Karl Marx's mother is supposed to have wistfully remarked how much better it would have been if her son had only *made* some capital instead of just writing about it. But she did not disown him.[53] Nor did the father of David Dubinsky, who grew up in Lodz, where the Jewish working class toiled from twelve to fifteen hours a day. At thirteen young Dubinsky quit school; at fourteen he was a master baker in his father's shop. At fifteen, as secretary of the baker's union, he struck, helping to shut down his own father's shop. But when the police came to jail him, Dubinsky's father gave him money and a *tallis* (prayer shawl).[54]

"Marx said it: Abolish such families!" cries the grandfather in exasperation in Clifford Odets' *Awake and Sing* (1935), as Jacob's relatives inflict wounds on one another trying to survive the Depression. But exactly how such families spawned radicals is a matter of dispute, and the theories are incompatible. For Glazer, Kenneth Keniston, Richard Flacks, and Seymour Martin Lipset, radical offspring wished to extend and make operational what liberal parents did not realize fully. The children thus discovered a fresh way to obey the Fifth Commandment. These analysts of the New Left stress the continuity in the values of the two generations, with the younger generation attempting to live up to the pieties of social justice with which they had been inculcated. The positive support of such parents gave the young confidence in the rightness of their values and in the propriety of their engagement in the civil rights movement, in anti-poverty campaigns, in battles for student power, and in their resistance to American military intervention in Vietnam. Impressionistic evidence is given along this line in Midge Decter's *Liberal Parents, Radical Children*. Social science supports this view in the monographs of Keniston and Flacks, admittedly based on small samples but clearly dependent upon psychological investigation of Jewish as well as non-Jewish New Leftists, plus the new and far more comprehensive *Roots of Radicalism*.[55]

An alternative – and indeed contradictory – reading of the relation of parents to radical children is most forcefully advanced in Lewis Feuer's *The Conflict of Generations*. His relevant evidence is drawn not only from Jewish student radicals in the United States but from Russian Jewish students in the late nineteenth century as well. Feuer emphasizes not only the self-destructiveness and nihilism of student movements but – in the special case of the Jews – a disgust with the fathers' 'passive recipience of persecution', leading them to identify with the sufferings and struggles of other groups, notably blacks. He explains the Jewish predilection for student radicalism in the 1920s and 1930s by referring to the nature of American culture itself and by the eagerness of the second generation to distance themselves from the ways of their immigrant parents. The cycle of irrational rejection was to be perpetuated in the 1960s.[56]

The evidence for the 'de-authorization of the fathers' is partly autobiographical (accounts by Morris Raphael Cohen and John Gates). But *The Conflict of Generations* leaves unclear why it is aberrant to repudiate the culture of, say, immigrant parents (even when – as in the case of Cohen – such rejection was neither violent or suicidal), and why it would have been rational to duplicate the values and lifestyles of the elders. Nor does Feuer acknowledge the possibility that many young Jews identified with and supported the civil rights movement because racial segregation and discrimination constituted a far more shameful violation of the promise of American democracy than any other social problem, including anti-Semitism. Certainly some signs of Oedipal strains should be mentioned which buttress Feuer's case, as in the wilder rantings of Jerry Rubin, as in Bob Dylan's advice to parents in changing times not to criticize what they could not understand, as in the admission of a former 1960s student in Lawrence Kasdan's retrospective film, *The Big Chill* (1983): "Half the stuff I did, I did to piss off my parents. And it worked." In Feiffer's satire *Grownups*, the Jewish parents are so obtrusive and inescapable that when the adult daughter whimsically proposes killing them, her brother demurs: "That would be only a short-range solution." Here too Hoffman's autobiography is revealing about its filial conflicts.[57]

Nevertheless the Feuer thesis is in general unconvincing. Even if he is right about the self-destructive irrationality of student movements in the 1960s, no links are firmly forged to Oedipal relations to parents. Social scientists and others who have investigated young radicals' attitudes toward their parents disclose neither such hostility, nor disgust with the passivity of their parents. Far from subsuming their own ethnicity in a romantic identification with the uprisings of oppressed peoples and minorities, many Jewish members and sympathizers quit the New Left after 1967, when it became hostile to Israel. Moreover Feuer's theory cannot explain the phenomenon of young radicals who insisted upon the compatibility of their politics with Jewishness and Judaism, a synthesis which flourished largely after publication of *The Conflict of Generations*.[58]

The traditional strength and cohesiveness of the Jewish family have undoubtedly been decisive in shaping the identities of the rebels who emerged from homes that – unlike their grandparents' English – were rarely broken. But bonds of continuity and affirmation appear to have been more powerful than the friction and disaffection which could also be detected. Why some families produced radicals, and most others liberals, nevertheless remains an enigma.

There is another theory that has been given almost no attention. It is advanced here only tentatively, because there are reportedly many places in purgatory reserved for those scholars who allow generalizations to outstrip their evidence. The explanation is given a fleeting glance in Fuchs' study (which is about liberalism, not radicalism) and in Glazer's reflections on student radicals (which is confined to America in the 1960s). In accounting for the propensity for radical ideologies and movements, scholars have given little, if any, weight to what the sociologist Talcott Parsons considered the most distinctive characteristic of the Jewish people: its intellectuality.[59] Many examples of the Jewish belief that happiness is the predicate of the contemplative life could be cited, but consider Maimonides' discussion of the world to come in his *Mishneh Torah*:

> Sages and prophets longed for the messianic age not in order that they should dominate the world and rule over the gentiles . . . but solely in order to be free to devote themselves to the Torah and divine wisdom without oppression and hindrance. . . . Therefore [the children of] Israel shall all be great scholars; they shall know hidden things and attain to the knowledge of God as far as it is within human reach.

In the twentieth century, in an age much denuded of such sacred dedication among scholars, is it too vast a distance from Maimonides to Trotsky's prediction that, under socialism, average citizens would scale the heights of Aristotle, Goethe, and Marx?[60]

Even the word 'intellectual' itself arose thanks to Jews; for it was coined during *l'affaire* in France to categorize the Dreyfusards, some of whom were the captain's co-religionists. If Jews have been disproportionately radicals, it may be because they have been disproportionately intellectuals. Randolph Bourne and Thorstein Veblen were among the first Americans to recognize the spectacular impact that Jewish intellectuals were making upon Western culture, but the remarks of Nikos Kazantzakis are even more to the point. "Ours is an age of revolution," the Greek writer said of the inter-war period: "That is, a Jewish age." Modern life had become fragmented and decomposed, and "the Jews have this supreme quality: to be restless, not to fit into the realities of the time; to struggle to escape; to consider every status quo and every idea a stifling prison. This spirit of the Jews shatters the equilibrium . . ."[61] They were often indifferent to or highly critical of the pieties of national honor, of blood and soil. Morris Hillquit, for example, wrote the Socialists' St Louis Declaration of 1917 that branded the Great War a crime. He was consistent in his opposition to militarism and "the false doctrine of national patriotism" even after Czarism had been overthrown and the world more plausibly made safe for democracy. Thus an editorialist for *The New York Times* could take a swipe at Hillquit, whose "singular genius" was to try "at once to betray the land of his birth and the land of his adoption". No wonder then that T. S. Eliot, at the same time as Kazantzakis, imagined a Christian society that would, for "reasons of race and religion," exclude large numbers of "free thinking Jews". From Eliot's perspective, such a minority would endanger the social stability and ideological coherence of his ideal order; and he warned against "a spirit of excessive tolerance".[62]

Revolutionary politics was a natural outlet, for Béla Kun and Georg Lukács in Hungary, for Gustav Landauer and Ernst Toller in Bavaria, for Rosa Luxemburg in Berlin, for Trotsky and Karl Radek and Julius Martov and Lev Kamenev and Grigori Zinoviev in Russia, for Michael Borodin in China, and innumerable others. In the United States, intellectuals were less driven to such extremism, but their presence gave the Jewish labor movement much of its distinctiveness. The early leaders of the ILGWU, for example, were almost invariably former students who, according to a labor historian, "found themselves in the garment industry as a matter of necessity, not of choice". But their experience in Russian revolutionary movements gave Benjamin Schlesinger, Julius Woolf, Saul Metz, Abraham Boroff and others a certain stature as well as militancy; and therefore "the Jewish workers turned instinctively to the intellectuals for help and were grateful for the leadership which they were able to provide". More than any other immigrant group, the Jews harbored intellectuals among their tired, huddled masses; and they fostered a radical spirit and outlook. Noam Chomsky, for instance, has recorded his own indebtedness to the "radical Jewish working class milieu" to which his family belonged: "it was a very unusual culture which I don't think exists any more, a mixture of a very high level of intense intellectual life, but at the same time it was really working class."[63]

That is why the Jews differed from other members of the proletariat, and remained sympathetic to radicalism even when nestled in the middle class. For in America in particular, radical movements have failed to appeal to their putative constituencies, have flunked the assignment of dialectical materialism. Nearly all the oppressed and impoverished have ignored the agitation of radicals, have listened only in moments of crisis, and have lapsed back into moderation or political indifference when episodic pressures were released. Because radicalism is an *ism* – that is, a modern ideology – it is likely to be sustained primarily by those groups pervaded by the presence of intellectuals. Consider a New York *Post* headline in 1950: "Einstein Red Faker, Should Be Deported, Rankin Screams."[64] In insinuating that independence of mind and political dissidence were the signature of Jews in particular, even a venomous crank like the Mississippi congressman was groping for a partial truth. And in a way Attorney General William Saxbe was right after all when he attributed the earlier strength of American Communism to the power of Jewish intellectuals yearning after strange gods that eventually failed.

Not only defense agencies should add, however, that most Jews have not been radicals. Even most Jewish intellectuals have not been radical. Such persons – and not serfs, or slaves, or peasants, or black sharecroppers – have nevertheless been overrepresented in revolutionary movements, almost as predictably as queens beat jacks in poker. In describing the homes of Jewish New Leftists, Glazer has singled out the importance of books, lectures and 'cultural style' in promoting a greater responsiveness to the social environment. He isolates certain measurable indices, such as college attendance and choices of fields of concentration, but did not speculate upon intellectuality as in itself a susceptibility to radical impulses. It is curious that Glazer himself should dwell on 'cultural style', because his own youthful radicalism was barely shaped by reading as such. His own family – which had to go on welfare in Harlem during the Great Depression – was so unfamiliar with his own vocation as a writer and an editor that his mother, once asked to describe his occupation, vaguely asserted that he was "in the pen business".[65] A theory that emphasizes such vocations and such inclinations does not require the ascription of intellectuality to the Judaic faith itself, as the source of a certain tropism toward radicalism. That is another advantage of the theory.

One final note should be sounded, in a time of uncertainty about a vital Jewish future and of quiescence on the left. Radical movements have attracted the best as well as the worst of humanity; but however equivocal the record of socialism in the annals of human freedom, the pain of the dispossessed ought to continue to disturb even those whose well-being is partly derived from the struggle, the travail, the martyrdom of earlier radicals. That heritage is, at its most sublime, difficult to sustain; and the Jewish intellectuals who once embraced such movements sometimes found a more enduring set of values and sancta in their own tradition. They may have appeared as deracinated as the earlier Jewish intellectuals of Alexandria in the days of Philo. But even then, at the dawn of the Diaspora, formal creeds proved less important than might have been suspected. The Hellenistic Jewish intellectuals found it easier, as Harry Wolfson observed, to surrender the God of their fathers than the cooking of their mothers, and hence they did not stray too far from Jewish neighborhoods, from the nexus of Jewish life itself.[66] Two millennia later, under conditions at least as hospitable for Jewish comfort and as risky for Judaic culture, the question remains whether this tiny and ancient people can still distinguish itself by the disproportionate numbers of sons and daughters it can produce who are famished for justice.

Among those who gave themselves to radical causes, many paid the price of passion, embroiling themselves more than others in folly and tragedy. Like all the other members of our species, however, they were subjugated to the ordinances of time and place and mortality, even as they dared to press against the limits that only social arrangements seemed to impose. They also dramatized the process of disenchantment, of putting into secular form longings for redemption which were once expressed in a religious idiom. It may still be appropriate to relate their ambiguous pursuit of brotherhood to the hope of Natalia Victorovna Haldin, the character in Joseph Conrad's *Under Western Eyes* who "believe[d] that the future will be merciful to us all. Revolutionist and reactionary, victim and executioner, betrayer and betrayed, they shall all be pitied when the light breaks . . ."[67]

Brandeis University

NOTES

* I appreciate the criticism of an early draft of this essay by Alexander Orbach and Robert S. Wistrich.

[1] Henry Kissinger, *Years of Upheaval* (Boston: Little, Brown, 1982), pp. 661, 975.

[2] Peter Grose, *Israel in the Mind of America* (New York: Knopf, 1983), pp. 72, 82, 118.

[3] *New York Times* (April 4, 1974), p. 7; (April 5, 1974), p. 18; (April 6, 1974), p. 13; (April 7, 1974), IV, p. 3; (April 10, 1974), p. 20.

[4] Henry L. Feingold, 'The Jewish Radical in His American Habitat', *Judaism* **22** (Winter 1973), 92–105; Peter Dreier and Jack Nusan Porter, 'Jewish Radicalism in Transition', *Society* **12** (Jan.–Feb. 1975), 34–43; Louis Ruchames, 'Jewish Radicalism in the United States' in Peter I. Rose (ed.), *The Ghetto and Beyond: Essays on Jewish Life in America* (New York: Random House, 1969), pp. 228–252; Stephen J. Whitfield, 'The Radical Persuasion in American Jewish History', *Judaism* **32** (Spring 1983), 136–152.

[5] Lewis Feuer, *Spinoza and the Rise of Liberalism* (Boston: Beacon, 1958), p. 48.

[6] Michael Walzer, *The Revolution of the Saints: A Study in the Origins of Radical Politics* (New York: Atheneum, 1968), p. 47.

[7] Steven E. Aschheim, *Brothers and Strangers: The East European Jew in German and German Jewish Consciousness, 1800–1923* (Madison: University of Wisconsin Press, 1982), p. 33.

[8] Daniel Bell, *Marxian Socialism in the United States* (Princeton: Princeton University Press, 1967), p. 98; John Higham, *Send These to Me: Jews and Other Immigrants in Urban America* (New York: Atheneum, 1975), p. 89.

[9] John Laslett, *Labor and the Left: A Study of Socialist and Radical Influences in the American Labor Movement, 1881–1924* (New York: Basic Books, 1970), pp. 99–100, 117, 119, 120–123, 129, 133–135.

[10] Richard Drinnon, *Rebel in Paradise: A Biography of Emma Goldman* (Boston: Beacon, 1961), pp. 71, 263–264, 274–275, 285–286; Emma Goldman, *Living My Life* (New York: Dover, 1970), I, 184, 314, 370, and II, 552, 824, 841; S. N. Behrman, *The Worcester Account* (New York: Random House, 1954), p. 181.

[11] Drinnon, *Rebel in Paradise*, pp. 140–142, 213, 215, 223; Peggy Lamson, *Roger Baldwin: Founder of the American Civil Liberties Union* (Boston: Houghton Mifflin, 1976), pp. 58–62; *New York Times* (April 17, 1973), p. 38; Alix Kates Shulman, Introduction to *Red Emma Speaks: Selected Writings and Speeches by Emma Goldman* (New York: Vintage, 1972), pp. 4–5.

[12] Zechariah Chafee, *Free Speech in the United States* (Cambridge, Mass.: Harvard University Press, 1941), pp. 108–113, 125, 140, 247–252, 269–282, 318; Paul L. Murphy, *The Meaning of Freedom of Speech: First Amendment Freedoms from Wilson to FDR* (Westport, Conn.: Greenwood, 1972), pp. 79–80, 87–90.

[13] Chafee, *Free Speech in the United States*, pp. 362–366.

[14] Martin David Dubin, 'Rosika Schwimmer' in *Notable American Women: A Biographical Dictionary, 1607–1950* (Cambridge, Mass.: Belknap, 1971), III, pp. 246–249; Chafee, *Free Speech in the United States*, pp. 369–374; Holmes, J., in *U.S.* v. *Schwimmer* (1928), 279 U.S. 644, 653.

[15] Bell, *Marxian Socialism*, p. 154; Theodore Draper, *American Communism and Soviet*

Russia: The Formative Period (New York: Viking, 1960), pp. 224–226, 433–434; *New York Times* (April 28, 1977), II, p. 8.

[16] Oscar Handlin, *A Continuing Task: The American Jewish Joint Distribution Committee, 1914–1964* (New York: Random House, 1964), pp. 44–46, 56–59; Henry L. Feingold, *Zion in America* (New York: Twayne, 1974), pp. 279; Draper, *American Communism and Soviet Russia*, p. 191; Irving Howe, *World of Our Fathers* (New York: Simon and Schuster, 1976), pp. 330–341.

[17] Daniel Aaron, *Writers on the Left* (New York: Avon, 1965), pp. 213–215, 343, 376, 437, 464; Earl Latham, *The Communist Controversy in Washington: From the New Deal to McCarthy* (New York: Atheneum, 1969), pp. 107.

[18] Eric F. Bentley (ed.), *Thirty Years of Treason* (New York: Viking, 1971), pp. 602, 604, 605, 608.

[19] Ernest Renan, *Oeuvres Complètes* (Paris: Calmann-Lévy, 1947), IV, p. 471.

[20] Bonnie Lyons, 'Lillian Hellman: "The First Jewish Nun on Prytania Street" ' in Sarah Blacher Cohen (ed.), *From Hester Street to Hollywood: The Jewish-American Stage and Screen* (Bloomington: Indiana University Press, 1983), pp. 106–122; Lillian Hellman, *An Unfinished Woman: A Memoir* (New York: Bantam, 1970), pp. 132–133; Lillian Hellman, Foreword to Lev Kopelev, *To Be Preserved Forever* (Philadelphia: J. B. Lippincott, 1977), unpaginated.

[21] Kwame Nkrumah, *Autobiography* (London: Panaf, 1973), pp. 36–37; Peter Goldman, *The Death and Life of Malcolm X* (New York: Harper and Row, 1974), pp. 175n–176n, 254–255.

[22] Robert S. Wistrich, *Revolutionary Jews from Marx to Trotsky* (London: Harrap, 1976), p. 187.

[23] Leonard B. Rosenberg, 'The "Failure" of the Socialist Party of America', *Review of Politics* **31** (July 1969), 329–352; Aileen S. Kraditor, *The Radical Persuasion, 1890–1917: Aspects of the Intellectual History and the Historiography of Three American Radical Organizations* (Baton Rouge: Louisiana State University Press, 1981), pp. 217–218.

[24] Samuel Gompers, *Seventy Years of Life and Labor*, rev. and ed. by Philip Taft and John A. Sessions (New York: E. P. Dutton, 1957), pp. 74–81, 213–215; Morton Keller, *Affairs of State* (Cambridge, Mass.: Harvard University Press, 1977), p. 397.

[25] Higham, *Send These to Me*, p. 98.

[26] *New York Times* (June 2, 1969), p. 45; Joseph Freeman, *An American Testament: A Narrative of Rebels and Romantics* (New York: Holt, Rinehart, 1936), pp. 58–60.

[27] Bentley, ed., *Thirty Years of Treason*, pp. 626–627, 630–631; *New York Times* (Feb. 10, 1980), II, pp. 1, 4; Martin Gottfried, *Broadway Musicals* (New York: Harry N. Abrams, 1979), p. 103.

[28] Copland quoted in Murray Polner (ed.), *American Jewish Biographies* (New York: Lakeville Press, 1982), p. 73.

[29] Steven Heller, *Man Bites Man: Two Decades of Satiric Art* (London: Hutchinson, 1981), p. 105.

[30] Victor S. Navasky, *Kennedy Justice* (New York: Atheneum, 1971), pp. 141–150; Eric Breindel, 'King's Communist Associates', *New Republic* **190** (Jan. 30, 1984), p. 14.

[31] Jacques E. Levy, *Cesar Chavez: Autobiography of La Causa* (New York: W. W. Norton, 1975), pp. 100, 117, 125–126, 146–147; Joan Baez, *Daybreak* (New York: Dial, 1968), pp. 56–67; Jeremy Larner, 'The Negro in the South: "Half a Loaf Is Ashes" ', *New Leader* **43** (Sept. 12, 1960), 11.

[32] Jack Newfield, *A Prophetic Minority* (New York: Signet, 1967), pp. 78, 79.
[33] Arthur M. Schlesinger, Jr, *A Thousand Days: John F. Kennedy in the White House* (London: Andre Deutsch, 1965), pp. 664–669.
[34] Mark I. Pinsky, 'Red, White and Jew', *Moment* **5** (Nov. 1980), p. 19.
[35] Stanley Rothman and S. Robert Lichter, *Roots of Radicalism: Jews, Christians, and the New Left* (New York: Oxford University Press, 1982), pp. 81–82; Nora Sayre, *Sixties Going on Seventies* (New York: Arbor House, 1973), pp. 78–92.
[36] 'A Yippie Comes in from the Damp', *Time* **116** (Sept. 15, 1980), 22.
[37] Abbie Hoffman, *Soon to Be a Major Motion Picture* (New York: Perigree, 1980), pp. 111–117, *passim*; John Murray Cuddihy, *The Ordeal of Civility: Freud, Marx, Lévi-Strauss and the Jewish Struggle with Modernity* (New York: Basic Books, 1974), pp. 190–197; Kenneth Dauber, 'The Problem of Poe', *Georgia Review* **32** (Fall 1978), 653.
[38] Hoffman, *Soon to Be a Major Motion Picture*, pp. 3, 9, 13–16.
[39] Norman Podhoretz, *Breaking Ranks: A Political Memoir* (New York: Harper and Row, 1979), pp. 302, 329–336, 344.
[40] Roger L. Simon, *The Big Fix* (New York: Pocket Books, 1974), pg. 19; 'From Jews to Jaws: A Chat with Richard Dreyfuss', *Davka* **5** (Fall 1975), 34–35; Alan Lelchuk, *American Mischief* (New York: Farrar, Straus and Giroux, 1973), pp. 3, 280–281, 286–293.
[41] Lawrence H. Fuchs, *The Political Behavior of American Jews* (Glencoe, I11.: Free Press, 1956), pp. 171–203.
[42] Charles Liebman, *The Ambivalent American Jew* (Philadelphia: Jewish Publication Society of America, 1973), pp. 143–144; Arthur Liebnam, *Jews and the Left* (New York: John Wiley, 1979), pp. 7–11.
[43] Liebman, *The Ambivalent American Jew*, p. 140; Albert Einstein, *Ideas and Opinions* (New York: Crown, 1954), pp. 185–187; Nathan Glazer, *American Judaism* (Chicago: University of Chicago Press, 1971), p. 141.
[44] Gershom Scholem, *Sabbatai Sevi: The Mystical Messiah, 1616–1676* (Princeton: Princeton University Press, 1973), pg. xi; David Biale, *Gerschom Scholem: Kabbalah and Counter-History* (Cambridge: Harvard University Press, 1979), pp. 111–113.
[45] Marc Karson, 'Catholic Anti-Socialism', in John H. M. Laslett and Seymour Martin Lipset (eds), *Failure of a Dream?: Essay in the History of American Socialism* (Garden City, N.Y.: Doubleday, 1974), pp. 164–166, 179–180; Laslett, *Labor and the Left*, p. 136, n.5.
[46] Robert Michels, *Political Parties* (New York: Free Press, 1962), pp. 247–248.
[47] Werner Cohn, 'The Politics of American Jews', in Marshall Sklare (ed.), *The Jews: Social Patterns of an American Group* (New York: Free Press, 1958), pp. 614–626; W. D. Rubinstein, *The Left, the Right and the Jews* (New York: Universe, 1982), pp. 9, 227, 230; Nathan Glazer, *Remembering the Answers: Essays on the American Student Revolt* (New York: Basic Books, 1970), p. 243.
[48] J. L. Talmon, *Israel Among the Nations* (New York: Macmillan, 1971), pp. 24–25.
[49] 'Playboy Interview: Saul Alinsky', *Playboy*, **19** (March 1972), 64.
[50] Glazer, *Remembering the Answers*, pp. 232, 236; Mordecai Richler, *The Apprenticeship of Duddy Kravitz* (New York: Ballantine, 1974), pp. 49–50.
[51] Dennis McNally, *Desolate Angel: Jack Kerouac, the Beat Generation and America* (New York: Random House, 1979), pp. 156, 185–186, 221, 280, 293–294; Allen Ginsberg, *Kaddish and Other Poems, 1958–1960* (San Francisco: City Lights, 1961), p. 31; Ted

Morgan, *On Becoming American* (Boston: Houghton Mifflin, 1978), p. 224.
[52] Robert and Michael Meeropol, *We Are Your Sons: The Legacy of Ethel and Julius* Rosenberg (Boston: Houghton Mifflin, 1975), pp. 276–281, 285–293, 296–335; *New York Times* (May 19, 1968), p. 1.
[53] Lewis S. Feuer, *Marx and the Intellectuals: A Set of Post-Ideological Essays* (Garden City, N. Y.: Doubleday Anchor, 1969), p. 34.
[54] Howe, *World of Our Fathers*, p. 352.
[55] Clifford Odets, *Six Plays* (New York: Modern Library, 1939), p. 55; Kenneth Keniston, *Youth and Dissent: The Rise of a New Opposition* (New York: Harcourt, Brace, Jovanovich, 1971), pp. 273–277.
[56] Lewis S. Feuer, *The Conflict of Generations: The Character and Significance of Student Movements* (New York: Basic Books, 1969), pp. 154–161, 423–431.
[57] Hoffman, *Major Motion Picture*, pp. 3–4, 15.
[58] See Jack Nusan Porter and Peter Dreier (eds), *Jewish Radicalism: A Selected Anthology* (New York: Grove Press, 1973); Alan Mintz and James A. Sleeper (eds), *The New Jews* (New York: Vintage, 1971); Arthur Waskow, *The Bush Is Burning!: Radical Judaism Faces the Pharoahs of the Modern Superstate* (New York: Macmillan, 1971); and *Response* Magazine (1967–).
[59] Fuchs, *Political Behavior*, pp. 178–180, 185, 189; Glazer, *Remembering the Answers*, pp. 236–238; Talcott Parsons, 'The Sociology of Anti-Semitism', in Isacque Graeber and S. H. Britt (eds), *Jews in a Gentile World: The Problem of Anti-Semitism* (New York: Macmillan, 1942), p. 106.
[60] Scholem, *Sabbatai Sevi*, pp. 12–13 Leon Trotsky, *Literature and Revolution* (Ann Arbor: University of Michigan Press, 1960), p. 256.
[61] Nikos Kazantzakis, *Journeying: Travels in Italy, Egypt, Sinai, Jerusalem and Cyprus* (Boston: Little, Brown, 1975), pp. 179–180.
[62] Howe, *World of Our Fathers*, p. 319; T. S. Eliot, *After Strange Gods: A Primer of Modern Heresy* (New York: Harcourt, Brace, 1934), p. 20.
[63] Laslett, *Labor and the Left*, pp. 99–100; Moses Rischin, *The Promised City: New York's Jews, 1870–1914* (New York: Harper and Row, 1970), pp. 175–176; Chomsky quoted in Polner, (ed.), *American Jewish Biographies*, p. 64.
[64] Edwin R. Bayley, *Joe McCarthy and the Press* (New York: Pantheon, 1982), p. 3.
[65] Glazer, *Remembering the Answers*, pp. 236–238; Leonard Dinnerstein and Gene Koppel, *Nathan Glazer: A Different Kind of Liberal* (Tucson: University of Arizona, 1973), pp. 53–54.
[66] Harry A. Wolfson, *Philo: Foundations of Religious Philosophy in Judaism, Christianity, and Islam* (Cambridge, Mass.: Harvard University Press, 1947), I, p. 83.
[67] Joseph Conrad, *Under Western Eyes* (London: Penguin, 1957), p. 291.

DENNIS H. WRONG

THE CONCEPT OF ALIENATION REVISITED

In reconsidering the concept of alienation, my aim is not to revive it nor to suggest new meanings and uses for it, but rather to throw a few more clods of earth on its coffin. The concept has tended to pass out of contemporary discourse largely as a result of its excessive use in the 1950s and 1960s when it became popular and soon acquired a vague and shapeless character, serving as a sort of verbal talisman connoting virtually any state of psychological discomfort or malaise. Yet the very popularity of the term depended on its suggesting something more than mere personal discontent, which is, after all, a pretty commonplace condition. 'Alienation' was meant to convey either a sense of metaphysical melancholy about the human condition or a socio-political diagnosis of, and protest against, the failures and limitations of the social order.

The latter emphasis was paramount in the 1960s when the term was fashionable among social scientists, political intellectuals and radical youth. To assert one's 'alienation' became a badge of honor, a credential attesting to superior moral sensitivity – in short, something to boast of rather than to complain about. This has never been true of such strictly subjective or psychological states as 'unhappiness', not even as expressed in the more sophisticated synonyms favored by the educated, like 'frustration', 'depression', 'anxiety', or 'neurosis'. When alienation was imputed to, or reportedly observed among, ordinary citizens, as allegedly exemplifying their attitude to politics or capitalism or American life in general, the implication was usually clear that these collective entities rather than the alienated individuals themselves, were to blame for the situation. Thus, even in its mosts clichéd uses the term was never quite reduced to individual psychology but retained a latent sociological and even political meaning.

I have no wish to survey yet again the uses and abuses of the concept, nor to explore its origins in Christian theology or its venerable practical meanings, shorn of wider philosophical or sociological significance, in jurisprudence and early psychiatry. Nor do I want to review once more the concept's intellectual history in Hegelian philosophy and Marxism:

S. Hook, W.L. O'Neill and R. O'Toole (eds.), Philosophy, History and Social Action, 461–473.

the complex conceptual links among alienation, objectification, estrangement and reification – *Entäusserung, Entfremdung, Verdinglichung, et al.* – in the German Romantic and Idealist traditions. Neither am I concerned to offer another account of the specifically sociological adaptations of the concept, its differences from and frequent erroneous equation with anomie, let alone various efforts to convert it into an empirically testable or even 'measurable' phenomenon. All of these things have been done before quite adequately by competent and intelligent scholars.[1]

Anyone, of course, remains free to define alienation as he or she pleases. But there is one conception that is most faithful to the German philosophical tradition that gave birth to the concept, faithful, despite variations, to its history from Hegel through Feuerbach to Marx, more indirectly in the writings of Nietzsche and the turn-of-the-century German sociologists, Tönnies, Simmel, Weber and Sombart, and in the later works of Lukács, the Frankfurt School and Sartre. As Peter Berger (1967, p. 85) defines it:

> Alienation is the process whereby the dialectical relationship between the individual and his world is lost to consciousness. The individual 'forgets' that this world was and continues to be co-produced by him . . . The essential difference between the socio-cultural world and the world of nature is obscured – namely, the difference that men have made the first, but not the second.

In another book, co-authored with Thomas Luckmann, Berger offered a definition of 'reification' that is almost identical, word for word, with the definition of alienation that I have quoted (Berger and Luckmann, 1966). Lukács, of course, is the source of the conception of reification as a dominant form of social consciousness, although he did not, as is often ignorantly alleged, invent the term. It had long meant – and is still so defined in English-language dictionaries – the mental habit of regarding abstract concepts as possessing a real or 'thing-like' existence in the external world independently of the mind that formulated them. Alfred North Whitehead labelled this habit "the fallacy of misplaced concreteness" and his vivid phrase was invoked by Talcott Parsons in his criticism of utilitarianism in *The Structure of Social Action*.

Lukács' achievement was to give reification a wider and collective meaning, signifying the disposition of most people most of the time, particularly, of course, social classes to perceive the social and historical world as fixed, unalterable and external to their lived experience. Lukács, in effect, rediscovered the Hegelian idea of alienation and recognized that it was implicit in some of Marx's later writings despite his own ignorance of the, as yet, unpublished Paris Manuscripts.

Whether one calls it alienation or reification, this conception differs from the sociological usages in which, as Richard Schacht remarks in his thorough survey of the various meanings of the concept, ". . . alienation is conceived mostly in terms of the presence of certain attitudes and feelings . . . Thus there is no question of alienated individuals who are not aware of their alienation."[2] Berger strenuously objects to what he calls the 'psychologization' of the concept, but his own version, which, following Lukács, he explicitly links to the idea of 'false consciousness', is psychological in referring to cognition even though it excludes feelings and attitudes. To be sure, it is the *absence* rather than the presence of something in a person's awareness or consciousness that is central to Berger's (and Lukács') conception.

The concept of alienation has different possible normative implications, which re-enforces the futility of prescribing a uniform definition for it. Although Berger, with alienation especially in mind, has argued that "the integration of some Marxian concepts into sociological theory" is "very important", (Berger, 1966) he also concedes that his own "use of the concept has 'right' rather than 'left' implications" (Berger, 1967, p. 197, n.5.) His conservatism is evident in his insistence that alienation and reification reflect mankind's universal, even socio-biological, need for order or *nomos*, a view that correctly sees alienation and anomie as opposites rather than like most sociologists treating them as virtually equivalent.[3] But one need not emulate Berger in regarding alienation as a necessary bulwark against anomie. Nor, for that matter, Lukács in regarding it as an obstacle to be overcome by a new world-transforming historical subject. Berger's own definition of alienation has other implications and it is these that I wish to stress, playing them off against the broader psychological and political resonances that inescapably cling to the term in order to point to some surprising and even paradoxical conclusions.

The affinity between alienation as a denial that the social and historical world is a human creation and the political outlook of the Left, grounded in the original ethos of the Enlightenment, is obvious enough. That 'men make their own history' is at the heart of the Left's world-view: men can therefore overcome the historically given, the burden of the past, and recreate the social world in the image of secular ideals validated by rational moral and political principles. It follows that the most alienated people of all are primitive men who do not sharply separate the natural from the cultural, seeing both as pervasively shaped by magical and animistic forces. Characterizations of primitive societies as changeless or 'historyless', as 'cold' social formations in contrast to the more incoherent and internally contradictory 'hot' societies immersed in 'historicity', or an ineluctable historical consciousness, are consistent with this view. One of the clearest and least philosophically pretentious formulations of the difference is that of the anthropologist Robert Redfield, who maintains that the absence of the idea of reform, of deliberate attempts to alter received customs and institutions, is perhaps the crucial dividing line between primitive – he calls them 'precivilized' – societies and civilization (Redfield, 1953).

Redfield goes on to observe that the conscious reform of their institutions is exceedingly rare even in civilized societies, noting its infrequency until 'quite modern times' in both Western and Chinese history. Beliefs in the divine origin and legitimation of the social order have obviously inhibited purposeful efforts to change it. Ludwig Feuerbach regarded religion as the prototype of alienation: the projection of human qualities onto imaginary beings who were then regarded as independent, external powers to whom man must abase himself. The overcoming of alienation required the rejection of religious beliefs and authorities in the name of a materialistic and secular humanism.

The young Marx extended the idea of alienation to the state and the economy, ultimately concluding that economic alienation was the foundation of both religious and political alienation. Alienation as a form of consciousness was dissolved into the social reality of subjection to capitalism and the class domination it imposed. Anxious to repudiate the abstract and mentalistic overtones of the Hegelian and Feuerbachian conceptions, it is hardly surprising that Marx virtually abandoned the term in his later works in which the exploitation and eventual revolt of the working class become the major themes. Some sociologists who have revived the concept have faintly echoed this in making alienation nearly synonymous with *powerlessness*, although, in contrast to Berger and Lukács, awareness of being powerless is central to their definitions. It seems, incidentally, quite superfluous to invoke 'alienation' to signify a sense of subordination, oppression or victimization when we have so many other less metaphysically loaded words meaning the same thing.

What are the features of modern societies that are regarded as alienating and themselves the products of alienation? One could hardly hope to offer an exhaustive list, but the following surely includes the major candidates: the division of labor, the rule of the bourgeoisie, capitalism, the market, private property, the tyranny of the majority, the big city, bureaucracy, the centralized state, mass society, commercial or popular culture, the impersonal norms of science and technology, rule by 'technocrats' dominated by 'instrumental rationality' – I don't think I have left out anything important.

Most people are indeed often overawed by, and feel estranged from, the massive 'facticity' of these pervasive social realities. But do they really regard them as forces that are not 'man-made', as something more, or less, than human products and creations? Might not the very opposite be the truth – namely, that it is precisely the correct perception of their man-made nature that accounts for their power to alienate? If so, perhaps the causes of alienation as estrangement lie not in what the Hegelian–Marxists tradition calls alienation or reification, nor in Durkheim's quite dissimilar notion of anomie, but in what was accurately described by Max Weber, in the phrase he borrowed from Schiller, as the "disenchantment of the world".

But before discussing this possibility, an immensely influential apparent exception to it needs to be considered. The example of the market seems to contradict the suggestion that all of the allegedly alienating social phenomena listed above are generally perceived as human products. Marx's famous passage in *Capital* on 'commodity fetishism', as has often been noted, echoes his youthful theorizing about alienation, although he does not repeat the term itself. Lukács invoked this passage as the root and prime exemplar of reification, which in generalized form became perhaps the central concept in the intellectual tradition of post-Lukácsian Western Marxism, including the Frankfurt 'critical theorists'.

One hesitates to assert bluntly that so illustrious a thinker as Marx was confused, so let us just say that the discussion of commodity fetishism is at the very least dense, apparently conflating several different meanings. (I first read it, incidentally, in high school before I had even heard of sociology and I did not feel then that I fully understood it, as perhaps I still don't).

Marx wrote that commodity production – that is, production of objects in order to sell them – creates a situation in which the 'social relation' between producers is indirect and takes the form of a relation between their material products that compete for sale on the market. "There is", he goes on to observe, "a definite social relation between men, that assumes, in their eyes, the fantastic form of a relation between things." He then alludes to the Feuerbachian view of religion, referring to "the mist-enveloped regions of the religious world [in which] the productions of the human brain appear as independent beings endowed with life." "So it is", he concludes, "in the world of commodities with the products of men's hands."

But Marx has given us no more than a description of the market where, at least in large-scale societies, the producers do not interact directly but influence each other only indirectly through the results of the exchange of their products mediated by money (and, often enough, by middlemen). Where is there any fantasy or mystification in this familiar circumstance? Marx, to be sure, did not like the market, nor money as the medium that made its expansion possible, and expressed his preference for "production by freely associated men . . . consciously regulated by them in accordance with a settled plan". As much as Durkheim or Parsons, Marx glorified consensus over conflict, competition, and individual choice, treating 'social integration' as the desirable human norm rather than mere 'system integration'. He differed only in thinking that capitalism and commodity production were insuperable obstacles to its attainment. Socialists before and after Marx have long identified socialism itself with production 'for use' rather than 'for profit' (or 'exchange-value'). One notes, however, that in contemporary Yugoslavia and Hungary, and in Czechoslovakia before Soviet tanks put an end to it, Marxists in power speak of, and have tried partially to implement, 'market socialism', a term which they do not regard as an oxymoron.

Marx contended more specifically that the price or 'exchange-value' of a commodity is perceived as an inherent property of the physical object rather than as a social meaning reflecting an actual social relation under capitalist conditions of production and exchange. He provides no evidence that this is so; the purported resemblances to primitive fetishism, or to Feuerbach's deified projections of the human mind, are at best no more than analogies. I am reminded of the old tale that children growing up in big cities – children, note, not adults – believe that milk is somehow produced by the milkman who bottles and delivers it. Marx, to be sure, ends the famous chapter by quoting several second-rate and now forgotten economists who weirdly claimed that 'value' (meaning exchange-value) was an actual property of things in contrast to their use-value, which is relative to human purposes. But this is simply an instance of the reification of theoretical concepts in the pre-Lukácsian sense; there is no reason to believe that the ordinary citizen, equipped only with his or her commonsense, saw things that way anymore than he or she saw milk as 'naturally' coming in bottles rather than from cows. Marx reveals himself here as the quintessential bourgeois intellectual in all of his bookishness and inveterate scholasticism. Today what Frank Parkin calls 'professorial Marxism' is, of course, a flourishing enterprise centered on Talmudic debates over texts to which nobody else is listening, certainly not the 'proletariat'. Perhaps Marxism has, in the West at least, come home again, or, to paraphrase T. S. Eliot, "in its end is its beginning".

Marx's major achievement was his insistence that the principles of classical political economy had to be firmly located in their social and historical context, as opposed to the claims of the early economists that they were timeless natural laws comparable to, indeed reflecting and partly modelled after, Newtonian mechanics or Darwinian natural selection. The core of Marx's Marxism was his assertion of the dependence of the laws of the market on capitalist economic institutions that are historical creations capable of being transformed through their own operation which prepares the ground for eventual social revolution. The connection with the older philosophical idea of alienation lies in the rejection of the assimilation of the workings of capitalism to immutable natural forces analogous to the omnipotent deities of religion in their independence from and control over human aims and actions. Here is the rationale for the more generalized attack by 'critical theorists' on the belief that the socio-cultural world is governed by laws of a deterministic 'nature-like' character, the basis of their ceaseless polemics against 'positivism' and 'objectivism' in social thought – which they manage to find almost everywhere, including in Marx himself as well as in various later versions of Marxism.

To recall the modern social phenomena held responsible for alienation – capitalism, bureaucracy, technocracy, etc. – does anyone today really consider these to be reflections of unalterable natural laws of the same order as the motions of the planets or the law of gravity itself? Just as I previously suggested that nobody deifies them, I doubt that many people see them as anything but man-made, 'socially constructed' entities. Even contemporary economists, for whom the workings of the free market remain the major organizing principle, the master 'paradigm', if you like, of their discipline, no longer reify their concepts in the manner of the classical economists criticized by Marx. Modern economists regard the free market, and *homo economicus* himself, as conceptual constructs or models that lend themselves to econometric manipulation permitting wide and varied inferences, but not as descriptions of the world. The professional economist's fundamental "equilibrium price-auction model", as Lester Thurow (1983) calls it in his recent assault upon it, remains open to criticism, but only to second-order criticism (in Ernest Gellner's sense), directed against the model's artificiality and remoteness *vis-à-vis* the real world, despite its seductive adaptability to mathematization, rather than to the charge that its builders naively think they are describing that world. Thus 'commodity fetishism' as alienation no longer applies even to the abstract thought processes of economists.

Marx believed that political economy was becoming the 'dominant ideology' (Abercrombie *et al.*, 1980) of so-called bourgeois societies. He was mistaken, for neither classical economics nor the social Darwinism, increasingly popular during his own most active years, ever fully achieved that status, not even for the bourgeoisie itself. Nor did sociology, for which Comte, and even Durkheim over a generation later, had such high hopes, although Marxism itself became a coercively imposed dominant ideology for those non-bourgeois societies that trace their legitimation back to the October Revolution of 1917. Contemporary sociologists, however, remain curiously blind to the extent to which the leading assumptions of their discipline have entered into the popular consciousness over the past fifty years.

Raymond Aron (1984, p. 76) has written: "To our customs and beliefs, the very ones we hold sacred, sociology ruthlessly attaches the adjective 'arbitrary'." He quotes an acute observation by Marcel Mauss, from the generation of French sociologists between Aron's own and Durkheim's, that "All social phenomena are, to some degree, the work of a collective will, and whoever says human will, says choice between different possible options" (Aron, 1984, p. 71). These statements suggest that the view of the social order as arbitrary or optional presented by sociologists may itself be a source of estrangement. Not only does such a view disenchant the world, ridding it of humanized gods and spirits, but it also deprives it of the aura of iron necessity and inevitability associated with the reign of natural laws. Modern men like all men have been thrust into an external and constraining society that is not of their own making, but they are more acutely aware than earlier generations that his society was made by other men rather than by God or nature; it often appears to them, accordingly, as a world of makeshift, even historically accidental social arrangements by which they feel ensnared and which they can readily imagine as being otherwise. The pathos of their situation is not so much that they reify the institutions and power structures impinging upon them, ascribing to them an existence and weight independent of human activity and choice, as that they are *unable* to reify them, seeing them instead as purely contingent human creations to which they must accommodate themselves. If alienation is the failure to remember the humanly produced and socially constructed nature of society, perhaps the very overcoming of such alienation is not the solution to but rather a cause of alienation as estrangement, homelessness, meaninglessness – the conception of it generally favored by American sociologists.

One remembers incongruities in the outlook of the youthful political and 'counter-cultural' protest movements of the 1960s that made 'alienation' into a catchword. Marxist slogans were combined with American Indian headbands, the chanting of mantras, and events like the famous effort, or pretended effort, to 'levitate' the Pentagon. It often seemed as if the desired alternative to commodity fetishism was the real, original fetishism of tribal cults so looked down upon by Marx from the heights of his nineteenth-century progressivism. Primitive peoples, far from representing the epitome of alienation because of their failure to see their culture and institutions as man-made and reformable, were admired as typifying the lack of alienation for this very reason, for possessing a deep feeling for the wholeness of the cosmos harmoniously embracing both nature and culture. No wonder 'alienation' became a discredited term under the weight of so many contradictory meanings!

Except for the numbers of people and the publicity involved, little was new in the religiosity and neo-primitivism of the 1960s counterculture. Fifty years earlier, Max Weber had noted "the need of some modern intellectuals to furnish their souls with, so to speak, guaranteed genuine antiques". "They play", he added, "at decorating a sort of domestic chapel with small sacred images from all over the world, or they produce surrogates through all sorts of psychic experiences to which they ascribe the dignity of mystic holiness, which they peddle in the book market" (Gerth and Mills, 1946). What was novel, though by no means entirely unprecedented, about the 1960s was that such attitudes were expressed under the auspices of the political Left and often included Marxist phrases like 'demystification' when the very opposite, 're-mystification' or the 're-enchantment' of a disenchanted world, appeared to be what the protesters really wanted.

Hannah Arendt is one recent thinker who understood that alienation as estrangement may, for modern man, result from his very awareness of living in an almost entirely man-made world rather than from the lack of such awareness, as argued by the Hegelian-Marxist school. "The modern age, with its growing world-alienation," she wrote, "has led to a situation where man, wherever he goes, encounters only himself. All the processes of the earth and the universe have revealed themselves either as man-made or as potentially man-made" (Arendt, 1968). Like the theorists of the Frankfurt School whose 'technophobia' she shared, she assigned responsibility for this result to the triumphs of science and technology which have "devoured . . . the solid objectivity of the given". As our major interpreter of totalitarianism, Arendt achieved a deeper understanding of it than the Frankfurt theorists in seeing it as an expression of the modern awareness that historical facts like all facts are 'theory-dependent', thus creating the possibility of 'making' – or 'making over' – history and at the same time as a panicky, remystifying, willed flight from the very alienation this awareness produces. The resulting synthesis is what gives totalitarianism its peculiarly modern character, that is, discontinuous with the past, including all forms of earlier authoritarian, tyrannical and exploitative domination.

Arendt grounded her understanding of totalitarianism concretely in European history and did not, like the Frankfurt School writers, regard it as the culmination, the inevitable end-product, of everything she disliked about modernity. She saw it rather as, at most, a latent possibility, a danger that might but need not recur, in the mass industrial societies of the modern world. Her archaism, her 'Hellenic nostalgia', was superior as a source of insight to the soured and embittered futurism of the Frankfurt theorists. (It is still possible to feel, nevertheless, that her recoil from modernity was excessive.)

The predictable response of latter-day critical theorists to the assertion that alienation is caused by the realization that social institutions are humanly created expedients rather than by the absence of such a realization is to reaffirm the possibility of an authentic and unified collective will to dereify what exists and reconstruct it anew. As a critic of critical theory has remarked,

> The Critical Theory of Society generalizes Marx's revolutionary dialectic into the assertion that, since there is nothing in our society but our activity and its ephemeral products, it is all totally plastic and liable to instant revolution at almost any moment we wish – if only we would wish" (McInnes, 1972).

Ah, there's the rub – "if only we would wish!" We are not, after all, talking about Periclean Athens, a few thousand residents of Parisian *arrondissements*, or even the Red Belt surrounding Berlin in the 1920s, but of hundreds of millions of people of diverse origins and outlooks dependent for survival itself on a technically complex network of goods and services that is increasingly international and intercontinental in scope. Marx had the realism to identify a tangible collective agent in the proletariat. He was wrong about its destiny, but a return to the Young Hegelian apotheosis of Humanity itself as a fictive collective actor is even less convincing than it was in the early nineteenth century when there were plausible excuses for identifying world history with European and Western history.

I shall close by trying to relate the previous discussion to sociological theory today, or at least to the three dominating figures who presently occupy the front row in its pantheon, the Holy Trinity of Marx, Weber and Durkheim.

Although in other respects their views were diametrically opposed – for example, on the relation of the individual to society – Marx and Durkheim were alike in divinizing society, not, of course, in its existing form, but in the ideal form they hoped it would assume in the future partly as a result of the influence of their own teachings. In Durkheim this is explicit: the sacred beings and objects men have worshipped in the past were in reality nothing but representations of society itself – he might just as well have said that they were 'alienated' or 'reified' projections of social relations. When we lose our reverence for the mighty and benign power that regulates us, we fall into anomie and endanger not only our sanity but our very life itself.

To Marx, the future prospect of a fully 'socialized humanity' gave meaning to history. Marxism, as Leszek Kolakowski (1978) has written, is a philosophical expression of the "self-deification of mankind". For what would be the most, if not the only, sanctified object of devotion in this future about which Marx notoriously told us so little? Clearly, it would be the collective forms of life that so many generations had labored to produce in the dark and bloody past.

Max Weber knew better. For him, as for Marx and Durkheim, the gods had been routed by the progress of knowledge, although he remained acutely sensitive to how "the ghost of dead religious beliefs . . . prowls about in our lives" (Weber, 1930). But he saw, in 1905, that "the rosy blush of the Enlightenment . . . seems also to be irretrievably fading". It is worth recalling, however, that over a century earlier the greatest thinker of the Enlightenment and the founder of the German Idealist philosophy that gave birth to the concept of alienation, Immanuel Kant, said that "Out of the crooked timber of humanity, no straight thing can ever be made."

New York University

NOTES

[1] Such as Richard Schacht, *Alienation*, Garden City, N.Y.: Doubleday, 1970; and Joachim Israel, *Alienation from Marx to Modern Sociology: A Sociological Analysis*, Boston: Allyn and Bacon, 1971.

[2] Schacht, *ibid.*, p. 154.

[3] For a brilliant and searching discussion that sees them as opposite dispositions rooted in human nature itself, see Gary B. Thom, *The Human Nature of Social Discontent: Alienation, Anomie, Ambivalence*. Totowa, N.J.: Rowman and Allanheld, 1983.

REFERENCES

Abercrombie, Nicholas. Hill, Stephen and Turner, Bryan S. 1980. *The Dominant Ideology Thesis*. London: George Allen and Unwin.

Arendt, Hannah 1968. *Between Past and Future*. New York: The Viking Press, p. 89.

Aron, Raymond. 1984. *Politics and History*. New Brunswick, N.J.: Transaction Books.

Berger, Peter L. 1966. 'Response', *New Left Review*, No. 35 (Jan.–Feb.), p. 76.

Berger, Peter L. 1967. *The Sacred Canopy*. Garden City, N.Y.: Doubleday Anchor Books, p. 85.

Berger, Peter L. and Luckmann, Thomas. 1966. *The Social Construction of Reality*. Garden City, N.Y.: Doubleday Anchor Books, p. 89.

Gerth, H. H. and Mills, C. Wright 1946. *From Max Weber: Essays in Sociology*. New York: Oxford University Press, p. 154.

Kolakowski, Leszek. 1978. *Main Currents of Marxism*, Vol. III. Oxford: The Clarendon Press, p. 530

McInnes, Neil. 1972. *The Western Marxists*. New York: Library Press, p. 176.

Redfield, Robert. 1953. *The Primitive World and Its Transformations*. Ithaca, N.Y.: Cornell University Press, pp. 111–138.

Thurow, Lester C. 1983. *Dangerous Currents: The State of Economics*. New York: Random House.

Weber, Max. 1930. *The Protestant Ethic and the Spirit of Capitalism*. New York: Scribner, p. 182.

CONTRIBUTORS

WILLIAM J. BAUMOL is Professor of Economics by joint appointment at Princeton University and New York University. He is the author of many books including, most recently, *Productivity Growth and U.S. Competitiveness* (with K. McLennan), *Superfairness: Applications and Theory*, and *Microtheory: Applications and Origins*.

WILLIAM BREIT is E. M. Stevens Distinguished Professor of Economics at Trinity University, San Antonio, Texas. A prolific writer of fiction (two mystery novels) and non-fiction, his most recent books are *Antitrust Penalty Reform: An Economic Survey* and *Lives of the Laureates*.

WILLIAM H. DRAY is Professor Emeritus of History and Philosophy at the University of Ottawa. He is the author of numerous articles and books including *Philosophy of History* and *Perspectives on History*.

ABRAHAM EDEL is Research Professor of Philosophy at the University of Pennsylvania. Among his most recent books are *Aristotle and His Philosophy* and *Interpreting Education*.

ROBERT G. GILPIN, JR. is Eisenhower Professor of International Affairs, Professor of Politics and International Affairs, and Faculty Associate of the Center of International Studies, Princeton University. Among the books he has authored are *American Scientists and Nuclear Weapons Policy* and *War and Change in World Politics*.

ELI GINZBERG is A. Barton Hepburn Professor Emerituis of Economics and Director of the Conservation of Human Resources, Columbia University. He is the author of about 100 books, mostly in the field of human resources and health economics.

CHARLES HANLY is Professor of Philosophy at the University of Toronto, a practising psychoanalyst and a training analyst in the Toronto Psychoanalytic Institute. He is the author of *Existentialism and*

S. Hook, W.L. O'Neill and R. O'Toole (eds.), Philosophy, History and Social Action, 475–478.

Psychoanalysis as well as numerous articles in philosophical and psychoanalytic publications.

SIDNEY HOOK is Emeritus Professor of Philosophy, New York University, and currently a Senior Research Fellow at the Hoover Institution for the Study of War, Revolution and Peace, Stanford.

IRVING LOUIS HOROWITZ is Hannah Arendt distinguished professor of sociology and political science at Rutgers University, where he also serves as editor-in-chief of Transaction/SOCIETY. He is the author of many books, among the most recent of which are *C. Wright Mills: An American Utopian* and *Winners and Losers*.

MILTON R. KONVITZ is Professor of Industrial and Labor Relations and Emeritus Professor of Law at Cornell University. He is chairman of the Editorial Board of *Midstream* and co-editor of *Jewish Social Studies*. Among other books he is the author of *Religious Liberties and Conscience* and *Judaism and the American Idea*.

PAUL KURTZ is Professor of Philosophy at the State University of New York at Buffalo. He is editor of *Free Inquiry* magazine, founding chairman of the Committee for the Scientific Investigation of Claims of the Paranormal, and editor of Prometheus Books. Among his recent books are *The Transcendental Temptation: A Critique of Religions and the Paranormal* and *A Skeptics Handbook of Parapsychology*.

THELMA Z. LAVINE is Robinson Professor in Philosophy and the Humanities at George Mason University. She is the author of many articles, book chapters, and television scripts, and is co-author of *History and Anti-History in Philosophy*.

GEORGE S. N. LUCKYJ is Professor Emeritus of Russian and Ukrainian literature at the University of Toronto.

CZESLAW MILOSZ is Professor Emeritus of Russian Language and Literature at the University of California, Berkeley. He is the author of many books of poetry and non-fiction. In 1980 he won the Nobel Prize for Literature.

WILLIAM L. O'NEILL is Professor of History at Rutgers University. His most recent books are *A Better World: The Great Schism: Stalinism and the Intellectuals* and *American High: The Years of Confidence, 1945–1960*.

ROGER O'TOOLE is Professor of Sociology at the University of Toronto. He is the author of *The Precipitous Path* and *Religion: Classic Sociological Approaches*.

RICHARD H. POPKIN is Emeritus Professor of Philosophy, Washington University, St. Louis and Adjunct Professor of History and Philosophy, University of California, Los Angeles. He is the author of *The History of Scepticism from Erasmus to Spinoza*, and *Isaac La Peyrere, His Life, His Work and His Influence*, among many studies in the history of philosophy and Jewish intellectual history.

SIDNEY RATNER is Emeritus Professor of History, Rutgers University. Among his books the most recent are *Taxation and Democracy in America* and *The Tariff in American History*.

VERA COOPER RUBIN is an astronomer who has specialized in observations of motions of stars in galaxies, and motions of galaxies in the universe. She is a Staff Member of the Carnegie Institution of Washington. She was the first woman permitted to observe at Palomar Observatory, and is a member of the U.S. National Academy of Sciences.

HAROLD Z. SCHIFFRIN is Professor of Chinese Studies and Sociology at the Hebrew University of Jerusalem and Academic Chairman of the Harry S. Truman Research Institute for the Advancement of Peace. Among his publications are *Sun Yat-Sen and the Origins of the Chinese Revolution* and *Sun-Yat Sen: Reluctant Revolutionary*.

HAROLD ISSADORE SHARLIN has a B.S. in electrical engineering and a Ph.D. in American economic history. Sharlin taught the history of science and technology for twenty-five years before going to Washington, D.C., in 1976, where he is a consultant on science policy.

JOSEF ŠKVORECKÝ is a Canadian who writes fiction in Czech. He is also a professor of English and Film at Erindale College, University of Toronto. His latest novels are *The Engineer of Human Souls* and *Dvorak In Love*.

STEPHEN J. WHITFIELD is the Max Richter Professor of American Civilization at Brandeis University. His most recent book is *A Critical American: The Politics of Dwight Macdonald*.

DENNIS H. WRONG is Professor of Sociology at New York University, and a member of the editorial boards of *Dissent* and *Partisan Review*. Among other works he is the author of *Skeptical Sociology* and *Power: Its Forms, Bases, and Uses*.

INDEX OF NAMES

BOSTON STUDIES IN THE PHILOSOPHY OF SCIENCE

Editors:

ROBERT S. COHEN and MARX W. WARTOFSKY

(Boston University)

1. Marx W. Wartofsky (ed.), *Proceedings of the Boston Colloquium for the Philosophy of Science 1961–1962.* 1963.
2. Robert S. Cohen and Marx W. Wartofsky (eds.), *In Honor of Philipp Frank.* 1965.
3. Robert S. Cohen and Marx W. Wartofsky (eds.), *Proceedings of the Boston Colloquium for the Philosophy of Science 1964–1966. In Memory of Norwood Russell Hanson.* 1967.
4. Robert S. Cohen and Marx W. Wartofsky (eds.), *Proceedings of the Boston Colloquium for the Philosophy of Science 1966–1968.* 1969.
5. Robert S. Cohen and Marx W. Wartofsky (eds.), *Proceedings of the Boston Colloquium for the Philosophy of Science 1966–1968.* 1969.
6. Robert S. Cohen and Raymond J. Seeger (eds.), *Ernst Mach: Physicist and Philosopher.* 1970.
7. Milic Capek, *Bergson and Modern Physics.* 1971.
8. Roger C. Buck and Robert S. Cohen (eds.), *PSA 1970. In Memory of Rudolf Carnap.* 1971.
9. A. A. Zinov'ev, *Foundations of the Logical Theory of Scientific Knowledge (Complex Logic).* (Revised and enlarged English edition with an appendix by G. A. Smirnov, E. A. Sidorenka, A. M. Fedina, and L. A. Bobrova). 1973.
10. Ladislav Tondl, *Scientific Procedures.* 1973.
11. R. J. Seeger and Robert S. Cohen (eds.), *Philosophical Foundations of Science.* 1974.
12. Adolf Grünbaum, *Philosophical Problems of Space and Time.* (Second, enlarged edition). 1973.
13. Robert S. Cohen and Marx W. Wartofsky (eds.), *Logical and Epistemological Studies in Contemporary Physics.* 1973.
14. Robert S. Cohen and Marx W. Wartofsky (eds.), *Methodological and Historical Essays in the Natural and Social Sciences. Proceedings of the Boston Colloquium for the Philosophy of Science 1969–1972.* 1974.
15. Robert S. Cohen, J. J. Stachel, and Marx W. Wartofsky (eds.), *For Dirk Struik. Scientific, Historical and Political Essays in Honor of Dirk Struik.* 1974.
16. Norman Geschwind, *Selected Papers on Language and the Brain.* 1974.
17. B. G. Kuznetsov, *Reason and Being: Studies in Classical Rationalism and Non-Classical Science.* 1987.
18. Peter Mittelstaedt, *Philosophical Problems of Modern Physics.* 1976.
19. Henry Mehlberg, *Time, Causality, and the Quantum Theory* (2 vols.). 1980.
20. Kenneth F. Schaffner and Robert S. Cohen (eds.), *Proceedings of the 1972 Biennial Meeting, Philosophy of Science Association.* 1974.
21. R. S. Cohen and J. J. Stachel (eds.), *Selected Papers of Léon Rosenfeld.* 1978.
22. Milic Čapek (ed.), *The Concepts of Space and Time. Their Structure and Their Development.* 1976.

23. Marjorie Grene, *The Understanding of Nature. Essays in the Philosophy of Biology.* 1974.
24. Don Ihde, *Technics and Praxis. A Philosophy of Technology.* 1978.
25. Jaakko Hintikka and Unto Remes. *The Method of Analysis. Its Geometrical Origin and Its General Significance.* 1974.
26. John Emery Murdoch and Edith Dudley Sylla, *The Cultural Context of Medieval Learning.* 1975.
27. Marjorie Grene and Everett Mendelsohn (eds.), *Topics in the Philosophy of Biology.* 1976.
28. Joseph Agassi, *Science in Flux.* 1975.
29. Jerzy J. Wiatr (ed.), *Polish Essays in the Methodology of the Social Sciences.* 1979.
30. Peter Janich, *Protophysics of Time.* 1985.
31. Robert S. Cohen and Marx W. Wartofsky (eds.), *Language, Logic, and Method.* 1983.
32. R. S. Cohen, C. A. Hooker, A. C. Michalos, and J. W. van Evra (eds.), *PSA 1974: Proceedings of the 1974 Biennial Meeting of the Philosophy of Science Association.* 1976.
33. Gerald Holton and William Blanpied (eds.), *Science and Its Public: The Changing Relationship.* 1976.
34. Mirko D. Grmek (ed.), *On Scientific Discovery.* 1980.
35. Stefan Amsterdamski, *Between Experience and Metaphysics. Philosophical Problems of the Evolution of Science.* 1975.
36. Mihailo Marković and Gajo Petrović (eds.), *Praxis. Yugoslav Essays in the Philosophy and Methodology of the Social Sciences.* 1979.
37. Hermann von Helmholtz, *Epistemological Writings. The Paul Hertz/Moritz Schlick Centenary Edition of 1921 with Notes and Commentary by the Editors.* (Newly translated by Malcolm F. Lowe. Edited, with an Introduction and Bibliography, by Robert S. Cohen and Yehuda Elkana). 1977.
38. R. M. Martin, *Pragmatics, Truth, and Language.* 1979.
39. R. S. Cohen, P. K. Feyerabend, and M. W. Wartofsky (eds.), *Essays in Memory of Imre Lakatos.* 1976.
40. B. M. Kedrov and V. Sadovsky. *Current Soviet Studies in the Philosophy of Science.*
41. M. Raphael, *Theorie des Geistigen Schaffens auf Marxistischer Grundlage.*
42. Humberto R. Maturana and Francisco J. Varela, *Autopoiesis and Cognition. The Realization of the Living.* 1980.
43. A. Kasher (ed.), *Language in Focus: Foundations, Methods and Systems. Essays Dedicated to Yehoshua Bar-Hillel.* 1976.
44. Trân Duc Thao, *Investigations into the Origin of Language and Consciousness.* (Translated by Daniel J. Herman and Robert L. Armstrong; edited by Carolyn R. Fawcett and Robert S. Cohen). 1984.
45. A. Ishmimoto (ed.), *Japanese Studies in the History and Philosophy of Science.*
46. Peter L. Kapitza, *Experiment, Theory, Practice.* 1980.
47. Maria L. Dalla Chiara (ed.), *Italian Studies in the Philosophy of Science.* 1980.
48. Marx W. Wartofsky, *Models: Representation and the Scientific Understanding.* 1979.
49. Trân Duc Thao, *Phenomenology and Dialectical Materialism.* 1985.
50. Yehuda Fried and Joseph Agassi, *Paranoia: A Study in Diagnosis.* 1976.
51. Kurt H. Wolff, *Surrender and Catch: Experience and Inquiry Today.* 1976.
52. Karel Kosík, *Dialectics of the Concrete.* 1976.

53. Nelson Goodman, *The Structure of Appearance.* (Third edition). 1977.
54. Herbert A. Simon, *Models of Discovery and Other Topics in the Methods of Science.* 1977.
55. Morris Lazerowitz, *The Language of Philosophy. Freud and Wittgenstein.* 1977.
56. Thomas Nickles (ed.), *Scientific Discovery, Logic, and Rationality.* 1980.
57. Joseph Margolis, *Persons and Minds. The Prospects of Nonreductive Materialism.* 1977.
58. G. Radnitzky and G. Andersson (eds.), *Progress and Rationality in Science,* 1978.
59. Gerard Radnitzky and Gunnar Andersson (eds.), *The Structure and Development of Science.* 1979.
60. Thomas Nickles (ed.), *Scientific Discovery: Case Studies.* 1980.
61. Maurice A. Finocchiaro, *Galileo and the Art of Reasoning.* 1980.
62. William A. Wallace, *Prelude to Galileo.* 1981.
63. Friedrich Rapp, *Analytical Philosophy of Technology.* 1981.
64. Robert S. Cohen and Marx W. Wartofsky (eds.), *Hegel and the Sciences.* 1984.
65. Joseph Agassi, *Science and Society.* 1981.
66. Ladislav Tondl, *Problems of Semantics.* 1981.
67. Joseph Agassi and Robert S. Cohen (eds.), *Scientific Philosophy Today.* 1982.
68. Wuadysuaw Krajewski (ed.), *Polish Essays in the Philosophy of the Natural Sciences.* 1982.
69. James H. Fetzer, *Scientific Knowledge.* 1981.
70. Stephen Grossberg, *Studies of Mind and Brain.* 1982.
71. Robert S. Cohen and Marx W. Wartofsky (eds.), *Epistemology, Methodology, and the Social Sciences.* 1983.
72. Karel Berka, *Measurement.* 1983.
73. G. L. Pandit, *The Structure and Growth of Scientific Knowledge.* 1983.
74. A. A. Zinov'ev, *Logical Physics.* 1983.
75. Gilles-Gaston Granger, *Formal Thought and the Sciences of Man.* 1983.
76. R. S. Cohen and L. Laudan (eds.), *Physics, Philosophy and Psychoanalysis.* 1983.
77. G. Böhme et al., *Finalization in Science,* ed. by W. Schäfer. 1983.
78. D. Shapere, *Reason and the Search for Knowledge.* 1983.
79. G. Andersson, *Rationality in Science and Politics.* 1984.
80. P. T. Durbin and F. Rapp, *Philosophy and Technology.* 1984.
81. M. Marković, *Dialectical Theory of Meaning.* 1984.
82. R. S. Cohen and M. W. Wartofsky, *Physical Sciences and History of Physics.* 1984.
83. E. Meyerson, *The Relativistic Deduction.* 1985.
84. R. S. Cohen and M. W. Wartofsky, *Methodology, Metaphysics and the History of Sciences.* 1984.
85. György Tamás, *The Logic of Categories.* 1985.
86. Sergio L. de C. Fernandes, *Foundations of Objective Knowledge.* 1985.
87. Robert S. Cohen and Thomas Schnelle (eds.), *Cognition and Fact.* 1985.
88. Gideon Freudenthal, *Atom and Individual in the Age of Newton.* 1985.
89. A. Donagan, A. N. Perovich, Jr., and M. V. Wedin (eds.), *Human Nature and Natural Knowledge.* 1985.
90. C. Mitcham and A. Huning (eds.), *Philosophy and Technology II.* 1986.
91. M. Grene and D. Nails (eds.), *Spinoza and the Sciences.* 1986.
92. S. P. Turner, *The Search for a Methodology of Social Science.* 1986.
93. I. C. Jarvie, *Thinking About Society: Theory and Practice.* 1986.
94. Edna Ullmann-Margalit (ed.), *The Kaleidoscope of Science.* 1986.

95. Edna Ullmann-Margalit (ed.), *The Prism of Science*. 1986.
96. G. Markus, *Language and Production*. 1986.
97. F. Amrine, F. J. Zucker, and H. Wheeler (eds.), *Goethe and the Sciences: A Reappraisal*. 1987.
98. Joseph C. Pitt and Marcella Pera (eds.), *Rational Changes in Science*.
99. O. Costa de Beauregard, *Time, the Physical Magnitude*. 1987.
100. Abner Shimony and Debra Nails (eds.), *Naturalistic Epistemology: A Symposium of Two Decades*. 1987.
101. Nathan Rotenstreich, *Time and Meaning in History*. 1987.
102. David B. Zilberman (ed.), *The Birth of Meaning in Hindu Thought*. 1987.
103. Thomas F. Glick (ed.), *The Comparative Reception of Relativity*. 1987.
104. Zellig Harris *et al.*, *The Form of Information in Science*. 1987
105. Frederick Burwick, *Approaches to Organic Form: Permutations in Science and Culture*. 1987.